SP Pvt. Ltd.

Algae Biofuel

Editors:

S.K. Bhatnagar
Department of Cell Biology
College of Biotechnology, S.V.P. University
of Agriculture & Technology Meerut - 250 110

Atul Saxena
Growdiesel Climate Care Council
B-196, Surajmal Vihar
Delhi - 110 092

&

Stefan Kraan
Ocean Harvest Technology
N 17 Business Park, Milltown, Tuam,
Co. Galway, Ireland

2011

Studium Press (India) Pvt. Ltd.

Algae Biofuel

ISBN: 978-93-80012-44-5

Published by:

Studium Press (India) Pvt. Ltd.
4735/22, 2nd Floor, Prakash Deep Building
(Near Delhi Medical Association),
Ansari Road, Darya Ganj, New Delhi-110 002
Tel.: 23240257, 65150447; Fax: 91-11-23240273;
jngovil@gmail.com; jngovil@hotmail.com

Printed at Magic International Pvt. Ltd. Greater Noida, UP

सत्यमेव जयते

डा. एस. अय्यप्पन

सचिव एवं महानिदेशक

DR. S. AYYAPPAN

SECRETARY & DIRECTOR-GENERAL

भारत सरकार
कृषि अनुसंधान और शिक्षा विभाग एवं
भारतीय कृषि अनुसंधान परिषद
कृषि मंत्रालय, कृषि भवन, नई दिल्ली 110 001

GOVERNMENT OF INDIA
DEPARTMENT OF AGRICULTURAL RESEARCH & EDUCATION
AND
INDIAN COUNCIL OF AGRICULTURAL RESEARCH
MINISTRY OF AGRICULTURE, KRISHI BHAVAN, NEW DELHI 110 001
Tel.: 23382629; Fax: 91-11-23384773
E-mail: ayyappans@icar.org.in; ayyapans@yahoo.co.uk

FOREWORD

Conservation of environment, water, soil and energy have emerged as main challenges before every one of us. Presently, at global level energy requirement are being met by fossil fuels which are limited natural resources. Scientists throughout the world are looking for various alternate fuel sources like thermal, solar, hydro and nuclear based energy. In the past few years, oil producing crops like maize, jatropha, soybean etc. have been identified but all these crops need fertile or rarely a degraded land for cultivation. The need of the hour is to find plant materials which are environment friendly and may be grown in barren land or are capable of utilizing waste water for multiplication. Recently, it has been explored that various algal strains have potential to give high oil yields which can sequester carbon dioxide and safeguard the environment.

The editors of ALGAE BIOFUEL have brought forth the issues pertaining to green energy together in this compilation. The articles included in this compendium have been written by renowned scientists in the field of algae through which all major issues pertaining to biofuel production, algae cultivation, harvesting, carbon sequestration, environmental cleaning, bioremediation, deoiled algal cake as animal feed etc. have been covered.

I am confident that this book will be used by the agricultural scientists, students and researchers to initiate algae based green fuel production.

The attempt made by the editors deserves appreciation for bringing out this publication on algae biofuel from India.

Dated the 20th August, 2010
New Delhi

(S. Ayyappan)

LIST OF CONTRIBUTORS

Abdul Nazar, A. K., National Institute of Ocean Technology, Palikaranai, Chennai, Tamil Nadu, India -601302

Abrol, Deepika, Department of Biotechnology, Bareilly College, Bareilly 245 001 India Email : dabrol2000@yahoo.com

Adey Walter, Department of Botany, Smithsonian Institute, U.S.A. Email : adey.walter@nmnh.si.edu

Ahluwalia, A.S., Department of Botany, Panjab University, Chandigarh-160014. India. Email : phykos@pu.ac.in

Ashish Bhatnagar, Algae Biofuels & Biomolecules Initiative, Department of Microbiology, Maharshi Dayanand Saraswati University, Ajmer 305 009 Rajasthan India Email : bhatnagarashis@gmail.com

Bannon Jefferey, Department of Botany, Smithsonian Institute, U.S.A. Email : jbenemann@aol.com

Bhatnagar. S. K., Department of Cell Biology, College of Biotechnology, Sardar Vallabhbhai Patel University of Agriculture & Technology, MEERUT 250 110 India Email : subodhbhatnagar18@gmail.com, drskb2000@yahoo.com

Bhosale. Rahul Department of Biotechnology & Bioinformatics, University of Pune, Pune-411007, Maharashtra, India

Chaugule B. B.: Department of Botany, University of Pune, Pune-411007, Maharashtra, India

Das KC , Biorefining and Carbon Cycling Program, Faculty of Biological and Agricultural Engineering, The University of Georgia, Athens, GA 30605

Dhar, Dolly Wattal, Centre for Conservation and Utilisation of Blue Green Algae Indian Agricultural Research Institute, New Delhi- 110 012

Dube, Lokesh Chandra, Emergent Ventures India Pvt. Limited, Gurgaon, India E-mail: lokesh@emergent-ventures.com

Dubois-Calero Nathalie, AirScience Biofuel, 1751 Richardson, Montreal, QC, Canada H3K 1G6

Gluck, Steven J., The Dow Chemical Company, Environmental Technology Centre, Freeport, TX USA

Gogoi H. K., Defence Research Laboratory, Tezpur, Assam-784001, India

Gopinath A., Internal Combustion Engineering Division, Department of Mechanical Engineering, Anna University, India, Chennai-25.

Horsman Mark, Department of Chemical Engineering, University of Ottawa, 161 Louis Pasteur St., Ottawa, ON, Canada K1N 6N5

Jitendra, Department of Botany, Panjab University, Chandigarh-160014. India. E-mail: ahlawat.jitender@gmail.com

Kalita M. C., Environment Biotechnology Laboratory, Department of Biotechnology, Gauhati University, Guwahati - 781014, Assam, INDIA

Kathiroli S., National Institute of Ocean Technology, Palikaranai, Chennai, Tamil Nadu, India -601302

Kaur Simrat, Defence Research Laboratory, Tezpur, Assam-784001, India

Khan Shakeel A., Division of Environmental Sciences, NRL Building, Pusa Campus, Indian Agricultural Research Institute, New Delhi-110012.India E- mail : shakeel_iari@yahoo.com, shakeel_env@iari.res.in

Kirubagaran R., National Institute of Ocean Technology, Palikaranai, Chennai, Tamil Nadu, India -601302

Kraan Stefan, Scientific Director and Co-founder Ocean Harvest Technology,N17 Business Park, Milltown, Tuam, Co. Galway, Ireland Email : stefan@oceanharvest.ie

Kumar Ankit, Emergent Ventures India Pvt. Limited, Gurgaon, India

Lan Christopher Q, Department of Chemical Engineering, University of Ottawa, 161 Louis Pasteur St., Ottawa, ON, Canada K1N 6N5 E-mail: christopher.lan@uottawa.ca.

Li Yanqun, College of Life Science, South China Normal University, Guangzhou, P.R. China

Mary Leema J., Andaman and Nicobar centre for Ocean Science and Technology, NIOT, R & D Complex, Port Blair, A&N Islands, India - 744103 Email: leemaroy2002@Yahoo.com

Mishra Sandhya, Marine Biotechnology and Ecology Discipline, Central Salt & Marine Chemicals Research Institute, CSIR, G. B. Marg, Bhavnagar - 364 002, India. E-mail: smishra@csmcri.org , smishracsmcri@rediffmail.com

Monica Bhatnagar, Algae Biofuels & Biomolecules Initiative, Department of Microbiology, Maharshi Dayanand Saraswati University, Ajmer 305 009 Rajasthan India

Nagarajan, G. Internal Combustion Engineering Division, Department of Mechanical Engineering, Anna University, India, Chennai-25.

Pabbi, Sunil, Centre for Conservation and Utilisation of Blue Green AlgaeIndian Agricultural Research Institute, New Delhi- 110 012 Email : Sunil.pabbi@gmail.com

Perriman, Damien, The Dow Chemical Company, Environmental Technology Centre, Freeport, TX USA Email : dperriman@verdezyne.com

Puhan, Sukumar, Anna University, Chennai - 6000 25, Tamil Nadu, India Email: sp_anna2006@yahoo.co.in

Purkayastha J. Defence Research Laboratory Tezpur, Post Bag No.-2, Tezpur-784001, Assam. E-mail: jubilee_p@rediffmail.com

Raghvendar Singh, College of Biotechnology, Sardar Vallabhbhai Patel University of Agriculture and Technology, Meerut-250110 India E-mail: raghavendar@gmail.com

Rashmi, Department of Pharmaceutical Technology (Biotechnology), National Institute of Pharmaceutical and Education Research, S.A.S. Nagar, Mohali-160062. India.

Rout Jaishree, Department of Ecology & Environmental Science, Assam University, Silchar-788011, Assam, India Email : rouyjaya@rediffmail.com

Saxena, Atul, CEO, Growdiesel Climate Care Council, B-196, Surajmal Vihar, Delhi 110 092 India Email : info@growdieselmail.com

Senthil Chinnasamy, Naruveli Ventures, No.1 Karunanidhi Second Street, Kottur, Chennai 600085

Singh, Renu, Division of Environmental Sciences, Indian Agricultural Research Institute, New Delhi-110 012 Email : renu_icar@yahoo.com

Sood, Anjuli, Department of Botany, University of Delhi, Delhi-110007. India. E-mail: anjulisood@gmail.com

Srivastava, R. B., Director, Defence Research Laboratory, Tezpur, Assam-784001, India.

Talukdar, Jayant, Microalgae Biotechnology Division, Environment Biotechnology Laboratory, Department of Biotechnology, Gauhati University, Guwahati - 781014, Assam, INDIA E-mail: jayanta_talukdar@yahoo.com

Wu Nan, Department of Chemical Engineering, University of Ottawa, 161 Louis Pasteur St., Ottawa, ON, Canada K1N 6N5

ACKNOWLEDGEMENTS

When entire world is looking at the alternative feedstock for energy production which should be eco-friendly, farmer friendly, economical having commercial use of its post extraction produce, the preparation of a compendium on Algae Biofuel had been a challenging task for the editors because lot of work has been initiated on this feedstock in western countries and India is preparing to launch it.

The editors wish to acknowledge the contributions made by renowned authors of various chapters who provided tremendous input to the theme. Authors from India and abroad have placed the data together in such a way that it is easily understandable to all readers including the entrepreneurs. We are extremely thankful to Prof. S Ayyappan, Director General, ICAR and Secretary DARE for generously going through the entire manuscript for writing the FOREWORD. Editors are grateful to Prof M P Yadav, Ex-V.C. and Prof Arvind Kumar Bakshi, V.C., Sardar Vallabhbhai Patel University of Agriculture & Technology, Meerut for their sustained co-operation and encouragements. We are grateful to Prof M Borowitzka and Prof M D Guiry who motivated us for this compendium during the visit of senior editor to National University of Ireland, Galway. We express our thanks to Dr Raghvendar Singh, Dr Deepika Abrol and Dr A S Ahluwalia for providing useful inputs while preparing the manuscript. Acknowledgements are due to M/S Studium Press (India) for taking up the task of bringing out a book on a very useful thrust area at an appropriate time. The editors are grateful to all those who are directly or indirectly involved with the publication of this book but could not get specific mention.

Editors

PREFACE

With the population explosion and expanding urbanization, this planet is facing drastic reduction in natural resources. Proportion of population and fertile soil is shrinking, water resources are receding, level of ground water is suspending and the forest litter is vanishing. Under these circumstances, our traditional sources of energy are going to last for few years and the next generation will be posed to a dreadful future. Visualizing these facts, entire world had been looking for alternative sources of energy which might replace the fossil fuel in the ages to come. Agricultural crops like palm, soybean, sweet corn, sugarcane, maize, Jatropha have been used as bio-fuel feedstock but majority of them cultivate on fertile land with few exceptions. Food grain security and reduced agricultural land does not permit us to proceed for the cultivation of land based plant feedstock for bio-fuel production despite the fact that all feedstock convert solar energy into chemical energy and store it as oil, carbohydrate and proteins. In order to reduce or minimize the use of fertile land, scientists are looking for a better feedstock which might grow on barren land, utilize least natural resource, can capture maximum solar energy and can produce maximum bio-fuel resource within a short span of time.

Algae, a group of highly diversified, autotrophic macro and microorganisms has emerged as a potent bio-fuel producer not only because its few strains have enough amount of fatty acids but due to its eco-friendly metabolism, easy cultivation, carbon sequestration tendency, versatile growth conditions and easy harvesting. Algae can utilize variety of habitats including factory effluents, sewage water, domestic waste water etc. for its cultivation. Simultaneously it can fix Carbon dioxide much faster to produce carbohydrates, fatty acids, proteins and vitamins etc. Algae being highly diversified crop can grow in variable environmental conditions.

Keeping in view these characteristics, biotechnologists and environmentalists focused on the cultivation and harvesting of indigenous and region specific algal strains which are rich in fatty acid. For this purpose, well designed close photo-bioreactors have been developed by western countries with the view to utilize power plant generated flue gases which is not only expensive but has certain drawbacks like the formation of algal blooms and contamination. Developing countries are searching for a farmer friendly module by using open, closed or hybrid ponds, commercially viable oil extraction system, and utilization of algal flakes as poultry and animal food etc.

Present book has been framed with the intention of providing consolidated information on various facets of utilizing algae as the most potent bio-fuel feedstock. All 21 chapters have been contributed by renowned scientists from various countries who are working in this discipline and are closely associated with algal researches.

The editors are of the view that algal cultivation model should be farmer friendly, region specific and economical so that it can be adopted by the farmers for their livelihood. The global scientific community is concentrating on the development of high fatty acid algal strains which might grow in variable environmental conditions. Present compendium is a novel attempt of creating awareness among the scientists, researchers and industries about algae bio-fuel. We are confident that this compendium will be a valuable reference source for the students, faculties, researchers, industrialists, environmentalists and the scientific community or every strata. The editors express their sincere gratitude to the learned authors for their valuable contributions and inputs in this volume. We are thankful to M/S Studium Press LLC, Houston, USA for publishing this book in time.

July 4, 2010 **EDITORS**

ABOUT THE EDITORS

Prof. S.K. Bhatnagar started his career as Lecturer of Botany at Bareilly College, Bareilly in 1977, Prof. Bhatnagar established Society for Plant Research in 1988 and started an International Research Journal VEGETOS which is being indexed by all major database. Prof Bhatnagar obtained Ph. D. in 1980, founded the department of Biotechnology at Bareilly College during 2003, carried out 10 research projects, produced 21 Ph. Ds, published more than 85 research papers in national and international journal, edited 7 research books, organized 6 national level conferences, delivered keynotes, invited lectures and Chaired sessions in various conferences.

Presently, he is working as Professor and Head, Department of Cell Biology, College of Biotechnology, S.V.P. University of Agriculture & Technology, Meerut. Prof. Bhatnagar visited various countries as INSA-COSTED fellow and as visiting scientists under INDO-GREEK program. He organized a mini symposium in 11th International Phycological Conference, NUI, Galway, Ireland and convened 1st Algae Biofuel Summit at New Delhi. He is actively engaged in promoting the cultivation of algal strains as oil feedstock, algal strains as the source of animal nutrient, enhancement of milk quality in cattles by using alternate food source, biodiversity conservation, transgenesis and micropropagation etc.

Mr Atul Saxena is the founder Chief Executive Officer of Growdiesel Climate Care Council. He is a technologist and turnkey project specialist with over 20 years of experience of executing large agro based projects. He launched Growdiesel on July 2nd, 2001 which is a renewable energy exploration and production organization dedicated to the cause of environment, energy & empowerment of masses. Growdiesel has also developed a proprietory catalyst free technology for production of Biodiesel from multiple feedstocks.

Mr. Saxena is engaged in initiation of multiple projects for development technologies to create Growdiesel BioCrude™, a renewable liquid fuel, which can replace fossil/petroleum crude oil in existing refineries (without any modifications) to produce carbon-neutral transportation fuels. He developed a low cost enclosed Photobioreactor system to produce Algae for Biofuels. Mr.Saxena is currently working on a project of converting Sunlight, non-potable water, all kinds of wastes and CO_2 to sustainable carbon negative Growdiesel BioCrude™ that can be converted to high quality liquid biofuels as Hi-cetane Green Diesel and Aviation turbine fuel". for rapid adoption by farmers and entrepreneurs.

Dr Stefan Kraan graduated with M. Sc. in marine biology from the University of Groningen, The Netherlands. He moved to Ireland to start Ph. D. on phylogenetics and aquaculture of edible brown seeweed *Alaria esculenta* at the National University of Ireland, Galway in 1998. He became Manager of the Irish Seaweed Industry organization in 1998 and finished his Ph. D. in 2001. Dr Kraan established the Irish Seaweed Centre (ISC; www.irishseaweed.com) in 2001, a virtual institute for seaweed-based research and development, which was launched in 2001. He was Manager of the Irish Seaweed Centre of the Martin Ryan Institute of the National University of Ireland, Galway for 10 years and National Seaweed Research Coordinator for the Irish Marine Institute for 5 years. Currently, Dr Kraan is Co-Founder and Scientific Director of Ocean Harvest Technology a company dedicated to aquaculture research (www.oceanharvest.ie) and Ocean Fuel Ltd, a company dedicated to develop large scale biomass cultivation programs for the bioethanol industry. His main fields of interests are aquaculture of seaweeds, sustainable development of algal resources, industrial applications of seaweeds and usage of seaweeds in biotechnology and biomedicine.

CONTENTS

1

ALGAE-BASED BIOFUELS: A REVIEW OF CHALLENGES AND OPPORTUNITIES FOR DEVELOPING COUNTRIES

Food and Agriculture Organization (FAO)[1*]

Abstract

Algae Based Biofuel (ABB) pathways need to be determined on case-by-case basis. Depending on the land based cultivation systems, different ABBs and bioenergy carriers can be produced. These land based systems can be run by using combustion gas, salt water and wastewater. Conditions like solar irradiation and temperature influence ABB design and they are suitable for many land and water types. ABB's economic viability is still in its nascent phase and the algae cultivation and processing systems require a high capital input than agriculture. Biomass output is potentially high, but it is still under development. ABBs have several significant sustainability opportunities, often unique to algae which include the use of land with a low economic and ecological value, no use of fresh water, reduced emission of several GHGs and the treatment of waste streams. In developing countries like India, ABB production holds future promise like the development of new industries, creation of jobs, GDP and energy independence. Developing countries are geographically interesting for algae cultivation due to favourable climatic conditions, cheap labour and underdeveloped land. All ABB concepts require significant capital investment along with a high level of expertise pertaining to productivity, operation, maintenance as well as processing. Despite this a knowledge gap still exists for several critical

1 Food and Agriculture Organization of the United Nations (FAO), Aquatic Biofuels, Environment, Climate Change and Bioenergy Division, Viale delle Terme di Caracalla 00153 Rome, Italy www.fao.org/bioenergy/aquaticbiofuels

* *Corresponding author*: E-mail: aquaticbiofuels@fao.org

factors. Technically, ABB might consider the land based systems which include a closed system involving high capital, technology, technical skill, research and maintenance requirement than open systems. The open systems suffer huge water loss due to evaporation. It is advisable to cultivate locally occurring algal species with a renewable nutrient source. Co-production of food or feed will increase economic viability and therefore it has to be considered as a strict requirement in the short term. The sea based systems have large potential in territorial waters of developing countries which can be labour intensive. In developing countries it might create more jobs but this has a small economic impact due to low wages. However, co-production of food or feed can increase economic viability and a synergy with fish cultivation will have high potential.

Key words: Algae, Biofuel, Bioenergy

Introduction

Algae have recently received a lot of attention as a new biomass source for the production of renewable energy. Some of the main characteristics which set algae apart from other biomass sources are that algae (can) have a high biomass yield per unit of light and area, can have a high oil or starch content, do not require agricultural land, fresh water is not essential and nutrients can be supplied by wastewater and CO_2 by combustion gas. The first distinction that needs to be made is between macroalgae (or seaweed) versus microalgae. Microalgae have many different species with widely varying compositions and live as single cells or colonies without any specialization. Although this makes their cultivation easier and more controllable, their small size makes subsequent harvesting more complicated. Macroalgae are less versatile, there are far fewer options of species to cultivate and there is only one main viable technology for producing renewable energy: anaerobic digestion to produce biogas. Both groups will be considered in this report, but as there is more research, practical experience, culture and there are more fuel options from microalgae, these will have a bigger share in the report. In Chapter 2, the different technological components that make up Algae Based Biofuels (ABB) are discussed: algae cultivation technology; processing to biofuel options; locations and carbon; light and nutrient inputs. Both land based and sea based applications are discussed.

In Chapter 3, ABB sustainability is investigated in depth. First, existing biofuel sustainability standards are analysed for applicability, followed by a thorough analysis of the opportunities and risks of ABB sustainability. Secondly, sustainability is discussed in the context of potential and threats for developing countries.

Concepts for Bio-Energy from Algae

Algae have a clear potential to be used as a source for the production of renewable energy. In order to utilize algae for this purpose, the most common systems for cultivating them are investigated. Since there are many species of algae with varying characteristics, a diversity of options for the production of algae-based energy have been analysed. Subsequently, the inputs and conditions needed for growing algae are examined. The chapter is concludes with a formulation of complete concepts for producing a renewable energy carrier from algae, which will be used for in-depth analysis later in the report.

Algae Culture Systems

Culture systems are very different between macroalgae (seaweed) and microalgae. Because of their small (µm) size, microalgae have to be cultivated in a system designed for that purpose (placed on land or floating on water), while seaweed can be grown directly in the open sea. The first mention of seaweed culture dates back to 1690, in Japan (Tamura in Buck and Buchholz, 2004). Japan and China are still the main producers of cultured seaweed. Seaweed is mainly used as a food product, either eaten directly, or used in many processed foods as stabilizers or emulsifiers. Besides culturing seaweed, part of the current seaweed production comes from harvesting natural populations or collecting beach-cast seaweed. Besides the disturbance of the ecosystem by these practices, they are clearly unsustainable for application on a very large scale. Therefore growing macroalgae in a dedicated cultivation system is worth considering. For microalgae, the development of dedicated culture systems only started in the 1950s when algae were investigated as an alternative protein source for the increasing world population. Later, algae were researched for the interesting compounds they produce, to convert CO_2 to O_2 during space travel and for remediation of wastewater. The energy crisis in the 1970s initiated the research on algae as a source of renewable energy. For algae to grow, a few relatively simple conditions have to be met: light, carbon source, water, nutrients and a suitably controlled temperature. Many different culture systems that meet these requirements have been developed over the years, however, meeting these conditions for scaled systems is difficult. One important prerequisite to grow algae commercially for energy production is the need for largescale systems which can range from very simple open air systems on- or offshore which expose the algae to the environment, to highly controllable, optimized but more expensive closed systems. The necessary technology for developing profitable algae-based fuel generation is still in various states of development and the final configuration is yet to be determined and demonstrated at the industrial scale.

Land based open culture systems

The simplest open air algae cultivation systems are shallow, unstirred ponds. The sizes range from a few m2 to 250 ha (Fig. 1a, b). CO_2 is the carbon source for algae. Its dissolution from air into water limits the growth rate, making the yield per hectare relatively low. Other negative influences are the slow diffusion of nutrients and flotation and sedimentation of dead and living algae, limiting the usage of available sunlight. This can be prevented by some form of agitation, which in practice is done in circular ponds with a mechanical arm stirring in a circular motion (Centre-Pivot ponds, Fig. 1c), or more commonly in so-called raceway ponds (Fig. 1d), in which a paddle wheel (Fig. 1e) forces a circulating water flow through a long narrow pond. Blowing gas bubbles through the medium provides both agitation and (part of the required) CO_2. Air, compressed CO_2 or CO_2- containing combustion gases can be applied. The major bottlenecks of these open systems are that there is almost no possibility for temperature control (unless a source of cheap surplus heat is available) and that they are very susceptible to invasion of algal predators, parasitic algae or other algal strains that grow better at the applied conditions and therefore out-compete the desired species. Only a limited amount of species is dominant enough to maintain itself in an open system (Carlsson *et al.*, 2007; Chisti, 2007; Pulz, 2001; Rodolfi *et al.*, 2009).

Land based closed cultivation systems

Temperature, gas exchange and competition problems can be alleviated through closing an open system by covering it with transparent material or a greenhouse, but this is expensive for large surfaces. Another simple, but inexpensive example is using polyethylene bags or sleeves (Fig. 2a) for batch culture. Sizes go up to 1,000 litres, but

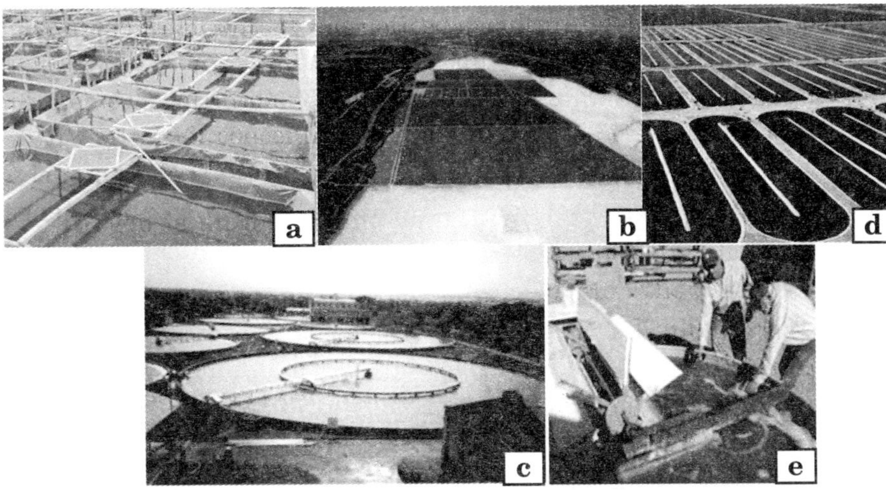

Fig. 1. Examples of open cultivation systems **a.** Small pond for *Spirulina* culture, Asia (Wikipedia, 2007); **b.** *Dunaliella salina* ponds of Cognis, Western Australia (BEAM); **c.** Centre-Pivot ponds for the culture of *Chlorella* in Taiwan (BEAM) **d.** Open raceway-type culture ponds of Earthrise in California, US(Spirulina.org.uk); **e.** Pedal wheel of a raceway pond (NMSU)

sensitivity to environmental conditions and short life expectancy make this system inappropriate for outside use. Several more advanced systems have been developed based on more durable transparent materials: glass, polyethylene and polycarbonate. These reactors offer continuous operation, a high level of controllability and elevated biomass concentrations, which results in lower space requirements and lower harvesting costs per tonne of algae. One example is the bubble column (Fig. 2b), a vertical tubular reactor. Scalability of this system is limited since, when putting several systems close to each other, they will cast a shadow on each other (Fig. 2c). Using a reactor consisting of long horizontal tubes eliminates this problem1 (tubular reactor, Fig. 2d-f). However, this has its own scaling problem: algae will consume nutrients and CO_2 while producing O_2 (which could inhibit2 algal growth at elevated concentrations), so growth conditions deteriorate further along the tube. Up-scaling can be achieved by installing individual modules with optimized size vs. tube length ratios. To make optimal use of surface area receiving solar irradiation, a flat photobioreactor can be applied (Fig. 2g). This system can potentially yield a much higher biomass concentration, but is still under developments. Difficulties are the complicated flow regime inside the reactor and scalability, although the latter has been greatly improved by a design called the green wall panel (Rodolfi *et al.*, 2009). Fig. 2h shows flat photobioreactors in a solar panel-like set-up. There are many variations and innovations on the previously described closed systems.

1. It should not be easily concluded that simply replacing vertical tubes with horizontal tubes eliminates the problem of shadowing. Overcoming the problem in both instances is a function of their spacing.
2. This will depend on the residence time of the algae in the tube.
3. Like vertical or horizontal tubes, flat-pane reactors are not exempt from the problem of shadowing when arranged in arrays.

Offshore cultivation systems

Historically, seaweed cultivation techniques have been developed based on the local circumstances, often using shallow, protected coastal areas that are safe, easily accessible and allow for easy immobilization of the culture system to the seabed. Generally the techniques used are labour intensive and therefore are restricted to regions with the lowest incomes. Harvesting natural macroalgae populations is also common practice in selected areas. For renewable energy production, macroalgae production is not based on harvesting natural populations and therefore focuses on growing seaweed species that can attach to underwater ropes or similar support structures. Furthermore, available locations not intensively used for other purposes such as shipping routes, port areas, recreation, etc. will often be in deeper seas with rougher conditions. Important examples of cultivation systems adaptable to these conditions are those based on vertical ropes, which allow the cultivated seaweed to catch all available light until the maximum light penetration depth; systems based on horizontal lines, which minimizes the amount of rope material needed per unit of area, or hybrid systems combining horizontal and vertical lines. In all cases the systems can be floating, anchored to the sea or both. Problems of damage to rope structures and washed off biomass have been reported (Chynoweth, 2002), so a cultivation system that prevents these problems needs to be designed. During experiments at sea (Buck and Buchholz, 2004), using rings (diameter of 5 m, surface of 19.6 m^2 and 80-100 m substrate rope) with ropes as abase for seaweed to attach to, gave the best results, especially under high flow or heavy weather conditions. These rings can be attached to each other and/or the seabed and can include a slow-release fertilizer. The main problem of this system is that the rings need to be harvested individually, making cost-price reduction through economy of scale more difficult. For seaweed cultivation on a large scale (1000s ha), a cultivation system that is simple, low-cost, low maintenance, and has a high light capture, productivity, resilience to climatic conditions, durability and life expectancy, while allowing easy harvesting and replantation is a great challenge and will require a lot of research and development. When selecting a location for seaweed cultivation, several considerations have to be made. Besides temperature, nutrient and light consideration (which will be mentioned later) and previously mentioned competition with other functions of the sea, distance from shore (or distance from a suitable harbour) are some of the important criteria, as they imply energy and time spent on transportation.

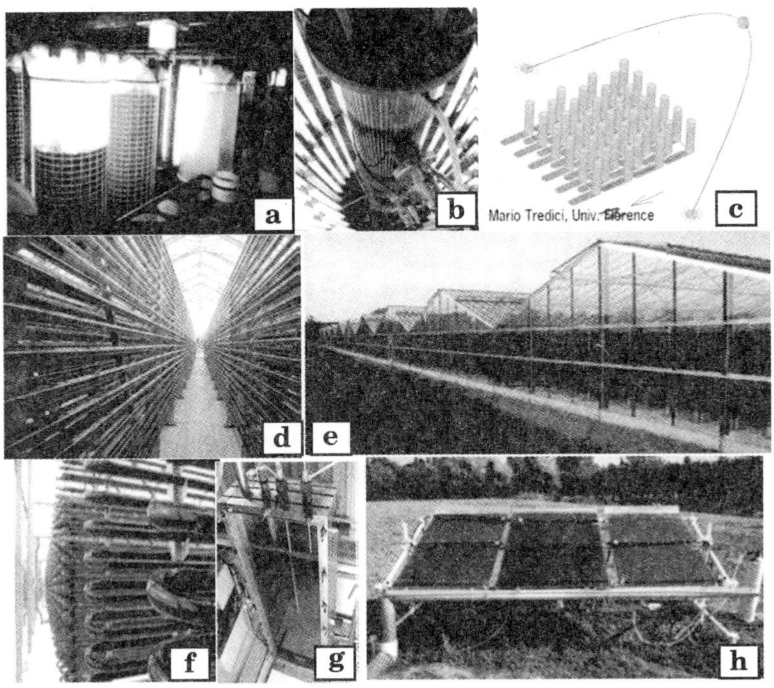

Fig. 2. Examples of closed cultivation systems **a.** 'Big Bag' culture of microalgae (BEAM); **b.** Bubble column reactor (Tredici in Wijffels, 2007) **c.** Field of bubble columns induce shading (Wijffels, 2007); **d-f.** Tubular reactor system (Bioprodukte-steinberg.de); **g.** Experimental flat photobioreactor (Wijffels, 2007); **h.** Experimental alveolar-panel photobioreactor (Tredici and Materassi, 1992)

Fresh seaweed contains around 90 percent water and thus a high amount of mass with a relatively low energy content. The water content can be reduced at the harvest location, pressure filtration will remove around 20 percent of the water, alternative options need to be investigated and energy spent on dewatering versus transport have to be optimized. Furthermore, the (necessity of) treatment of the released press water needs to be investigated (Reith *et al.*, 2005). Another criterion is the availability of existing offshore infrastructure. Offshore oil or gas platforms provide an anchoring point, boat and helicopter landing, personnel accommodation and in some cases pipes to shore that could be converted to pump seaweed or biogas to shore, or even CO_2 or nutrient rich waste water to the cultivation location. Also, offshore wind turbine parks contain a considerable amount of unused space between the turbines (distance between turbines is seven times the rotor blade length) which is restricted to maintenance vessels. Turbines provide anchoring points and have and individual boat landing, and in case seaweed is (partially) converted into electricity on-site, or if electricity is needed for a first processing step, a grid connection is available. Most considerations also apply to installations that capture wave energy, with the added benefit of milder sea conditions "behind" the wave energy devices.

Culture systems appropriate for energy production through algae

Energy is a low-value product. ("High" oil prices are tens of •cents per litre, algal biomass for health-foods and cosmetics can cost •1 000s per kilo.) This means that algaculture should be as cheap as possible in order to make conversion into energy carriers economically feasible. Current production systems of micro- and macroalgae supply knowledge and experience on cultivation, but do not supply biomass for a production price that can compete with other sources of (renewable and non-renewable) energy. Systems have to be optimized for a minimal financial and energy input, economy of scale has to be utilized to reduce production prices and additional revenue from other sources than purely energy from algae have to be obtained, through co-production or treatment of waste streams.

Within the open systems, raceway ponds provide a much better yield than more extensive systems, while keeping capital investment generally low. In scientific literature, this system is sometimes mentioned as the only open system practicable for large-scale production (Carvalho *et al.*, 2006; Chisti, 2007). An important condition however, is the availability of a microalga that can remain dominant year round and has good energy-yielding properties. In case this proves difficult or required land is costly, a horizontal tubular system has the best characteristics. Although much more expensive, it offers a much higher yield per hectare and controllability, while up-scaling is easier than with other closed systems. While a few large-scale commercial raceway examples exist, the biggest closed systems cover a few hectares. For either option, no commercial example of energy production from algae exists. In recent years, many claims have been made on possible productivities (and often oil contents) that approach or even cross the theoretical maximum. Often these are the result of extrapolating preliminary laboratory test and experiments of hours or days to dry weight or oil per hectare per year claims, often suspected to have been made to achieve financial gain. This has given rise to both euphoria over the potential of algae for renewable energy on the one hand and disappointment and distrust by the well informed on the other. This has resulted in an exponential growth of interest in the topic but also at times a negative perception surrounding commercial efforts. The fact is that no commercial algal energy producer yet exists, that high-yielding terrestrial crops will not easily be surpassed by algae in terms of energy content (but the area available for algaculture surpasses that of terrestrial crops by far), that a positive energy balance of energy output versus energy input for operation and the production of the cultivation system, and the subsequent financial returns to cover the process costs and the initial investment for the cultivation system are very challenging.

Table 1. Comparison of open versus closed land based systems

Parameter or issue	***Open ponds and raceways***	***Photobioreactors (PBR)***
Required space	High	For PBR itself low
Water loss	Very high, may also causesalt precipitation	Low
CO_2-loss	High, depending on pond depth	Low
Oxygen concentration	Usually low enough because of continuous spontaneous outgassing	Build-up in closed system requires gas exchange devices (O_2 must be removed to prevent inhibition of photosynthesis and photo oxidative damage)
Temperature	Highly variable, some control possible by pond depth	Cooling often required (by spraying water on PBR or immersing tubes in cooling baths)

Table 1. *(Contd...)*

Table 1. *(Contd...)*

Parameter or issue	***Open ponds and raceways***	***Photobioreactors (PBR)***
Shear	Usually low (gentle mixing)	Usually high (fast and turbulent flows required for good mixing, pumping through gas exchange devices)
Cleaning	No issue	Required (wall-growth and dirt reduce light intensity), but causes abrasion, limiting PBR lifetime
Contamination risk	High (limiting the numberof species that can be grown)	Low (Medium to Low)
Biomass quality	Variable	Reproducible
Biomass concentration	Low, between 0.1 and 0.5g/l	High, generally between 0.5 and 8 g/l
Production flexibility	Only few species possible, difficult to switch	High, switching possible
Process control and reproducibility	Limited (flow speed, mixing, temperature only by pond depth)	Possible within certain tolerances
Weather dependence	High (light intensity, temperature, rainfall)	Medium (light intensity, cooling required)
Start-up	6–8 weeks	2–4 weeks
Capital costs	High ~ US$100 000 per hectare	Very high ~ US$250 000 to 1000 000 per hectare (PBR plus supportingsystems)
Operating costs	Low (paddle wheel, CO_2 addition)	Higher (CO_2 addition, oxygen removal, cooling, cleaning, maintenance)
Harvesting cost	High, species dependent	Lower due to high biomass concentration and better control over species and conditions
Current commercial applications	5 000 (8 to 10 000) t of algal biomass per year	Limited to processes forhigh added value-compounds or algae usedin food and cosmetics

Source: (Pulz, 2001 adapted in Carlsson, 2007)

Algal Products

Since there are so many different algal species, algae as a group can produce a wide variety of products. This section investigates the algal energy products which are relevant from a renewable energy perspective; these will be treated in the paragraphs below.

Complete Cell Biomass

It is possible to dry algal biomass and combust it directly to produce heat and electricity, or use high-temperature high-pressure processes like pyrolysis, gasification and hydro-thermal upgrading (HTU) to produce fuel gas or fuel oil respectively. These technologies require dry biomass. Drying requires a great deal of energy, which has a strong negative effect on the energy balance and capital costs of required equipment (Wijffels, 2007) (drying with solar heat would compete for solar light with algae production). Thermochemical liquefaction is a high temperature, high-pressure treatment in which a wet biomass stream can be applied (Banerjee *et al.*, 2002; Dote *et al.*, 1994; Tsukahara and Sawayama, 2005), but this technology is still under development and is likely to require at least five years before it can be commercially applied (Meuleman, 2007). A biochemical way to process the whole biomass is anaerobic digestion. This produces biogas from the wet stream and requires much less energy input than the thermochemical options. There is 55-75 percent methane in biogas (Mes *et al.*, 2003), which can be combusted to produce heat and/or electricity, or upgraded to replace natural gas. Some experiments showed that microalgae with an intact cell wall are quite resilient against fermentation, causing the valuable energy to remain locked inside the cells. In many cases a pre-treatment step to break these cell walls will be necessary. Anaerobic digestion is a well developed, robust technology with many commercial examples This technology is applied to waste streams containing organic compounds, with a very low feedstock price (Reith, 2004). Considering the constraints of the other whole-cell treatment technologies, anaerobic digestion appears the most favourable. With this option it is relatively easy to recover the nutrients as the nutrient-free energy carrier is in the gas phase. Also, this technology can be utilized as a second step after extracting compounds with a higher value or to treat substandard batches of algae grown for other uses.

Unique products

There are many initiatives on energy generation from algae, some ideas have been around for many decades, some are currently at the pilot-stage, but so far there is no commercial implementation. Algaculture however, is performed worldwide to produce products with a higher economical value than energy. Sometimes the entire alga is the product, but often compounds are extracted which are very difficult or impossible to produce in other ways. Some examples of these so-called "unique products" include food, food-additives and health-food, feed for fish, shrimp and shellfish, colorants and omega-3-fatty acids (Molina Grima *et al.*, 2003; Reith, 2004). The prices of these unique products range considerably: cheapest is healthfood Spirulina from Myanmar for 8•/kg, food for aquatic organisms Nannochloropsis costs 500•/kg, Vitamin A precursor ß-carotene costs 1 000 •/kg and the strongest known anti-oxidant astaxanthin costs up to 10 000 •/kg (Wijffels, 2007), at the extreme end are 13C labelled fatty acids, its price per kg is US$38 000 000 (the company Spectra Stable Isotopes produces this commercially, but only about 400 g per year) (Fernandez *et al.*, 2003). Unique products are not treated further in this report, but the experience with culture systems for these algae is successfully used for modelling energy generation.

Lipids and biodiesel

Lipids are one of the main components of microalgae; depending on the species and growth conditions 2–60 percent of total cell dry matter (Wijffels, 2006), as membrane components, storage products, metabolites and storages of energy. These lipids can be used as a liquid fuel in adapted engines as Straight Vegetable Oil (SVO). Tri-glycerides and free fatty acids, a fraction of the total lipid content, can be converted into biodiesel. In comparison with SVO, algal oil is unsaturated to a larger degree making it less appropriate for direct combustion in sensitive engines.

In order to efficiently produce biodiesel from algae, strains have to be selected with a high growth rate and oil content. If an open culture system is used, the selected strain must have the ability to remain dominant under the applied conditions. Because of environmental conditions such as temperature, this means in practice that using a locally occurring strain is preferable in most cases (Sheehan *et al.*, 1998). In a closed photo-bioreactor, competition from other algae can be prevented to some extent and optimal growth conditions can be more easily maintained. Lipid accumulation in algae usually occurs during periods of environmental stress, culture under nutrient-deficient conditions is most often referred to. This implies a tradeoff; rapid growth but low lipid content under nutrient sufficient conditions, decrease or near-zero growth but lipid increase under nutrient-deficient conditions.

To the contrary, a fairly unique result by Rodolfi *et al.,* (2009) showed an almost consistent productivity and almost doubling of the lipid content to 60 percent after switching to nutrient-deficient conditions in a outdoor pilot reactor under natural light.

From all energy carriers produced from algae, biodiesel has received the most attention and is the only initiative which is on the border of pilot-scale and full-scale deployment.

Carbohydrates and ethanol

Bioethanol can be used as a biofuel which can replace part of the fossil-derived petrol. Currently bioethanol is produced by fermenting sugars, which in the case of corn are derived from hydrolyzing starch. Algae species starch contents over 50 percent have been reported. With new technologies, cellulose and hemicelluloses can be hydrolysed to sugars (Hamelinck *et al.*, 2005), creating the possibility of converting an even larger part of algal dry matter to ethanol. Algae have some beneficial characteristics compared to woody biomass, the traditional target for this technology. Most notable is the absence of lignin in algae, making its removal needed for woody material redundant. Furthermore, algae composition is generally much more uniform and consistent than biomass from terrestrial plants, because algae lack specific functional parts such as roots and leaves. Algal cell walls are largely made up of polysaccharides, which can be hydrolyzed to sugar.

Another algae-specific technology for ethanol production is being developed, in which green algae are genetically modified to produce ethanol from sunlight and CO_2 (Deng and Coleman, 1999). Ethanol production from or by algae has very interesting prospects, but is currently only in the preliminary phase of research. More development is needed to analyse a full-scale production system.

Hydrocarbons

One species of algae, *Botryococcus braunii* is well known for its ability to produce hydrocarbons which have been loosely described as equivalent to the "gas-oil fraction of crude oil." (Hillen *et al.*, 1982). Like petroleum, these hydrocarbons can be turned into gasoline, kerosene and diesel. While other algal species usually contain less than 1 percent hydrocarbons, in *B. braunii* they typically occupy 20-60 percent of its dry matter, with a reported maximum of >80 percent (Wijffels, 2006). Depending on the strain, these hydrocarbons are either C30 to C37 alkenes or C23 to C33 odd numbered alkenes (Ranga Rao and Ravishankar, 2007). These hydrocarbons are mainly accumulated on the outside of the cell, making extraction easier than when the cell wall has to be passed to reach the organics inside the cell (Wijffels, 2006). *B. braunii* lives in freshwater, but can also adapt to large range of (sea) salt concentrations. At present, the highest known salt concentration that a *Botryococcus* species can survive is 3 M NaCl, the optimum salinity being around 0.2 M NaCl (Qin, 2005) (seawater contains about 0.6 M NaCl (Dickson and Goyet, 1994)). Salinity manipulation may be used as a tool to yield algal biomass containing the desired lipid composition. Other factors affecting *B. braunii* growth and hydrocarbon production include availability of nitrogen and phosphate, light intensity and pH (Qin, 2005). *B. braunii*'s main disadvantage is that it grows very slowly: it's doubling time is 72 h (Sheehan *et al.*, 1998), and two days underlaboratory conditions (Qin, 2005). This is >20 times slower than fast-growing algae, therefore only low-investment growth systems like raceway ponds are interesting (Banerjee *et al.*, 2002). In such a system *B. braunii* would have to compete with natural occurring algae. Using saline water could give *B. braunii* a strong competitive position. There are parts in the world with brackish water or salty groundwater, which makes the land unusable for agriculture. Specifically for these regions, further investigation on large-scale *B. braunii* culture is certainly warranted, but in the scope of this research not treated further.

Hydrogen

As an energy carrier, hydrogen offers great promise as a fuel of the future, since it can be applied in mobile applications with only water as exhaust product and no NOx emissions when used in a fuel cell. One major bottleneck for the full-scale implementation of hydrogen-based technology is the absence of a large-scale sustainable production method for hydrogen. Currently, hydrogen gas is produced by the process of steam reformation of fossil fuels. Large-scale electrolysis of water is also possible, but this production method costs more electricity than can be generated from the hydrogen it yields. Biological hydrogen production is possible; several bacteria can extract hydrogen from carbohydrates in the dark, a group called purple non-sulphur bacteria can use energy from light to extract more hydrogen gas (H_2) from a wider range of substrates, while green sulphur bacteria can make H_2 from H_2S or $S_2O_3{}^{2-}$. These options are only interesting if a waste water with these compounds is available (Rupprecht *et al.*, 2006). Other algae can make hydrogen directly from sunlight and water, although only in the complete absence of oxygen. In practice, this means that hydrogen formation is only possible under conditions that either cost a great deal of energy, or prevent storing solar energy (Kapdan and Kargi, 2006), and a closed culture system is required. At the moment, it is only possible to produce a fraction of the theoretical maximum of 20 g H2/m2/d, making bulk-scale hydrogen production by algae not yet viable. For this to change in the future, more knowledge of the organisms that can produce hydrogen (only a few have been investigated) and the required conditions is necessary, as well as optimization of the biological route of solar energy to hydrogen, through genetic modification. If these improvements prove to be possible, this would constitute a profitable and renewable hydrogen production (Melis and Happe, 2001).

Conclusions

The choice for the most suitable energy carrier to be produced from algae is most clear in the case of seaweed. As visualized in Fig. 3, only the utilization of the entire biomass is an option, because conditions in the open sea cannot be controlled as they can on land, and therefore specifically stimulating the production of *e.g.* alkanes, lipids or hydrogen are not possible. In Reith *et al.* (2005), an economic analyses of treating macroalgae with anaerobic digestion, ethanol fermentation, HTU and super critical gasification resulted in the conclusion that only anaerobic digestion allowed for a feedstock price to be paid, the other options were not economically viable even if the seaweed biomass was free. Therefore seaweed is assumed to be converted through anaerobic digestion, if possible after extraction of compounds with a higher market value.

Of the algae-to-energy options reviewed in this paragraph, hydrogen production by algae is mostly far from commercial implementation. Although yield improvement options are being investigated, a breakthrough is not likely to occur in the next decade.

Using algae for ethanol production is in such an early stage that not much can be concluded yet about its strengths and weaknesses and is therefore not investigated further in this report. Nevertheless, its future development deserves attention.

The main product of *Botryococcus braunii* resembles compounds from fossil fuel, which offers exciting possibilities. A culture system used to produce *B. braunii* would cost the same to build and operate as with a fast-growing alga, but the yield would be 20 times less. There are possible scenarios where this would be acceptable, as this seems a relatively low-tech and robust option.

Algae for biodiesel is generally the favoured algae-for-energy option and has been researched the most. (Chisti, 2008a; Chiu *et al.*, 2009; Li *et al.*, 2008a; Li *et al.*, 2008b; Liu *et al.*, 2008; Meng *et al.*, 2009; Mulbry *et al.*, 2008; Rodolfi *et al.*, 2009; Sazdanoff, 2006; Sheehan *et al.*, 1998). Both open and closed land based cultivation systems appear suitable for this option. The conversion of the extracted lipids to biodiesel is relatively easy, and the product price can easily be compared with fossil fuel prices. Most commercially aimed pilot installations also chose this pathway. Since nutrient-limitation is often used as a lipid stimulation strategy, this technology requires strict nutrient input control, therefore using manure or wastewater as a nutrient source may be relatively complicated.

Algae Cultivation Inputs

Besides acquiring the product of interest, growing algae causes a flow of other inputs and outputs. Growing algae for the purpose of energy production requires analysis of these streams to be able to choose the most economical and environmentally friendly options.

CO_2 capture

As with all photosynthetic organisms, algae use CO_2 as a carbon source. No growth can occur in the absence of CO_2, and an insufficient supply of CO_2 is often the limiting factor in productivity. Based on the average chemical composition of algal biomass, approximately 1.8 tonnes of CO_2 are needed to grow 1 tonne of biomass. Natural dissolution of CO_2 from the

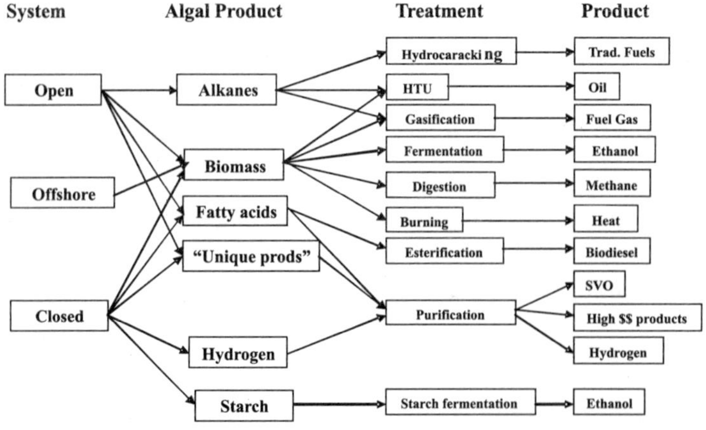

Fig. 3. Overview of algae to energy options

air into the water is not enough. This could be improved by bubbling air through the water, but, since air contains about 0.0383 percent of CO_2, all of the CO_2 in about 37 000 m^3 air is needed for 1 ton of dry algae.

Other options are using pure CO_2, which is rather expensive, or a waste source of CO_2 like flue gas from a boiler. Flue gas typically contains about 4 percent to 15 percent of CO_2 and is free or even produces revenue if a financial scheme for the prevention of greenhouse gas emissions is available. The only cost is the supply from the source to the cultivation system, which can be significant depending on the distance and water depth. CO_2 is only needed during daylight, at night it is actually produced, since algae also produce CO_2 for respiration, like all aerobic organisms. (A possible alternative is to dissolve CO_2 from flue gas to the maximum concentration at night, and add this water to the cultivation system during the day). In open systems not all of the supplied CO_2 will be absorbed, so flue gas is added in excess. In principle all available flue gas can be added to the algae pond. Part of the CO_2, NOx and SOx will dissolve, the rest will enter the atmosphere as happens conventionally.

After studying the culture of algae with flue gas, Negoro *et al.* reported that the SO_2 and NOx in the flue gas inhibited algal growth, but a few years later they reported there was barely any difference with growth on pure CO_2 (Negoro *et al.*, 1991; Negoro *et al.*, 1993). More recently, others have also confirmed that flue gas can be used to grow algae without harmful effects (Brown 1996, Hauck *et al.* 1996 and Doucha *et al.* 2005) and at least one commercial algae cultivator on Hawaii is using CO_2 from a small power plant (Pedroni *et al.*, 2001). In addition, dissolved NOx can be used by algae as a nitrogen source. The amount of flue gas needed per hectare will differ per species of algae, and will also vary throughout the day with light intensity and temperature, thus needs to be optimized for each specific application. High dissolved concentrations of CO_2 (and also SO_2) will affect the pH, thus need to be controlled or buffered. Solubility of CO_2 in a salt water system (higher pH) is higher than when sweet water is used. Also, extra CO_2 can be sequestered by growing algae that produce hard scales around theirs cells, made of calcium carbonate ($CaCO_3$) (Moheimani and Borowitzka, 2007).

Light

Algae rely on (sun) light for photosynthesis and thus growth. Light gets absorbed by the algae, so the higher the algae-concentration, the less deep light enters into the algae broth. Therefore, all algaculture systems are shallow and optimized to catch as much light as possible. Light is available in different quantities in different geological locations (see Fig. 4). Only a part (about 45 percent) of the total light spectrum is photosynthetically active radiation (PAR, ~400-700 nm), thus can be used by algae to capture CO_2, during photosynthesis, a process with a maximum efficiency of 27 percent multiplying these two factors gives the maximum theoretical conversion of light energy to chemical energy by photosynthesis: about 11 percent (Gao *et al.*, 2007). At night (or other dark conditions) photosynthesis cannot occur, so algae consume stored energy for respiration. Depending on the temperature and other conditions, up to 25 percent of the biomass produced during the day may be lost again at night (Chisti, 2007).

Algae have evolved under conditions where light is often limiting, therefore harvest as much of the available light as they can, but under good light conditions, this characteristic makes algae waste up to 60 percent of the absorbed irradiance as heat (Melis and Happe, 2001).

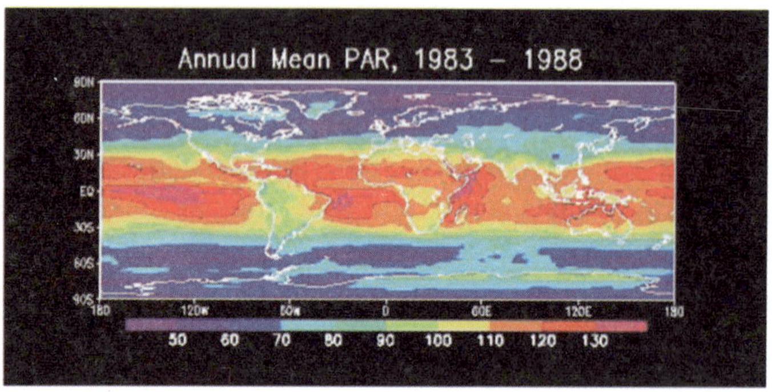

Fig. 4. Annual mean PAR levels (W/m2) (atmos.umd.edu)

Nutrient removal

Besides CO_2 and light, algae require nutrients to grow, nitrogen (N) and phosphorus (P) being the most important ones. These can be supplied in the form of agricultural fertilizer, which is simple, easily available but can be a significant cost factor (Braun and Reith, 1993; Chisti, 2008b). There are several options for cheaper sources of these nutrients. Aresta *et al.* (2005) mention wastewater effluent for example from a fishery, Olguín *et al.* (2003) describe a system where 84-96 percent of N and 72-87 percent of P was removed from the anaerobic effluent of piggery wastewater by growing algae, thereby reducing eutrophication in the environment. Another option is nutrient recycling within the process, depending on the treatment technology chosen. For example nutrient recycling after anaerobic digestion (Braun and Reith, 1993) or after gasification (Minowa and Sawayama, 1999).

The combination of nutrients and light will cause organisms capable of photosynthesis to produce chlorophyll in any part of the world. Because of it's reflection of green light, chlorophyll concentrations can be measured by satellite, which is a good indication of the best geographic locations for (unfertilized) algae growth. Fig. 5 shows a global map with chlorophyll distribution. With the help of GIS, Ecofys has plotted in white the coastal area between 0 and 25 kilometres from the coast, on locations with at least 5 mg/m3 chlorophyll. Even this very narrow strip with a high natural nutrient level account for an impressive 370 million ha worldwide (Total European union 432 million ha).

Temperature

In temperate and subtropical regions, algae have a growth season (for example, in the Netherlands from April to November); in the winter the outside temperature rops and algae only grow at a fraction of the summertime growth rate. In many industrial processes, heat is produced. Sometimes this heat is used somewhere else in the process, sold to neighbouring industries or used to heat neighbouring houses or other buildings. But often this surplus heat has no further use. This heat could be employed to optimize the temperature of the medium in which the algae grow.

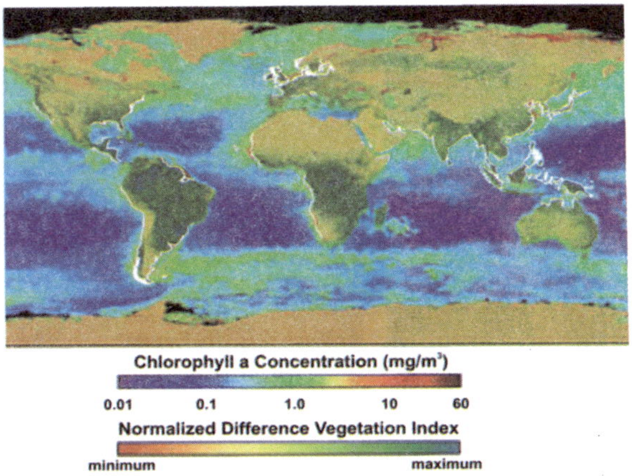

Fig. 5. Satellite data, showing worldwide photosynthesis: an indicator for fertility

Area closer than 25 km to shore and at least 5 mg/m3 chlorophyll are marked in white (total 370 ha) (Florentinus *et al.*, 2008).

Especially interesting is the combination with CO_2 of power plants, since power plants are important sources of this surplus heat. On the other hand, closed systems can get too hot and require cooling, which can be done with heat exchangers (Chisti, 2007) or spraying water on the outside (Chini Zittelli *et al.*, 1999).

Biomass Harvesting

When cultivating seaweed, every system requires a specific method of harvesting the biomass, but most commonly a specially developed harvesting vessel is used, which cuts the seaweed and hauls it inside (Reith *et al.*, 2005).

Microalgae concentrations always remain very low while growing, typically 0.02 percent to 0.05 percent dry matter in raceways and between 0.1 percent and 0.5 percent dry matter in tubular reactors (Tredici, 2009) (this means 1 tonne dry biomass has to be recovered from 200 m3 to 5 000 m^3 water). The size of algae is only a few micrometers. These two aspects make the harvesting and further concentration of algae difficult and therefore expensive. Harvesting has been claimed to contribute 20–30 percent to the total cost of producing the biomass (Grima *et al.*, 2003). In order to produce energy from algae as economically as possible, the cheapest way of concentrating the algal biomass to a water content that is low enough for oil pressing is essential.

The technically simplest option is the use of settling ponds. Once a day the settling pond is filled with a fully grown algae culture and drained at the end of that day, leaving a concentrated biomass volume at the bottom, which is stored for further processing (Benemann and Oswald, 1996). This way generally 85 percent (and up to 95 percent) of the algal biomass was found to be concentrated in the bottom of the settler (Sheehan *et al.*, 1998) at 3 percent dry matter (Sazdanoff, 2006) although this will depend on the species used. Settling pond require significant additional space (see Fig. 6).

Another way of separating algae from the water they grown in, is filtration. Many options have been described, including different materials, vacuum, pressured and rotating filtering. Some acceptable results have been obtained for colonial microalgae, but not for unicellular species (Benemann and Oswald, 1996; Grima *et al.,* 2003). Furthermore, filtration is a slow process (Sazdanoff, 2006), so a very large total capacity

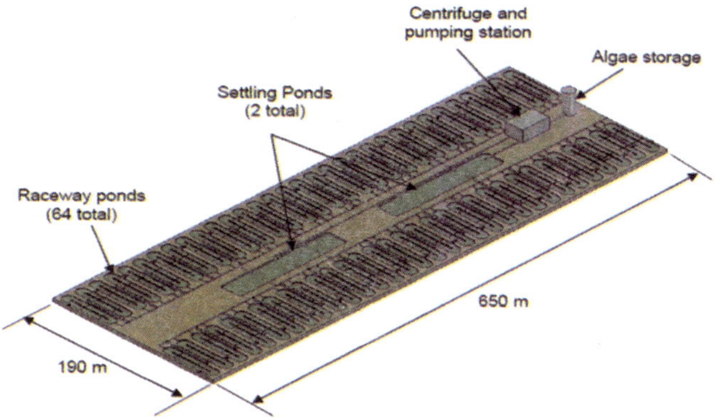

Fig. 6. Scaled model of algae farm with raceways and settling ponds (Sazdanoff, 2006)

system would be required to keep up with the production of a large algae farm. Filtration is not analysed further in this report.

Centrifugation is often used for the concentration of high-value algae, and generally considered expensive and electricity consuming. It is however the best known method of concentrating small unicellular algae (Grima *et al.*, 2003). Within the Aquatic Species Program the costs for centrifugation were estimated at 40 percent of production cost and 50 percent of investment cost. However, applying centrifugation as a secondary harvest method, to concentrate from 1-5 percent DM (dry matter) to 15-20 percent would reduce centrifugation costs at least 50 times (Benemann and Oswald, 1996). Benemann recommends in Sazdanoff's report (2006) to use centrifugation after pond settling, and a specific centrifuge (Fig. 7) with an acceptable energy consumption is mentioned.

Fig. 7. Alfa laval CH-36B GOF separator centrifuge

Other options include flocculation and killing the cells with ultrasound, which are discarded off-hand because the huge consumption of chemicals and energy respectively is very undesirable. To acquire a dry matter content of above 30%, some form of thermal technology is needed (Braun and Reith, 1993). Algae produce oils as a storage for energy. Thus, when removed from their reactor, may consume part of the stored energy for maintenance. According to Benemann and Oswald (1996) this does not cause a significant loss, as long as those conditions do not last longer than 24 h.

Sustainability of Energy from Algae

In this chapter, the sustainability of ABB will be analysed, firstly by reviewing the three main documents describing biofuel sustainability criteria: Roundtable Sustainable Biofuels, Renewable Energy Directive and the UK's Renewable Transport Fuel Obligation. Secondly, other algae specific sustainability issues not yet identified in these documents will be described and the suitability and environmental, economic and social sustainability of this technology in developing countries will be analysed.

Existing Biofuel Sustainability Standards

The discussion on the sustainability of biofuels has focused on terrestrial biomass based biofuels production, driven by the sustainability concerns arising from largescale feedstock production. The increasing interest in algae based biofuel (ABB) production may lead to a similar development that biofuel from terrestrial biomass is currently undergoing. Existing sustainability standards may be extended to properly cover ABB related sustainability issues.

In reaction to the concerns on the sustainability of biofuels, governments and organizations around the world have initiated policy developments that aim to secure the sustainable production of biofuels – respecting both environmental and social sustainability criteria. Certification schemes are envisaged to play an important role in practically demonstrating compliance with such sustainability criteria.

The first country to have such a system operational is the UK, where the Renewable Transport Fuel Obligation (RTFO) requires suppliers of biofuel to report on the sustainability and greenhouse gas performance of their biofuels. Currently, only the UK has a full operating system in place by means of the RTFO. More recently, sustainability criteria for biofuels have been developed at the level of the EU in the Renewable Energy Directive (RED). The RED requires producers to comply with a set of sustainability criteria for the biofuel to count towards the EU biofuels target. On the global level, the Roundtable on Sustainable Biofuels (RSB), a multistakeholders initiative of the UNEP in cooperation with the Ecole Polytechnique Federale de Lausanne, has developed a set of global sustainability standards for biofuels. The creation of a global certification scheme using a "meta-standard" approach is under discussion. The Algal Biomass Organization for instance is developing standards across a wide variety of topics for the algal industry.

Table 2. Sustainability issues addressed by the RSB, RED and RTFO Standard

Sustainability issues	***RSB***	***RED***	***RTFO***
GHG emissions	+	+	+
Carbon stock conservation	+	+	
Biodiversity	+	+	+
Environmental protection (soil, water and air)	+	b	+
Land rights	+	d	+
Labour conditions	+	c	+
Community relations	+	+	
Food security	+		
Competition with other non-food biomass applications	d		
Economic welfare	+		
Legal compliance	+	+	+
Indirect effects	+		

Notes:

(*a*) As a part of the GHG emissions reduction standard

(*b*) Only for biomass grown in the EU. For third countries, no standards but report obligation for economic operators, EU Member States and the European Commission – no minimum compliance requirement. Scope of reporting obligation for economic operators to be defined.

(*c*) Reporting obligation for the Commission and EU Member States on compliance with minimum requirement by countries – *e.g.* implementation of international conventions. Scope of reporting obligation for economic operators to be defined.

(*d*) Reporting obligation for the Commission and EU Member States. Scope of reporting obligation for economic operators to be defined. (Based on Dehue *et al.*, 2008a)

Table 2 presents an overview of the sustainability issues addressed by the RSB, the RED and the RTFO. Only the RED mentions algae, but only as part of a list of possible future second generation biofuel options (Dehue *et al.*, 2008b; EU, 2008; RSB, 2008). (Detailed analysis of the extent to which individual standards apply to ABB is outside of the scope of this report). In general, all of the issues addressed are to some degree relevant to algae based biofuels production. However, in the case of ABB not all issues are risks, in several points important opportunities are present, often unique to algae systems. The potential and likelihood of both risk and opportunities are often dependent on individual parts of an ABB concept, such as nutrient source, CO_2 source, location, conversion option etc. This chapter aims to give a broad overview of risks and opportunities of ABB from a sustainability point of view.

Algae Based Biofuel Specific Sustainability Issues

ABB has some unique features that can greatly reduce some of the sustainability problems faced by many terrestrial biofuel crops, for example little or no competition for agricultural land, or even positive effects, such as fertilizer production instead of consumption. However, integrating the full potential of these benefits influences other choices within an ABB concept. Choosing the most environmentally, economically and socially sustainable concept is very complex. It is dependent on many factors and will need to be done on a case-by-case basis. Important risks and opportunities to be considered are discussed below.

Land use: risks and opportunities

Direct land use changes (LUC) are caused when new areas (*e.g.* forest areas or degraded land) are taken into production to directly cover the additional feedstock demand. An indirect land-use change (ILUC) is caused when existing agricultural land is used to cover the feedstock demand of additional biofuel production. This will indirectly cause an expansion of the land use for biomass production to new areas (*e.g.* forestland and degraded land) when the previous users of the feedstock (*e.g.* food markets) do not reduce their feedstock demand. LUC and ILUC can have both positive and negative consequences on aspects such as biodiversity, carbon stocks and livelihoods (Dehue, 2008; Dehue and Meyer, 2009)

Opportunities for land based ABB production

Because algae are grown in water, the cultivation systems have much lower land quality requirements then agriculture. Soil fertility is not an issue at all. Land needs to be relatively level and sufficiently solid to build the cultivation systems on, which still allows a huge area of land to be used, such as deserts, infertile saline soils, polluted land and other land with low economic (and ecologic) value. As an example, Glenn *et al.* (1998) indicate that 43 percent of the earth's total land surface is arid or semi-arid and estimate that 15 percent of undeveloped land has sufficient access seawater (max 100 km), which amounts to 130 million ha. Furthermore, Bai *et al.* (2008) estimated that 24 percent of the earth's total land (about 3 500 million ha) has been degrading over the last 25 years. Part of the world's deserts further away from the sea may have saline ground water that could be used, or access to wastewater or sufficient fresh water. There is an enormous amount of land suitable for ABB production, that does not compete with agricultural land and avoids conversion of land with high carbon stocks. Much of the infertile land area overlaps receives high amounts of sunlight, compare the yellow land area in Fig. 5 with sunlight levels in Fig. 4.

Risks for land based ABB production

One of the possible problems of production on land is the enormous size of envisaged ABB production systems, with facilities in the order of 1,000 ha having been proposed (Wijffels, 2008). Plots of this size with low current economic and ecologic value can be scarce. Investigation of the maximum facility size and location is outside the scope of this report. Large-scale open systems can cause fog blankets and other local climatic changes. During heavy rain, closed systems built on hard surface may result in water disposal problems (BCIC, 2009) and in open systems this may lead to high nutrient, high biomass excess water. Small-scale systems are possible in the future with suitable high technology controls when considering co-production of bioenergy and higher value products. Independent of facility size, site preparation for a facility and connected infrastructure could have varying ecological impacts.

Opportunities for sea based ABB production

Considering cultivation at sea, it is well known that two thirds of the planet is covered by water. Clearly, not all this surface is equally suitable for seaweed culture.

Seaweed cultures will provide a safe place for young fish to grow up, which can contribute to the recovery of fish populations (Florentinus *et al.*, 2008; Reith *et al.*, 2005).

Synergy with offshore infrastructure

Ecofys work estimated the potential area for offshore wind turbines to be 550 million ha (considering water depths, shipping routes etc.) (Hoogwijk, 2004). The space between wind turbines could be used for seaweed culture, as it is off limits for shipping and supplies anchoring point and other possible advantages. Gas and oil platforms provide anchoring points, housing and existing fuel transport facilities. The area "behind" wave energy facilities will have milder sea conditions (Florentinus *et al.*, 2008).

Near shore high nutrient areas

Transportation costs will be of major influence in offshore concepts, but even only considering sea surface close to shore (which will contain a high nutrient concentration due to runoff of fertilizers and other anthropogenic and natural sources, as well as natural upwelling of nutrients from deeper sea layers) generates a potential cultivation area of about 370 million ha considering only 0–25 km distance to shore and at least 5 mg/m3 of chlorophyll a, an indicator of elevated nutrient - and other growth conditions. See Fig. 5 on page 24. (For several reasons this amount of chlorophyll does not indicates that there is sufficient nutrients for the required productivity [Tredici, 2009]).

Open ocean biological deserts

The dark blue/purple areas in Fig. 5 represent areas which are virtually devoid of nutrients, so called biological deserts. In these areas there is no competition with any economic activity and an ecosystem is practically non-existent. The potential area is over 5 billion ha (worldwide area of agricultural land 5.6 billion ha), but fertilization will need to be applied, however there is an option based on artificial upwelling (described in section 3.2.3) that has a high sustainability potential. Fig. 5 also shows that a strip on the equator with a higher nutrient concentration is separating the biological deserts. This is caused by natural upwelling of nutrients due to the earth's rotation and the zone has a relatively constant gulf stream and no significant seasonal changes, all good characteristics for algae cultivation.

Risks for sea based ABB production

Much of the spatial area at sea has existing uses, for example for nature, shipping, fishery, recreation, military training grounds and offshore infrastructure (Florentinus *et al.*, 2008). Finding cultivation space that has not been claimed for other uses may prove difficult. Conflicts may arise from changing existing use to seaweed production. Another important aspect to be mentioned is the potential damage to ecological or species balance in the ocean, when algae is mass cultivated in parts of the ocean where they were previously not present in large populations, along with the potential decline in oxygen concentration in the water at night which might kill other species.

Further, the design of sea-based cultivation systems has to guarantee the prevention of sea mammals and other wildlife getting entangled in it (Florentinus *et al.*, 2008; Reith *et al.*, 2005).

Greenhouse gases: risks and opportunities

CO_2: opportunities

CO_2 is the carbon source for photosynthesis, for algae and plants alike. The amount of CO_2 captured by algae varies per species, but in general about 1.8 tonne CO_2 is integrated in 1 tonne algal biomass (Chisti, 2007). This way, all carbon in ABB is directly sourced from Greenhouse Gas (GHG) CO_2 through algal biochemical conversion (other carbon sources are possible). The same is true for plant derived fuel. Algae have a few advantages over plant based CO_2 capture.

- Algae can grow directly on combustion gas (typically containing 4-15 percent CO_2), whereas plants take up CO_2 from the atmosphere (open air concentration 0.036 percent)
- Algae can contain large concentrations of the desired product, where as in plants this is often concentrated in the seeds, as in corn, soy and rapeseed.
- Algae are much more uniform than higher plants. Algae lack roots and leaves, microalgae are in many cases unicellular thus completely uniform, while seaweed may have very basic structural differentiation but all without hard, cellulosic material, therefore easily pumpable. This way the entire biomass can be processed, rather than just parts like seeds or roots.
- Algae growth is dependent on climatic conditions, and thus on the seasons, but does not have the annual growth cycle of sprouting in the spring and harvest in autumn.

CO_2 Risks

Overall algae have potentially higher annual yields and conversion to biofuels, thus more CO_2 capture per unit of area. CO_2 can originate directly from burning fossil fuel or ambient air, either way CO_2 captured in algal biomass is (temporarily) removed from the atmosphere. Ambient CO_2 is of such a low concentration that growth is limited. Extra CO_2 from combustion gas is cheap and desirable. However, gas cleaning is often necessary to remove components which are toxic to algae, the CO_2 (-containing gas) has to be transported and dissolved into the water phase, which will lower the pH, so energy input, chemicals, control equipment and gas cleaning facilities may be needed. Although the potential for CO_2 capture is huge, achieving an overall positive CO_2 balance is very challenging. Further decision criteria on whether or not to use combustion gas is outside the scope of this report. At sea, introducing CO_2 would lower the pH, which is undesirable unless controlled. However, if algae use the CO_2, pH rises again. The infrastructure needed for this would be very costly and energy intensive.

Further risks may arise if algae cultures establishment causes direct and indirect land use changes in high carbon stock areas.

Other greenhouse gases: opportunities

Besides CO_2, combustion gas contains other gases like NOx and often SO_2. Both have indirect greenhouse effects. Algae cultivation using these pollutants as nutrients, with high removal efficiencies, has been reported in literature. GHGs N_2O and CH_4 are emitted both naturally as well as induced by human activity. They are produced by micro-organisms and their formation requires an anaerobic or anoxic step. During daylight algae produce oxygen, preventing these steps from taking place, in effect reducing the emission of these gasses, both in natural systems, the sea as in manmade systems and wastewater treatment plants.

Other greenhouse gases risks

ABB has waste streams, GHG emission of these streams need to be prevented. At night algae consume oxygen, this may cause anaerobic conditions that can lead to the emission of N_2O and CH_4 gas, as explained in the previous section. Also, when cultivation conditions become toxic, mass algae death can occur, which in turn can result in GHG emissions.

Nutrients: risks and opportunities

Opportunities for land based ABB production

The amount and kind of nutrients needed for algal growth depends very heavily on the species, but as an indication for the most important nutrients, about 5 percent (typically is 7-8 percent) of algal dry matter consists of nitrogen (N) and 1 percent of phosphorus (P).

The most convenient form to supply nutrients is through chemical fertilizer, but from sustainability point of view this is undesirable since nitrate production has a very high energy input and phosphorus is mined, thus a fossil resource that will be depleted.

A second option commonly used in agriculture is manure. This source is also applicable to algae. Wastewater is often too diluted and polluted with pathogens to be used in agriculture, but application in algaculture is certainly possible. Algae cultivation systems are watery systems that produce oxygen (O_2) during daylight, therefore there is a lower chance of ammonia (NH_3), nitrous oxide (N_2O) and methane (CH_4) emissions than with land application. Per tonne of manure, algae systems are likely to have lower over GHG emissions than agriculture. However, the fact that not all components in manure and wastewater are easily biodegradable, can contain heavy metals and other micropollutants such as medicines and introduces a high concentration of bacteria and other organisms that may compete with target algae, using manure or wastewater as a nutrient source limits the type of cultivation system and algal species used significantly.

Another unique nutrient source is nitrogen oxides (NOx) and to a lesser extent sulphur dioxide (SO_2, has a large pH effect) from combustion gas. The gasses dissolve into the water phase (together with CO_2), where they are used for algae as a nutrient source.

Another option with significant sustainability benefits is the production of organic fertilizer. When applied in agriculture, the nutrients are released slowly which both benefits plant growth and reduces GHG emissions (Mulbry *et al.*, 2008). Furthermore, it offsets fertilizer use.

Risks of land based ABB production

When algae are grown for their oil or starch content, the leftover biomass will contain all the nutrients. It is very important that this is treated/ disposed of in a sustainable way. If not, nutrients could cause eutrophication of surface water, change the local ecosystem, leach to the groundwater, or get converted into GHGs by micro-organisms.

Ideally, leftover biomass can be treated as a co-product rather then a waste product, for example as protein rich feed for cattle, fish or even humans. This would reduce feed production needs with a positive effect on feedstocks and/or land availability for food production. Common wastewater treatment techniques can be applied, but these results in extra costs, loss of nutrients and possible increased GHG emissions. One more viable option is nutrient recycling. A common option is anaerobic digestion of leftover biomass. Not only will the nutrients be mostly concentrated in the liquid phase of the digestate, biogas is produced as well.

Opportunities of sea based ABB production

Seaweed cultivation relies on nutrients already present in the water, be it from natural sources or fertilizer runoff, wastewater outlets or other human activities. In fact, seaweed cultivation can be used to help remediate the anthropogenic flow of nutrients into the sea, which can cause serious eutrophication problems. Combining algaculture with fish cultivation allows for (partial) capture of the nutrients released to the environment though fish feed and faeces.

A more exotic algae concept has strong sustainability potential considering nutrients. When nutrients reach deeper sea layers, there is not much mixing and therefore these nutrients are basically removed from the nutrient cycle (unless brought to the surface again by natural upwelling). In the case of nitrogen, this loss is no significant problem because the atmosphere mainly consists of nitrogen gas. Phosphate however is produced in mines, which face depletion. As mentioned earlier, algae have the potential to capture nutrients and to be used as fertilizer. One concept involves artificially upwelling deep nutrient rich ocean water in biological desert areas, the nutrients are captured by seaweed, processed and serve as a new, renewable source of nutrients, of which phosphorus is the most interesting, since currently no large-scale alternatives are available.

Risks of sea based ABB production

For seaweed cultivation, fertilization should generally be avoided, as the nutrients will not be taken up completely by the algae, but will diffuse and thus pollute outside the cultivation area.

If the distance to shore is too big, seaweed can be processed offshore. The nutrients contained in the seaweed need to be disposed of properly, which can be a significant problem or costly offshore. In some cases, nutrients can be released back to the sea, but doing this widely and in low concentrations is difficult.

Artificially upwelling deep nutrient rich ocean water in biological desert areas could again potentially damage the ecological or species equilibrium.

Water consumption: risks and opportunities

Opportunities

One major advantage of algae cultivation over fuel crops and agriculture in general, is the ability of many algae to tolerate saline water. Fresh water is scarce in many parts of the world, and agriculture uses 70% of the total global fresh water consumption. This characteristic of algae allows a much greater area to be potentially utilized for algaculture. Besides seawater, other sources of salt water (usually with a lower salt concentration) are saline groundwater/aquifers and certain wastewaters.

Risks

Since seaweed is cultivated directly in seawater, water consumption is not a concern. However, in land-based cultivation systems, algae are harvested and separated from the watery medium in which they grew. This rest water will contain some organic components due to imperfect separation and leaked algal cell content. Ideally, this water is (at least partially) recycled, which may require additional treatment and energy input. Possible problems are the accumulation of unwanted compounds and proliferation of micro-organisms feeding on the organics in the recycled water. Non-recycled water should be disposed of properly, which can be more difficult due to the high salt content.

Land based open cultivation systems will suffer from significant water loss due to evaporation. In this process, water is removed but the salts are left behind. Replenishing this water loss with saline water will lead to accumulation of salt in the cultivation system, which sooner or later will reduce algal growth. Evaporative losses therefore have to be replenished with low salt water, even in systems with a high salt concentration. Alternative options are batch culture without water recycling, or in case the saline water supply has a lower salt concentration then the cultivation system, replacing part of the cultivation medium to dilute the salt concentration (blowdown, (Neenan *et al.*, 1986)).

Genetically modified organisms (GMOs): opportunities and risks

Opportunities

While many algal species show different interesting characteristics for biofuel production, there is still plenty of potential for modern biotechnology tools to increase their effectiveness. Two possible improvements have already received significant attention in research.

1. Reduction of antenna size. Under natural conditions, algae have to compete for light with surrounding photosynthetic organisms. Therefore the so-called antenna part of chlorophyll which receive light has evolved to catch more light than can be photosynthetically processed under optimal lighting conditions, wasting up to 60 percent of the received light energy (Melis and Happe, 2001). Scientists have been able to reduce the chlorophyll antenna size, resulting in more efficient use of high intensity light (Mitra and Melis, 2008).
2. Triggering of lipid production. As mentioned before, certain algae species accumulate large amounts of intracellular lipids, usually after being exposed to some form of stress. The exact biochemical pathway has received significant attention, in order to find out what triggers this accumulation. Genetic tools could exploit this trigger to modify algal species to produce more lipids and throughout the whole growth cycle (Sheehan *et al.*, 1998). No significant progress has been reported.

Risks

The application of genetically modified organisms (GMO) has received considerable protest and is subject to strict safety precautions. While there is in theory a large potential for advancement through genetic modification, large-scale application of genetically modified algae does not appear viable, because besides the general arguments against the use of GMOs, safety requirements will have a detrimental impact on both the economic and energetic output due to measures like sterilization of the entire cultivation system, and measures to prevent introduction of the modified species to the environment.

Sustainability Standards Specific to Developing Countries

Generally, few sustainability issues specific to developing countries are unique to ABB, most issues similar to terrestrial biofuel issues. As can be seen in Fig. 4, many developing countries are in the zones with the highest annual solar irradiances, so with the highest biomass production potential.

Opportunities

Development of ABB production can generate employment. A certain level of technical education is required during project deployment and, strategy and dealing with unforeseen circumstances, for operation and maintenance and biomass processing low-skilled workers can be employed. Independency of (foreign) energy can be achieved, as well as energy availability and access for the poor. This can be an important factor to allow economic development (FAO, 2008). Both income and other products such as food, feed, fertilizer, and base chemicals, can be generated, first for self-sufficiency, on the longer term as new export products as well.

A few sustainability issues of plant based biofuels have a high impact in developing countries. Important examples are food security and loss of agrobiodiversity (Rossi and Lambrou, 2008). ABB can avoid these issues by using non-arable land or sea, and even add to food security through co-production of (often highly nutritional and essential nutrient containing) food.

During the preparation of this report, an elaborate scan was made, in order to identify and describe current ABB projects in developing countries. With a few exceptions, it proved very difficult to find examples, and the ones that were found were almost all in the idea or very early start-up phase. This in contrast to developed countries, that may have suboptimal conditions for ABB, but tens if not hundreds of initiatives. Algae technologies have the potential to supply fuel, food and feed in developing countries, but there appears a lack of stimulus for the development of these concepts.

Risks

The biggest threat of the algae concept in developing countries stems from the fact that the scale envisaged for mass production is in the order of magnitude of 1,000 ha. There are obviously very few places on land suitable for algaculture at this scale, which are completely devoid of inhabitants and their economic activities, so great care has to be taken to prevent forced displacement of weaker social groups (small farmers and fishermen in particular).

Large-scale facilities require land, capital and technology, which small farmers traditionally have limited access to, making them less likely to adopt ABB technology. The risk of exclusion particularly high for women and female-headed households, due to persistent gender-based inequalities in most developing countries (Rossi and Lambrou, 2008). The high capital investments required and the high production potential in developing countries is likely to attract foreign investments, which provide economic stimulus, but will also lead to export of part of the revenues.

Developing countries typically cope with pressure of growing agriculture and urbanization, resulting in deforestation, land and water degradation and mechanisms preventing this are developing slowly. Care should be taken that ABB contributes to these issues as little as possible (FAO, 2008). Labour rights and environmental protection policies are generally enforced less strict in developing countries, so as with any industry, care should be taken with regard to these issues. The same could be considered for quality control for fuel and food.

Knowledge Gaps

Algae technology has enormous potential, not only for ABB, but also for food, feed, renewable chemicals and many other products that are critical for a more sustainable society. More research however, is relatively young and disperse and most importantly on a very small scale. To prove the viability of algae concepts, more information is needed on the economics of the process: optimized costs of the different inputs, but also the market value and market size of the outputs, not only fuels but also higher-value compounds. Especially in developing countries, if economical viability and robustness can be proven, many projects can be deployed rapidly, through microcredits or similar measures.

Not only economic sustainability is vital, but also environmental safety. Important tools to quantify this are energy and GHG balances. As there are many different concepts, generalized balances are difficult, but currently many factors required for a complete and consistent balance are unknown.

For both the economics and environmental balances, it is critical to have dependable and reproducible, long-term figures on productivity per unit area, as well as productivity of the desired compounds as a percentage of the total biomass. Much of the currently available data is from a very small scale, extrapolated from short-term experiments under optimal conditions, or even tainted by commercial and marketing benefits of high numbers. Also, it is often not clear if the total biomass is the measured unit, or only the organic fraction. This has a big effect, especially on algae and seaweed grown in seawater: some species contain over 40 percent of inorganic salts. In the case of biodiesel, the total lipid content is usually measured and reported, not the part that is extractable and usable for biodiesel. Other environmental effects that need to be investigated: effects on the local ecosystem of mass cultivation of productive algal species. Especially at sea, but also on land it has to be shown that this is a controllable risk. Methane and nitrous oxide emissions from different systems may have significant environmental impact, but may also be easily avoided, to the extend of our knowledge, no research has been done on this.

Many descriptions of algae project, mention the generation of carbon credits. Although algae can be used to adsorb CO_2 from combustion gasses, this CO_2 is not sequestered, as it becomes available again during the use of the algal end-product. The GHG reduction of algae concepts is the displacement of less sustainable alternatives, such as fossil fuels, chemical fertilizers, but also any other end product that may be displace by an algae-based alternative. In order to generate carbon credits from this, a baseline methodology is required, to determine how to calculate the exact amount of GHG emission prevented. To the extent of our knowledge, no such baseline methodology exists, nor is under development, and furthermore it is not possible to accurately estimate the cost of CO_2 abatement through algae.

Acknowledgements

In October 2008, the FAO Interdepartmental Working Group on Bioenergy recommended the preparation of this review paper on algae-based biofuels as a reference platform for possible future work of the Organization. Development of the review paper was coordinated by Alessandro Flammini (GBEP Secretariat) under the overall guidance of Mr Jeff Tschirley (Chair, FAO Bioenergy Working Group). The lead author of this paper is Sjors van Iersel (Ecofys) with supporting inputs from Liliana Gamba, Andrea Rossi, Sacha Alberici, Bart Dehue, Jasper van de Staaij and Alessandro Flammini.

We would like to acknowledge the valuable contributions of Prof. Mario Tredici (University of Florence), Lew Fulton (IEA), Soufien Taamallah (IEA), Joel Cuello (University of Arizona), Jim Sears (Algal

Biomass Organization), Mohammad Hasan (FAO, Fisheries Department) and other FAO colleagues, who provided inputs and support in the preparation of this paper. The information contained does not necessarily reflect the official views of the Organization. This review paper provides a joint contribution to the programme of work of the FAO Interdepartmental Working Group on Bioenergy and the Global Bioenergy Partnership (GBEP).

References

Aresta, M., Dibenedetto, A. and Barberio, G. (2005). Utilization of macro-algae for enhanced CO_2 fixation and biofuels production: Development of a computing software for an LCA study. *Fuel Process. Technol.* **86(14-15)**: 1679-1693. Atmos.umd.edu. Photosynthetically Active Radiation (PAR) as Observed from Satellites Retrieved 17-05-2007, from http://www.atmos.umd.edu/~srb/par/Figure01.htm.

Bai, Z.G., Dent, D.L., Olsson, L. and Schaepman, M.E. (2008). Global assessment of land degradation and improvement: Identification by remote sensing. GLADA report; 5. Rome/Wageningen, FAO/ISRIC.

Banerjee, A., Sharma, R., Chisti, Y. and Banerjee, U.C. (2002). *Botryococcus braunii*: A renewable source of hydrocarbons and other chemicals. *Crit. Rev. Biotechnol.* **22(3)**: 245-279.

BCIC (2009). Microalgae Technologies & Processes for Biofuels – Bioenergy Production in British Columbia. BEAM. Biotechnological and Environmental Applications of Microalgae. Retrieved 17-05-2007, from http://www.scieng.murdoch.edu.au/centres/algae/BEAMNet/BEAM-Net.htm.

Benemann, J.R. and Oswald, W.J. (1996). Systems and economic analysis of microalgae ponds for conversion of CO_2 to biomass. Final report: Size: p. 214. Bioprodukte-steinberg.de.Retrieved 17-05-2007, from http://www.bioproduktesteinberg.de.

Braun, A. and Reith, J. (1993). Algen in de Nederlandse energiehuishouding, in opdracht van het programma, Energiewinning uit Afval en Biomassa (EWAB) van Novem. Utrecht, Braun Consultants.

Buck, B.H. and Buchholz, C.M. (2004). The offshore-ring: A new system design for the open ocean aquaculture of macroalgae. *Journal of Applied Phycology* **16(5)**: 355-368.

Carlsson, A., Beilen van, J., Möller, R., Clayton, D. and Bowles, D.E. (2007). Microand macroalgae - utility for industrial applications Bioproducts, E.R. t. E.P. o. S. R. and Crops, f. N.-f., CNAP, University of York: 86.

Carvalho, A.P., Meireles, L.A. and Malcata, F.X. (2006). Microalgal reactors: A review of enclosed system designs and performances. *Biotechnol. Prog.* **22(6)**: 1490-1506.

Chini Zittelli, G., Lavista, F., Bastianini, A., Rodolfi, L., Vincenzini, M. and Tredici, M.R. (1999). Production of eicosapentaenoic acid by Nannochloropsis sp cultures in outdoor tubular photobioreactors. *J. Biotechnol.* **70(1-3)**: 299-312.

Chisti, Y. (2007). Biodiesel from microalgae. *Biotechnol. Adv.* **25(3)**: 294-306.

Chisti, Y. (2008a). Biodiesel from microalgae beats bioethanol. *Trends Biotechnol.* **26(3)**: 126-131.

Chisti, Y. (2008b). Response to Reijnders: Do biofuels from microalgae beat biofuels from terrestrial plants? *Trends Biotechnol.* **26(7)**: 351-352.

Chiu, S.Y., Kao, C.Y., Tsai, M.T., Ong, S.C., Chen, C.H. and Lin, C.S. (2009). Lipid accumulation and CO_2 utilization of *Nannochloropsis oculata* in response to CO_2 aeration. *Bioresour. Technol.* **100(2)**: 833-838.

Chynoweth, D.P. (2002). Review of Biomethane from Marine Biomass. Gainesville, Florida, USA, Department of Agricultural and Biological Engineering, University of Florida.

Dehue, B., Gamba, L. and Hettinga, W. (2008a). Benchmark of Colombian Sustainability Standard for Biofuels. Utrecht, Ecofys.

Dehue, B. (2008). Practical options to prevent indirect LUC in the RTFO. Draft 1.0. Ecofys. Utrecht, The Netherlands, Ecofys.

Dehue, B., Hamelinck, C., Reece, G., de Lint, S., Archer, R. and Garcia, E. (2008b). Sustainability Reporting Within the RTFO: Framework Report. Ecofys. Utrecht, The Netherlands, Ecofys.

Dehue, B. and Meyer, S. (2009). Practical definition and methodology to identify areas for responsible energy crop cultivation. Working version. Ecofys. Utrecht, The Netherlands, Ecofys.

Deng, M.D. and Coleman, J.R. (1999). Ethanol synthesis by genetic engineering in cyanobacteria. *Appl. Environ. Microbiol.* **65(2)**: 523-528.

Dickson, A.G. and Goyet, C. (1994). Handbook of methods for the analysis of the various parameters of the carbon dioxide system in sea water. Version 2: Size: p. 197.

Dote, Y., Sawayama, S., Inoue, S., Minowa, T. and Yokoyama, S. (1994). Recovery of liquid fuel from hydrocarbon-Rich Microalgae by thermochemical Liquefaction. *Fuel* **73(12)**: 1855-1857.

EU (2008). Provisional Text of the European Directive on the Promotion of the Use of Energy from Renewable Sources. Brussels.

FAO (2008). The State of Food and Agriculture - Biofuels - Prospects, Risks and Opportunities. Rome, Italy, FAO.

Fernandez, F.G.A., Alias, C.B., Lopez, M., Sevilla, J.M.F., Gonzalez, M.J.I., Gomez, R.N. and Grima, E.M. (2003). Assessment of the production of C-13 labeled compounds from phototrophic microalgae at laboratory scale. *Biomol. Eng.* **20(4-6)**: 149-162.

Florentinus, A., Hamelinck, C., Lint, S.D. and Iersel, S.V. (2008). Worldwide Potential of Aquatic Biomass. Utrecht, Ecofys.

Gao, K.S., Wu, Y.P., Li, G., Wu, H.Y., Villafane, V.E. and Helbling, E.W. (2007). Solar UV radiation drives CO_2 fixation in marine phytoplankton: A double edged sword." *Plant Physiol.* **144(1)**: 54-59.

Glenn, E.P., Brown, J.J. and O'Leary, J.W. (1998). Irrigating crops with seawater. *Scientific American* **279(2)**: 76-81.

Grima, E.M., Belarbi, E.H., Fernandez, F.G.A., Medina, A.R. and Chisti, Y. (2003). Recovery of microalgal biomass and metabolites: process options and economics. *Biotechnol. Adv.* **20(7-8)**: 491-515.

Hamelinck, C.N., van Hooijdonk, G. and Faaij, A.P.C. (2005). Ethanol from lignocellulosic biomass: techno-economic performance in short, middle- and long-term. Biomass Bioenerg. **28(4)**: 384-410.

Hillen, L.W., Pollard, G., Wake, L.V. and White, N. (1982). Hydrocracking of the oils of *Botryococcus-Braunii* to Transport Fuels. *Biotechnol. Bioeng.* **24(1)**: 193-205.

Hoogwijk, M. (2004). On the global and regional potential of renewable energy sources. Faculteit Scheikunde. Utrecht, Universiteit Utrecht. PhD Thesis.

Kapdan, I.K. and Kargi, F. (2006). Bio-hydrogen production from waste materials. *Enzyme Microb. Technol.* **38(5)**: 569-582.

Li, Y., Horsman, M., Wu, N., Lan, C.Q. and Dubois-Calero, N. (2008a). Biofuels from microalgae. *Biotechnol. Prog.* **24(4)**: 815-820.

Li, Y.Q., Horsman, M., Wang, B., Wu, N. and Lan, C.Q. (2008b). Effects of nitrogen sources on cell growth and lipid accumulation of green alga *Neochloris oleoabundans*. *Appl. Microbiol. Biotechnol.* **81(4)**: 629-636.

Liu, Z.Y., Wang, G.C. and Zhou, B.C. (2008). Effect of iron on growth and lipid accumulation in *Chlorella vulgaris*. *Bioresour. Technol.* **99(11)**: 4717-4722.

Melis, A. and Happe, T. (2001). Hydrogen production. Green algae as a source of energy. *Plant Physiol.* **127(3)**: 740-748.

Meng, X., Yang, J. M., Xu, X., Zhang, L., Nie, Q. J. and Xian, M. (2009). Biodiesel production from oleaginous microorganisms. *Renew. Energy* **34(1)**: 1-5.

Mes, T.Z.D.D., Stams, A.J.M., Reith, J.H. and Zeeman, G. (2003). Methane production by anaerobic digestion of wastewater and solid wastes. Biomethane & bio-hydrogen: status and perspectives of biological methane and hydrogen production. Reith, J.H., Wijffels, R.H. and Barten, H. Petten, Dutch Biological Hydrogen Foundation. pp. 58-102.

Meuleman, B. (2007). Personal communication on biomass conversion technology.

Minowa, T. and Sawayama, S. (1999). A novel microalgal system for energy production with nitrogen cycling. *Fuel* **78**: 1213.

Mitra, M. and Melis, A. (2008). Optical properties of microalgae for enhanced biofuels production. *Opt. Express* **16(26)**: 21807-21820.

Moheimani, N.R. and Borowitzka, M.A. (2007). Limits to productivity of the alga *Pleurochrysis carterae* (Haptophyta) grown in outdoor raceway ponds. *Biotechnol. Bioeng.* **96(1)**: 27-36.

Molina Grima, E.M., Belarbi, E.H., Fernandez, F.G.A., Medina, A.R. and Chisti, Y. (2003). Recovery of microalgal biomass and metabolites: process options and economics. *Biotechnol. Adv.* **20(7-8)**: 491-515.

Mulbry, W., Kondrad, S. and Buyer, J. (2008). Treatment of dairy and swine manure effluents using freshwater algae: fatty acid content and composition of algal biomass at different manure loading rates. *Journal of Applied Phycology* **20(6)**: 1079-1085.

Neenan, B., Feinberg, D., Hill, A., McIntosh, R. and Terry, K. (1986). Fuels from microalgae: Technology status, potential, and research requirements: Size: Pages: 171.

Negoro, M., Shioji, N., Miyamoto, K. and Miura, Y. (1991). Growth of Microalgae in High CO_2 Gas and Effects of Sox and Nox. *Appl. Biochem. Biotechnol.* **28-9**: 877-886.

Negoro, M., Hamasaki, A., Ikuta, Y., Makita, T., Hirayama, K. and Suzuki, S. (1993). Carbon-Dioxide Fixation by Microalgae Photosynthesis Using Actual Flue-Gas Discharged from a Boiler. *Appl. Biochem. Biotechnol.* **39**: 643-653.

NMSU. Scientists research best ways algae can fuel biodiesel production. Retrieved 17-05-2007, from http://www.nmsu.edu/~ucomm/Releases/2007/january/algae_biodiesel.htm.

Olguín, E.J., Galicia, S., Mercado, G. and Pérez, T. (2003). Annual productivity of *Spirulina* (Arthrospira) and nutrient removal in a pig wastewater recycling process under tropical conditions. *Journal of Applied Phycology* **15(2)**: 249.

Pedroni, P., Davidson, J., Beckert, H., Bergman, P. and Benemann, J. (2001). A Proposal to Establish an International Network on Biofixation of CO_2 and Greenhouse Gas Abatement with Microalgae.

Pulz, O. (2001). Photobioreactors: production systems for phototrophic microorganisms. *Appl. Microbiol. Biotechnol.* **57(3)**: 287-293.

Qin, J. (2005). Bio-hydrocarbons from algae: impacts of temperature, light and salinity on algae growth. Barton, Rirdc.

Ranga Rao, A. and Ravishankar, G.A. (2007). Influence of CO_2 on growth and hydrocarbon production in *Botryococcus braunii*. *J. Microbiol. Biotechnol.* **17(3)**: 414-419.

Reith, J.H. (2004). Duurzame co-productie van fijnchemicaliën en energie uit microalgen: openbaar eindrapport E.E.T. project K99005/398510-1010. Petten, Energieonderzoek Centrum Nederland.

Reith, J.H., Deurwaarder, E.P., Hemmes, K., Curvers, A.P.W.M., Kamermans, P., Brandenburg, W. and Zeeman, G. (2005). Bio-offshore: grootschalige teelt van zeewieren in combinatie met offshore windparken in de Noordzee. Petten [etc.], Energieonderzoek Centrum Nederland [etc.].

Rodolfi, L., Zittelli, G.C., Bassi, N., Padovani, G., Biondi, N., Bonini, G. and Tredici, M.R. (2009). Microalgae for oil: Strain selection, induction of lipid synthesis and outdoor mass cultivation in a low-cost photobioreactor. *Biotechnol. Bioeng.* **102(1)**: 100-112.

Rossi, A. and Lambrou, Y. (2009). Making Sustainable Biofuels Work for Smallholder Farmers and Rural Households - Issues and Perspectives, Food and Agriculture Organization of the UN (FAO).

RSB (2008). Global principles and criteria for sustainable biofuels production Version Zero, RSB.

Rupprecht, J., Hankamer, B., Mussgnug, J.H., Ananyev, G., Dismukes, C. and Kruse, O. (2006). Perspectives and advances of biological H2 production in microorganisms. *Appl. Microbiol. Biotechnol.* **72(3)**: 442-449.

Sazdanoff, N. (2006). Modeling and Simulation of the Algae to Biodiesel Fuel Cycle, College of Engineering, Department of Mechanical Engineering, The Ohio State University.

Sheehan, J., Dunahay, T., Benemann, J. and Roessler, P. (1998). Look Back at the U.S. Department of Energy's Aquatic Species Program: Biodiesel from Algae; Close-Out Report: Size: p. 325.

Spirulina.org.uk. Retrieved 17-05-2007, from http://spirulina.org.uk/images/spirulinaponds.jpg.

Tredici, M.R. and Materassi, R. (1992). From Open Ponds to Vertical Alveolar Panels - the Italian Experience in the Development of Reactors for the Mass Cultivation of Phototrophic Microorganisms. *Journal of Applied Phycology* **4(3)**: 221-231.

Tredici, M.R. (2009). Expert opinion.

Tsukahara, K. and Sawayama, S. (2005). Liquid fuel production using microalgae. *J. Jpn. Pet. Inst* **48(5)**: 251-259.

Wijffels, R. (2006). Energie via microbiologie: Status en toekomstperspectief voor Nederland. Utrecht, SenterNovem.

Wijffels, R. (2007). Presentation Microalgae for production of energy.

Wijffels, R.H. (2008). Potential of sponges and microalgae for marine biotechnology. *Trends Biotechnol.* **26(1)**: 26-31.

Wikipedia. (2007). Algaculture. Retrieved 1 April 2007, from http://en.wikipedia.org/w/index.php?title=Algaculture&oldid=119402781.

2

BIOFUELS FROM MICROALGAE

Yanqun Li[2], Mark Horsman[1], Nan Wu[1], Christopher Q. Lan[1*]
and Nathalie Dubois-Calero[3]

Abstract

Microalgae are a diverse group of prokaryotic and eukaryotic photosynthetic microorganisms that grow rapidly due to their simple structure. They can potentially be employed for the production of biofuels in an economically effective and environmentally sustainable manner. Microalgae have been investigated for the production of a number of different biofuels including biodiesel, bio-oil, bio-syngas, and bio-hydrogen. The production of these biofuels can be coupled with flue gas CO2 mitigation, wastewater treatment, and the production of high-value chemicals. Microalgal farming can also be carried out with seawater using marine microalgal species as the producers. Developments in microalgal cultivation and downstream processing (e.g., harvesting, drying, and thermochemical processing) are expected to further enhance the costeffectiveness of the biofuel from microalgae strategy.

Key words : Microalgae, Bio-fuel, Harvesting, Cyanobacteria, Prokaryotes, Eukaryotes

Introduction

A variety of biomasses from different sources, including forestry, agricultural, and aquatic sources have been investigated as the feedstock for the production of different biofuels including biodiesel, bio-ethanol, bio-hydrogen, bio-oil, and bio-gas. Techno-economic assessments

1 Department of Chemical Engineering, University of Ottawa, 161 Louis Pasteur St., Ottawa, ON, Canada K1N 6N5
2 College of Life Science, South China Normal University, Guangzhou, P.R. China
3 AirScience Biofuel, 1751 Richardson, Montreal, QC, Canada H3K 1G6
* *Corresponding author*: E-mail: christopher.lan@uottawa.ca.

indicated that cost-effectiveness of biofuel production was achievable (Bridgwater *et al.,* 2002). However, burning fuels derived from existing biomass has an environmental impact similar to the combustion of fossil fuels in terms of its impact to the carbon cycle (carbon balance), *i.e.*, conversion of fixed carbon into CO_2. In addition, depletion of certain existing biomasses (*e.g.*, wood) without appropriate compensation (*e.g.*, replanting) may result in massive biomass deficit, resulting in serious environmental problems (*e.g.*, deforestation). Conventional terrestrial plants are not very efficient in capturing solar energy. It was estimated that switch grass, the fastest-growing terrestrial crop, can convert solar energy to biomass energy at a yearly rate of no more than 1 W/m 2, less than 0.5% of the solar energy received at a typical mid-latitude region (200-300 W/m 2) (Lewis and Nocera, 2006, UN World Energy Assessment Report, 2003). On the other hand, studies have shown the photosynthetic efficiency of microalgae could well be in the range of 10-20% or higher (Richmond, 2000; Huntley and Redalje, 2007). Furthermore, recent studies showed that the extra N_2O entering the atmosphere as a result of using nitrogen fertilizers to produce crops for biofuels, when calculated in "CO_2-equivalent" global warming terms and compared with the quasi-cooling effect of "saving" emissions of fossil fuel derived CO_2, could contribute as much or more to global warming by N_2O emissions than cooling by fossil fuel savings (Crutzen, *et al.,* 2007). These concerns may be addressed by using fast growing microalgal species for biofuel production. For this review, we define microalgae as all unicellular and simple multicellular photosynthetic micro-organisms, be they prokaryotes (cyanobacteria) or eukaryotes (also called microalgae in a more narrow sense). They have high growth rates and photosynthetic efficiencies due to their simple structures. It is estimated that the biomass productivity of microalgae could be 50 times more than that of switchgrass, which is the fastest growing terrestrial plant (Nakamura, 2006; Demirbas, 2006). Biofuel production using microalgal farming offers the following advantages:

(1) The high growth rate of microalgae makes it possible to satisfy the massive demand on biofuels using limited land resources without causing potential biomass deficit.

(2) Microalgal cultivation consumes less water than land crops.

(3) The tolerance of microalgae to high CO_2 content in gas streams allows high-efficiency CO_2 mitigation.

(4) Nitrous oxide release could be minimized when microalgae are used for biofuel production.

(5) Microalgal farming could be potentially more cost effective than conventional farming.

On the other hand, one of the major disadvantages of microalgae for biofuel production is the low biomass concentration in the microalgal culture due to the limit of light penetration, which in combination with the small size of algal cells makes the harvest of algal biomasses relatively costly. The large water content of harvested algal biomass also means its drying would be an energy-consuming process. The higher capital costs of and the rather intensive care required by a microalgal farming facility compared to a conventional agricultural farm is another factor that impedes the commercial implementation of the biofuels from microalgae strategy. Nevertheless, these problems are expected to be overcome or minimized by technology development. Given the vast potential of microalgae as the most efficient primary producers of biomass, there is little doubt that they will eventually become one of the most important alternative energy sources.

Biofuels from Microalgae

Algae was initially examined as a potential replacement fuel source for fossil fuels in the 1970s amidst the gas scare (Barkley *et al.,* 1987), but prohibitive production costs and limitations discouraged the commercial development of algae-based fuel production. Subsequent studies, continued through the 1980s and heightened in the last 15 years, illustrate that research developments are enabling the commercial potential of microalgae to shift from aquaculture, fine chemicals, and health food (De la Noue, and De Pauw, 1988) to fuel production.

Biodiesel

Biodiesel is produced by a mono-alcoholic trans-esterification process, in which triglycerides reacts with a mono-alcohol (most commonly methanol or ethanol) with the catalysis of alkali, acids, or enzymes (Demirbas, 2007; Meher, *et al.,* 2006). It has combustion properties similar to those of diesel (Marchetti, *et al.,* 2007) and has been produced commercially or in backyard facilities to fuel vehicles. Significant technical advances have been achieved to optimize the trans-esterification process. For instance, Canadian researchers in the Department of Chemical Engineering at the University of Ottawa have developed a novel two-phase membrane reactor (Dube *et al.,* 2007), which exploits the immiscibility of canola oil in methanol to enable the separation of reaction products (biodiesel/glycerol) from the residual canola oil. The two-phase membrane reactor was particularly useful in removing unreacted canola oil from the product, yielding high purity biodiesel and shifting the reaction equilibrium to the product side. Nevertheless, one major challenge of biodiesel production is the high costs of feedstock. Currently, biodiesel production relies on animal fats and plant oils. This agricultural approach will eventually compete for land resource against food industry. For instance, it was estimated that to produce 5.54 Mtoe (million tons of oil equivalent) of biodiesel (1.72% of the 321 Mtoe estimated EU-25 (the 25 European Union countries) consumption for transportation fuels in 2003) would require 9.3 Mha of land for canola (rapeseed) and sunflower cultivation. This is equivalent to 150% of the current land used for these crops in EU-25, which is approximately 6.4 Mha (Kondili and Kaldellis, 2007). On the other hand, some microalgal species could accumulate lipids to a significant portion of their biomass (30-50% on dry weight basis), serving as a promising alternative source of lipids for biodiesel production (Miao and Wu, 2006, Xu *et al.,* 2006). It was estimated by Sheehan and his co-workers that microalgal farming using 200,000 ha of land (less than 0.1% climately suitable lands in the U.S. or 3.2% of the land currently used for the cultivation of sunflower and rapeseed in EU-25) would allow the production of a quad (*i.e.*, a quadrillion BTU) of fuel in the form of biodiesel (Sheehan *et al.,* 1998). To put this in perspective, one quad is approximately one-eighth of Canada's total energy consumption in 2004 (8543.3 _ 1015 J) (Energy Use Data Handbook, 1990 and 1998 to 2004). Yusuf Christi discussed the economics and quality constraints of biodiesel from microalgae in his recent review paper (Chisti, 2007). He pointed out that the cost of growing microalgae for biofuel production must be drastically reduced to compete *directly* with traditional energy sources. It is essential to consider the other roles algal cultures can play concurrently with biofuel production and the long term benefits this entails. It is interesting to notice that, even though the two major project sponsored by the U.S. government and the Japanese government concluded that algal oil was not economically feasible (Huntley and Redalje, 2007), the private sector has moved forward in building commercial facilities to produce biodiesel using algal oils. It was reported that a privately funded US$20 million program has engineered, built, and successfully operated for several years a commercial-scale (2 ha) modular production system coupling photo-bioreactors with open ponds in a two-stage process to produce *Haematococcus pluvialis* for biodiesel production with a annual averaged rate of achieved microbial oil production equivalent to 420 GJ ha^{-1} yr^{-1}, which exceeds the most optimistic estimates of biofuel production from plantations of terrestrial "energy crops." The maximum production rate achieved was equivalent to 1014 GJ ha^{-1} yr^{-1}. It was claimed that a rate of 3200 GJ ha^{-1} yr^{-1} is feasible using *Chlorella* under conditions that prevail in the existing production system, a rate possible to replace reliance on current fossil fuel usage equivalent to about 300 EJ yr^{-1} and eliminate fossil fuel emissions of CO_2 of about 6.5 gigatons of carbon (GtC) per year using only 7.3% of the surplus arable land projected to be available by 2050 (Huntley and Redalje, 2007). This rather optimistic estimation was based on a photosynthetic efficiency of 20% and a biomass productivity of around 70 g m^{-2} Day^{-1} of *Chlorella*. (Huntley and Redalje, 2007). It was also expected that some algal biodiesel processes, such as the one being developed at the University of Utah, would be cost competitive with regular diesel by 2009 (Seefeldt, 2007). Wu (Lv *et al.,* 2007) in China reported the production of biodiesel using *Chlorella protothecoides* at a scale of 11,000 L. However, they adopted a heterotrophic cultivation strategy, which does not necessarily fulfill the mandate of microalgal farming to convert solar energy to biofuel. Instead, organic carbons such as glucose are used for fuel production.

Bio-oil and Bio-syngas

When biomass is processed under high temperature at the absence of oxygen, products are produced as three phases: the vapor phase, the liquid phase, and the solid phase. The liquid phase is a complex mixture called bio-oil. The compositions of bio-oils vary significantly with the types of feedstock and processing conditions. For instance, the bio-oils produced from a commercial wood biomass feed (Lignocel HBS 150-500) originated from beech wood contain mainly phenols, alcohols, and carbonyls with concentrations varying significantly with conditions of prolysis (Iliopoulou *et al.,* 2007), while the bio-oils obtained via fast pyrolysis of rice husk comprise mainly formic acid (7.69%), â-hydroxybutyric acid (2.31%), toluene (5.00%), benzoic acid, 3-methyl (1.15%), 1,2-benzenedicarboxylic acid (1.22%), and other organic compounds (Ji, 2007). Extensive studies have been carried out on biomass conversion; several technologies such as an entrained flow reactor, circulating fluid bed gasifier (Prins *et al.,* 2006, Meier and Faix, 1999), vacuum pyrolysis (Pakdel and Roy, 1991), and vortex reactor (Meier and Faix, 1999) have been demonstrated to be effective. The overall energy to biomass ratio of a well-controlled pyrolytic process could be as high as 95.5% (Demirbas, 2001). These technologies can be classified into two categories: (1) pyrolysis, the primary product of which is pyrolytic liquids (bio-oils) and (2) gasification, with "syngas" as the primary product (Lv *et al.,* 2007, Prins *et al.,* 2006). Canadian companies have shown world leadership in this field. Ensyn Corp (EC) (http://www.ensyn.com/), a private company based in Ottawa, Ontario, developed one of the most successful biomass conversion technologies, Rapid Thermal Processing (RTP). RTP is a patented, state-of-the-art process that transforms carbon-based feedstocks, either wood "biomass" or petroleum hydrocarbons, into more valuable chemical and fuel products. Plascoenergy Group, another private company based in Ottawa, Ontario, has a proprietary Plasco Conversion System (PCS) that converts carbonaceous materials such as municipal solid wastes, into an energy-rich fuel or "syngas" and a commercially useful inert solid or "slag" (http://www.plascoenergygroup.com/). Bio-oils have been demonstrated to be suitable for power generation via both external combustion (*e.g.*, steam cycles, organic Rankine cycles, and Stirling engines) and internal combustion (*e.g.*, diesel engines and gas-turbine engines) or by co-firing with fossil diesel or natural gas (Czernik and Bridgwater, 2004; Bridgwater *et al.,* 1999; Chiaramonti *et al.,* 2007). Nevertheless, they have several undesirable features such as high oxygen content, low heat content, high viscosity at low temperature, and chemical instability (Czernik and Bridgwater, 2004, Bridgwater *et al.,* 1999, Chiaramonti, *et al.,* 2007) that impede their use as quality transportation fuels. To overcome this limitation, studies have been taken to upgrade the bio-oils to high quality fuels. For instance, production of high-grade transportation fuels from biomass has been demonstrated to be technically feasible by gasification and subsequent Fischer Tropsch synthesis (Lv *et al.,* 2007, Petrus and Noordermeer, 2006). Recent work by a group in China has demonstrated that hydrogen can be derived reliably by steam-reforming bio-oil (Wang, *et al.,* 2007). Most studies have so far focused on the use of conventional biomasses from forestry and agricultural sources (Demirbas, 2001). It was estimated that in year 2000, the majority of biomass energy was produced from wood and wood wastes (64%), followed by municipal solid waste (MSW) (24%), agricultural waste (5%) and landfill gases (5%) (Demirbas, 2001, Demirbasü, 2000). Recently, a few investigations have been carried out regarding the suitability of microalgal biomass for bio-oil production (Demirbas, 2006, Miao and Wu, 2004, Miao *et al.,* 2004). It was shown that, in general, microalgae bio-oils are of higher quality than bio-oil from wood (Demirbas, 2006).

Bio-hydrogen

Hydrogen is an important fuel with wide applications in fuel cells, liquefaction of coal, and upgrading of heavy oils (*e.g.*, bitumen). Hydrogen can be produced biologically by a variety of means, including the steamreformation of bio-oils (Wang *et al.*, 2000), dark and photo fermentation of organic materials (Kapdan, I. K.; Kargi, F. 2006), and photolysis of water catalyzed by special microalgal species (Kapdan and Kargi, 2006, Ran *et al.*, 2006).

Integrated Pollution Control and Biofuel Production using Microalgae

Microalgal Farming and CO_2 Mitigation

One of the key advantages of using microalgae for biofuel production lies in the ability of some microalgal species to tolerate high CO_2 content in feeding air streams (Chang and Yang, 2003), allowing efficient capturing of CO_2 from high-CO_2 streams such as flue gases and flaring gases (CO_2 content 5-15%) (Hsueh *et al.*, 2007). In comparison to terrestrialplants, which typical absorb CO_2 from the atmosphere containing only 0.03-0.06% CO_2, the benefit of microalgae is evident in terms of CO_2 mitigation. It was reported that using a outdoor cultivation of *Chlorella* sp. in a 55 m2 culture area photobioreactor, flue gas containing 6-8% by volume of CO_2, 10-50% CO_2 mitigation (flue gas decarbonization) was achievable and the residual NO_2 and NO in the flue gas was found not to affect algal growth (Doucha *et al.*, 2005). In such a facility, employment of appropriate flue/flaring gas pretreatment procedure and optimizing culture media is of critical importance (Lee *et al.*, 2000). A higher CO_2 mitigation rate between 50.1 (6.5% on cloudy days and 82.3 (12.5%on sunny days was reported by other researchers using different algal species (Vunjak-Novakovic *et al.*, 2005). Depending on the microalgal species and condition used in the facilities, algal biomass produced could be further processed for biodiesel, bio-oil, and bio-syngas production.

Microalgal Farming using Wastewater

In addition to the apparent benefit of combining micro algal biomass, and therefore biofuel, production and wastewater treatment, successful implementation of this strategy would allow the minimizing of the use of freshwater, another precious resource especially for dry or populous countries, for biofuel production. Extensive works have been conducted to explore the feasibility of using microalgae for wastewater treatment, especially for the removal of nitrogen and phosphorus from effluents (Mallick 2002, Aslan and Kapdan, 2006; Hernandez *et al.*, 2006; Abdel

Hameed, 2007; Lebeau and Robert, 2003),which would otherwise result in eutrophication if dumped into lakes and rivers (Galvez-Cloutier *et al.,* 2006). Ironically enough, it is algae in the lakes and rivers that cause this problem. It is simply a matter of allowing the consumption of nitrogen and phosphorus by microalgae in a controlled manner that benefits rather than deteriorates the environment. Levels of several contaminant heavy metals have also been shown to be reduced by the cultivation of microalgae, which is a subject discussed extensively by Mun˜oz and Guieysse (2006). A major concern associated with using wastewater for microalgae cultivation is contamination(Mun˜oz and Guieysse, 1988, 2006). This can be managed by using appropriate pretreatment technologies to remove sediment and to deactivate (sterilize) the wastewater (Tamer *et al.,* 2006).

Microalgal Farming using Marine Microalgae

Fresh wateris another natural resource besides variable land that may cap biofuel production. This concern is particular evident for populous countries such as China, India and dry coastal regions such as the Middle East. It is a novel idea to employ marine microalgae for CO_2 mitigation and biofuel production. Extensive studies have been carried out for the cultivation of different marine microalgae using a variety of cultivation systems including both open ponds (Blanco *et al.,* 2007) and various types of closed photobioreactors (Blanco *et al.,* 2007, Wu and Merchuk, 2004; Janssen *et al.,* 2001). A few examples of marine microalgal species that have been studied for microalgal farming include red marine alga *Porphyridium* sp. (Wu and Merchuk, 2004), N-fixing cyanobacterium *Anabaena* (Blanco *et al.,* 2007), macrophytic marine red alga *Agardhiella subulata* (Huang *et al.,* 2006), marine green alga *Dunaliella tertiolecta*(Janssen *et al.,* 2001), and marine phytoplankter *Tetraselmis suecica* (Laws and Berning, 1991).It is intriguing to notice that nitric oxide (NO) and carbondioxide (CO_2) were reported to be simultaneously eliminated from a model flue gas using a marine phytoplankter, *Tetraselmis suecica* (Laws and Berning, 1991) making it possible to combine biomass (biofuel)production with CO_2 and NOx mitigation. It is expected that, with the selection of appropriate marine microalgal strains, cost effective production of different biofuels should be achievable using seawater as the medium. Indeed, the 2-ha demonstration facility reported by Huntley and Redalje (Huntley and Redalje, 2007), which was discussed briefly in the introduction, utilized seawater as the medium and for temperature control.

Enhancement of Economic Feasibility of Biofuels from Microalgae

Biorefinery: The High-Value Coproduct Strategy

The term biorefinery was coined to describe the production of a wide range of chemicals and biofuels from biomasses by the integration of bioprocessing and appropriate low environmental impact chemical technologies in a cost-effective and environmentally sustainable manner (Chisti, 2007). Examples include the two phase conversion reaction of fructose to 5-hydroxymethylfurfural (Kamm, 2007), fermentative production of ethanol from sugars derived from cellulose and semi-cellulose (Ltd, A.B.C. Company news, 2007), and bio-oils and/or biosyngas by the pyrolysis/gasification of woods or other types of biomasses (Mohan *et al.,* 2006). Microalgae have the capacity of producing a vast array of high-value bioactive compounds that can be used as pharmaceutical compounds, health foods, and natural pigments (Oh *et al.,* 2003, Jiang, 2000). Some well-studied examples include acetylic acids, â-carotene (Huang *et al.,* 2006, Del Campo *et al.,* 2007), vitamin B (He *et al.,* 2005), keto carotenoid astaxanthin (Ip and Chen, 2005), polyunsaturated fatty acids (Wen and Chen, 2003, Jiang *et al.,* 2004), and lutein (Blanco *et al.,* 2007, Shi *et al.,* 2002) (Table 1). The economical feasibility of micro algal biofuel production should be significantly enhanced by a high-value co-product strategy, which would, conceptually, involve sequentially the cultivation of microalgae in a micro algal farming facility (CO_2 mitigation), extracting bioreactive products from harvested algal biomass, thermal processing (pyrolysis, liquefaction, or gasification), extracting high-value chemicals from the resulting liquid, vapor, and/or solid phases, and reforming/upgrading biofuels for different applications. The employment of a high-value co-product strategy through the integrated bio-refinery approach is expected to significantly enhance the overall cost-effectiveness of micro algal biofuel production.

Design of Advanced Photobioreactors

The choice of cultivation systems is another key aspect that significantly affectsthe efficiency and cost-effectiveness of a microalgal biofuelproduction process. This topic has been discussed extensively by a few authors (Chisti, 2007, Lee, 2001, Pulz, 2001, Carvalho *et al.,* 2006, Chaumont, 1993). Carvalho *et al.* (2006) explained several closed systems in detail. Lee (2001) discussed a few open systems and systematically compared them with closed systems over different

Table 1. Some high-value bio-products extracted from microalgae

Product group	***Applications***	***Examples (Producer)***
Phycobiliproteins carotenoids	Pigments, cosmatics, provitamins, pigmentation	Phycocyanin (*Spirulina platensis*) (Furuki, *et al.*, 2003), â- carotene (*Dunelliela salina*) (Borowitzka, 1991), astraxanthin and leutin (*Haematococcus pluvialis*) (Olaizola, 2003)
Polyunsaturated fatty acids (PUFAs)	Food additive, nutraceutics	Eicosapentaenoic acid (EPA) (*Chlorella minutissima*) (Cardozo *et al.*, 2007), docosahexaenoic acid (DHA), (*Schizochitrium* sp.) (Valencia *et al.*, 2007), arachidonic acid (AA), Parietochlorisincise (Bigogno *et al.*, 2002),
Vitamins	Nutrition	Biotin (*Euglena gracils*) (Baker *et al.*, 1981), á-tocopherol (Vitamin E), (*Euglena gracilis*[a]) (Survase *et al.*, 2006), ascorbic acid (Vitamin C), (*Prototheca moriformis*[a], *Chlorella* spp.[a]) (Bremus *et al.*, 2002)

[a] Heterotrophic growth

geographical regions. Pulz (2001) focused more on process parameters and suggested a number of open systems. Janssen *et al.* (2003) offered useful conceptual diagrams for some of the discussed closed systems and described new systems to be examined, including the use of optical fiber to enhance lighting. Even though the open pond systems seem to be favored for commercial cultivation of microalgae at present due to their low capital costs, closed systems offer better control over contamination, mass transfer, and other cultivation conditions. The combination of the closed photobioreactor and open pond combines the benefits of the two and has been demonstrated to be effective at a 2-ha scale (Huntley and Redalje, 2007).

Selection of Cost-Effective Technologies for Biomass Harvesting and Drying

Given the relatively low biomass concentration obtainable in microalgal cultivation systems due to the limit of light penetration (typically in the range of 1-5g/L) and the small size of microalgal cells (typically in the range of 2-20 μm in diameter), costs and energy consumption forbiomass harvesting are a significant concern needs to bead dressed properly. Different technologies, including chemical flocculation (Knuckey *et al.,* 2006), biological flocculation (Divakaran and Pillai, 2002), filtration (Molina *et al.,* 2003), centrifugation (Olaizola, 2003), and ultrasonic aggregation (Bosma *et al.,* 2003) have been investigated for microalgal biomass harvesting. In general, chemical and biological flocculation require only low operating costs; however, they have the disadvantage of requiring long processing period and having the risk of bioreactive product decomposition. On the other hand, filtration, centrifuge, and ultrasonic flocculation are more efficient but more costly. The selection of appropriate harvesting technology depends on the value of the target products, the biomass concentration, and the size of microalgal cells of interest. Biomass drying before further lipid/ bioproduct extraction and/or thermochemical processing is another step that needs to betaken into consideration. Sun drying is probably the cheapest drying method that has been employed for the processing of microalgal biomass (Millamena *et al.,* 1990, Prakash *et al.,* 1997). However, this method takes long drying time, requires large drying surface, and risks the loss of some bio reactive products. Low-pressure shelf drying is another low-cost drying technology that has been investigated (Prakash *et al.,* 1997). It is nevertheless also of low efficiency. More efficient but more costly drying technologies having been investigated for drying microalgae include drum drying (Prakash *et al.,* 1997), spray drying (Desmorieux and Decaen, 2006, Leach *et al.,* 1998), fluidized bed drying (Leach *et al.,* 1998), freeze drying (Millamena *et al.,* 1990), and refractance window dehydration technology (Nindo and Tang, 2007). It is important to find the balance between the drying efficiency and cost-effectiveness to maximize the net energy output of the fuels from micro algae strategy.

Conclusions

Microalgae are a diverse group of prokaryotic and eukaryotic photosynthetic microorganisms that can grow rapidly due to their simple structure. They have been investigated for the production of different biofuels including biodiesel, bio-oil, bio-syngas, and bio-hydrogen. Microalgal biofuel production is potentially sustainable. It is possible to produce adequate microalgal biofuels to satisfy the fast growing energy demand within there straints of land and water resources. Microalgal farming can be coupled with flue gas CO_2 mitigation and wastewater treatment. It can also be carried out with seawater as the medium, given that marine microalgal species are adopted, providing a feasible alternative for biofuel production to populous and dry coastal regions. Microalgae can produce a large variety of novel bioproducts with wide applications in medicine, food, and cosmetic industries. Combining microalgal farming and the production of biofuels using bio refinery strategy is expected to significantly enhance the overall cost-effectiveness of the biofuel from microalgae approach. Technological developments, including advances in photo bioreactor design, microalgal biomass harvesting, drying, and other downstream processing technologies are important areas that may lead to enhanced cost-effectiveness and therefore, effective commercial implementation of the biofuel from microalgae strategy.

Acknowledgments

Financial support from the Natural Science and Engineering Research Council (NSERC) of Canada and Air Science Technologies Inc., Montreal, Quebec, Canada, are gratefully acknowledged.

References

Abdel Hameed, M.S. (2007). Effect of algal density in bead, bead size and bead concentrations on wastewater nutrient removal. *Afr. J. Biotechnol.* **6(10)**: 1185-1191.

Aslan, S. and Kapdan, I.K. (2006). Batch kinetics of nitrogen and phosphorus removal from synthetic waste water by algae. *Ecol. Eng.* **28(1)**: 64-70.

Baker, E.R., McLaughlin, J.J.A. and Hutner, S.H. (1981). Water-soluble vitamins in cells and spent culture supernatants of *Poteriochromonas stipitata*, *Euglena gracilis*, and *Tetrahymena thermophila*. *Arch. Microbiol.* **129(4)**: 310-313.

Barkley, W.C.W., Lewin, R.A. and Cheng, L. (1987). *In*: *Development of Microalgal Systems for the Production of Liquid Fuels*, Socie´te´ pour l'Algologie Applique´e, Villeneuve d'Ascq, France. Sadler, T. *Ed.*; Elsevier Applied Science: Villeneuve d'Ascq, France.

Bigogno, C., Khozin-Goldberg, I., Boussiba, S., Vonshak, A. and Cohen, Z. (2002). Lipid and fatty acid composition of the green oleaginous alga *Parietochloris incisa*, the richest plant source of arachidonic acid. *Phytochemistry* **60(5)**: 497-503.

Blanco, A.M., Moreno, J., Del Campo, J.A., Rivas, J. and Guerrero, M.G. (2007). Outdoor cultivation of lutein-rich cells of *Muriellopsis* sp. in open ponds. *Appl. Microbiol. Biotechnol.* **73(6)**: 1259-1266.

Borowitzka, L.J. (1991). Development of Western Biotechnology's algal â-carotene plant. *Bioresour. Technol.* **38(2-3)**: 251-252.

Bosma, R., Van Spronsen, W.A., Tramper, J. and Wijffels, R.H. (2003). Ultrasound, A new separation technique to harvest microalgae. *J. Appl. Phycol.* **15(2-3)**: 143-153.

Bremus, C., Herrmann, U., Bringer-Meyer, S. and Sahm, H. (2006). The use of microorganisms in l-ascorbic acid production. *J. Biotechnol.* **124(1)**: 196-205.

Bridgwater, A.V., Meier, D. and Radlein, D. (1996). An overview of fast pyrolysis of biomass. *Org. Geochem.* **30(12)**: 1479-1493.

Bridgwater, A.V., Toft, A.J. and Brammer, J.G. (2002). A techno-economic comparison of power production by biomass fast pyrolysis with gasification and combustion. *Renewable Sustainable Energy Re* **6(3)**: 181-248.

Cardozo, K.H.M., Guaratini, T., Barros, M.P., Falcao, V.R., Tonon, A.P., Lopes, N.P., Campos, S., Torres, M.A., Souza, A.O., Colepicolo, P. and Pinto, E. (2007). Metabolites from algae with economical impact. *Compe. Biochem. Physiol., Part C: Toxicol. Pharmacol.* **146(1-2 SPEC. ISS.)**: 60-78.

Carvalho, A.P., Meireles, L.A. and Malcata, F.X. (2006). Microalgal reactors: A review of enclosed system designs and performances. *Biotechnol. Prog.* **22(6)**: 1490-1506.

Chang, E.H. and Yang, S.S. (2003). Some characteristics of microalgae isolated in Taiwan for biofixation of carbon dioxide. *Bot. Bull. Acad. Sin.*, **44(1)**: 43-52.

Chaumont, D. (1993). Biotechnology of algal biomass production: A review of systems for outdoor mass culture. *J. Appl. Phycol.* **5(6)**: 593-604.

Chiaramonti, D., Oasmaa, A. and Solantausta, Y. (2007). Power generation using fast pyrolysis liquids from biomass. *Renewable Sustainable Energy Rev.* **11(6)**: 1056-1086.

Chisti, Y. (2007). Biodiesel from microalgae. *Biotechnol. Adv.*, **25(3)**: 294-306.

Crutzen, P.J., Mosier, A.R., Smith, K.A. and Winiwarter, W. (2007). N_2O release from agro-biofuel production negates global warming reduction by replacing fossil fuels. *Atmos. Chem. Phys. Discuss.* **7(4)**: 11191-11205.

Czernik, S. and Bridgwater, A.V. (2004). Overview of applications of biomass fast pyrolysis oil. *Energy Fuels* **18(2)**: 590-598.

De la Noue, J. and De Pauw, N. (1988). The potential of microalgal biotechnology: a review of production and uses of microalgae. *Biotechnol. Adv.* **6**: 725-770.

Del Campo, J.A., Garcia-Gonzalez, M. and Guerrero, M.G. (2007). Outdoor cultivation of microalgae for carotenoid production: Current state and perspectives. *Appl. Microbiol. Biotechnol.* **74(6)**: 1163-1174.

Demirbas, A. (2001). Biomass resource facilities and biomass conversion processing for fuels and chemicals. *Energy Convers. Manage.* **42(11)**: 1357-1378.

Demirbas, A. (2006). Oily products from mosses and algae via pyrolysis. *Energy Sources, A* **28(10)**: 933-940.

Demirbas, A. (2007). Recent developments in biodiesel fuels. *Int. J. Green Energy*, **4(1)**: 15-26.

Demirbasü, A. (2000). Industry, B. r. f. e. a. c., Biomass resources for energy and chemical industry. *Energy, Educ., Sci. Technol.* **5**: 21-45.

Desmorieux, H. and Decaen, N. (2006). Convective drying of *Spirulina* in thin layer. *J. Food Eng.* **77(1)**: 64-70.

Divakaran, R. and Pillai, V.N.S. (2002). Flocculation of algae using chitosan. *J. Appl. Phycol.* **14(5)**: 419-422.

Doucha, J., Straka, F. and Livansky, K. (2005). Utilization of flue gas for cultivation of microalgae (*Chlorella* sp.) in an outdoor open thin layer photobioreactor. *J. Appl. Phycol.*, **17(5)**: 403-412.

Dube, M.A., Tremblay, A.Y. and Liu, J. (2007). Biodiesel production using a membrane reactor. *Bioresour. Technol.* **98(3)**: 639-647.

Energy Use Data Handbook, (1990, 1998-2004, 2006). Energy Publications, Office of Energy Efficiency, Natural Resources Canada: Ottawa.

Furuki, T., Maeda, S., Imajo, S., Hiroi, T., Amaya, T., Hirokawa, T., Ito, K. and Nozawa, H. (2003). Rapid and selective extraction of phycocyanin from *Spirulina platensis* with ultrasonic cell disruption. *J. Appl. Phycol.* **15(4)**: 319-324.

Galvez-Cloutier, R., Leroueil, S., Allier, D., Locat, J. and Arsenault, S.A. (2006). Combined method: Precipitation and capping, to attenuate eutrophication in Canadian lakes. *J. ASTM Int.* **3(6)**.

He, H.Z., Li, H.B. and Chen, F. (2005). Determination of vitamin B1 in sea water and microalgal fermentation media by high-performance liquid chromatography with fluorescence detection. *Anal. Bioanal. Chem.* **383(5)**: 875-879.

Hernandez, J.P., De-Bashan, L.E. and Bashan, Y. (2006). Starvation enhances phosphorus removal from wastewater by the microalga *Chlorella,* spp. co-immobilized with *Azospirillum brasilense. Enzyme Microb. Technol.* **38(1-2)**: 190-198.

Hsueh, H.T., Chu, H. and Yu, S.T. 2007. A batch study on the bio-fixation of carbon dioxide in the absorbed solution from a chemical wet scrubber by hot spring and marine algae. *Chemosphere,* **66(5)**: 878-886.

Huang, J.C., Chen, F. and Sandmann, G. (2006). Stress-related differential expression of multiple â-carotene ketolase genes in the unicellular green alga *Haematococcus pluvialis. J. Biotechnol.* **122(2)**: 176-185.

Huntley, M.E. and Redalje, D.G. (2007). CO_2 mitigation and renewable oil from photosynthetic microbes: A new appraisal. *Mitigation Adaptation Strategies Global Change* **12(4)**: 573-608.

Iliopoulou, E.F., Antonakou, E.V., Karakoulia, S.A., Vasalos, I.A., Lappas, A.A. and Triantafyllidis, K.S. (2007). Catalytic conversion of biomass pyrolysis products by mesoporous materials: Effect of steam stability and acidity of Al-MCM-41 catalysts. *Chem. Eng. J.* **134(1-3)**: 51-57.

Ip, P.F. and Chen, F. (2005). Employment of reactive oxygen species to enhance astaxanthin formation in *Chlorella zofingiensis* in heterotrophic culture. *Process Biochem.* **40(11)**: 3491-3496.

Janssen, M., Slenders, P., Tramper, J., Mur, L.R. and Wijffels, R.H. (2001). Photosynthetic efficiency of *Dunaliella tertiolecta* under shortlight/dark cycles. *Enzyme Microb. Technol.* **29(4-5)**: 298-305.

Janssen, M., Tramper, J., Mur, L.R. and Wijffels, R.H. (2003). Enclosed outdoor photobioreactors: Light regime, photosynthetic efficiency, scale-up, and future prospects. *Biotechnol. Bioeng.* **81(2)**: 193-210.

Jiang, F.C. and A.Y. (2000). *Algae and Their Biotechnological Potential.* Kluwer Academic Publishers: Dordrecht/Boston/London.

Jiang, Y., Fan, K.W., Wong, R.T.Y. and Chen, F. (2004). Fatty acid composition and squalene content of the marine microalga *Schizochytrium mangrovei. J. Agric. Food Chem.* **52(5)**: 1196-1200.

Ji-lu, Z. (2007). Bio-oil from fast pyrolysis of rice husk: Yields and related properties and improvement of the pyrolysis system. *J. Anal. Appl. Pyrolysis* **80(1)**: 30-35.

Kamm, B. (2007). Production of platform chemicals and synthesis gas from biomass. *Angew. Chem., Int. Ed.* **46(27)**: 5056-5058.

Kapdan, I.K. and Kargi, F. (2006). Bio-hydrogen production from waste materials. *Enzyme Microb. Technol.* **38(5)**: 569-582.

Knuckey, R.M., Brown, M.R., Robert, R. and Frampton, D.M.F. (2006). Production of microalgal concentrates by flocculation and their assessment as aquaculture feeds. *Aquacult. Eng.* **35(3)**: 300-313.

Kondili, E.M. and Kaldellis, J.K. (2007). Biofuel implementation in East Europe: Current status and future prospects. *Renewable Sustainable Energy Rev.*, **11(9)**: 2137-2151.

Laws, E.A. and Berning, J.L. (1991). A study of the energetics and economics of microalgal mass culture with the marine chlorophyte *Tetraselmis suecica*: Implications for use of power plant stack gases. *Biotechnol. Bioeng.* **37(10)**: 936-947.

Leach, G., Oliveira, G. and Morais, R. (1998). Spray-drying of *Dunaliella salina* to produce a â-carotene rich powder. *J. Ind. Microbiol. Biotechnol.* **20(2)**: 82-85.

Lebeau, T. and Robert, J.M. (2003). Diatom cultivation and biotechnologically relevant products. Part II: Current and putative products. *Appl. Microbiol. Biotechnol.* **60(6)**: 624-632.

Lee, J.N., Lee, J.S., Shin, C.S., Park, S.C. and Kim, S.W. (2000). Methods to enhance tolerances of Chlorella KR-1 to toxic compounds in fluegas. *Appl. Biochem. Biotechnol., A* **84-86**: 329-342.

Lee, Y.K. (2001). Microalgal mass culture systems and methods: Their limitation and potential. *J. Appl. Phycol.* **13(4)**: 307-315.

Lewis, N.S. and Nocera, D.G. (2006). Powering the planet: Chemical challenges in solar energy utilization. *Proc. Natl. Acad. Sci. U.S.A.*, **103(43)**: 15729-15735.

Ltd, A.B.C. Company news (2007). Biodiesel from algae makes debut. *Fuels Lubes Int.* **13(2)**: 28.

Lv, P., Yuan, Z., Wu, C., Ma, L., Chen, Y. and Tsubaki, N. (2007). Biosyngas production from biomass catalytic gasification. *Energy Convers. Manage.* **48(4)**: 1132-1139.

Mallick, N. (2002). Biotechnological potential of immobilized algae for waste water N, P and metal removal: A review. *Bio. Metals* **15(4)**: 377-390.

Marchetti, J.M., Miguel, V.U. and Errazu, A.F. (2007). Possible methods for biodiesel production. *Renewable Sustainable Energy Rev.* **11(6)**: 1300-1311.

Meher, L.C., Vidya Sagar, D. and Naik, S.N. (2006). Technical aspects of biodiesel production by transesterification-A review. *Renewable Sustainable Energy Rev.* **10(3)**: 248-268.

Meier, D. and Faix, O. (1999). State of the art of applied fast pyrolysis of lignocellulosic materials-A review. *Bioresour. Technol.* **68(1)**: 71-77.

Miao, X. and Wu, Q. (2004). High yield bio-oil production from fast pyrolysis by metabolic controlling of *Chlorella protothecoides*. *J. Biotechnol.* **110(1)**: 85-93.

Miao, X. and Wu, Q. (2006). Biodiesel production from heterotrophic microalgal oil. *Bioresour. Technol.* **97(6)**: 841-846.

Miao, X., Wu, Q. and Yang, C. (2004). Fast pyrolysis of microalgae to produce renewable fuels. *J. Anal. Appl. Pyrolysis*, **71(2)**: 855-863.

Millamena, O.M., Aujero, E.J. and Borlongan, I.G. (1990). Techniques on algae harvesting and preservation for use in culture and as larval food. *Aquacult. Eng.* **9(5)**: 295-304.

Mohan, D., Pittman, Jr. C.U. and Steele, P.H. (2006). Pyrolysis of wood/biomass for bio-oil: A critical review. *Energy Fuels* **20(3)**: 848-889.

Molina Grima, E., Belarbi, E.H., Acien Fernandez, F.G., Robles Medina, A. and Chisti, Y. (2003). Recovery of microalgal biomass and metabolites: Process options and economics. *Biotechnol. Adv.* **20(7-8)**: 491-515.

Mun˜oz, R. and Guieysse, B. (2006). Algal-bacterial processes for the treatment of hazardous contaminants: A review. *Water Res.* **40(15)**: 2799-2815.

Nakamura, D.N. (2006). Journally speaking: The mass appeal of biomass. *Oil Gas J.* **104(45)**: 15.

Nindo, C.I. and Tang, J. (2007). Refractance window dehydration technology: A novel contact drying method. *Drying Technol.* **25(1)**: 37-48.

Oh, H.M., Choi, A. and Mheen, T.I. (2003). High-value materials from microalgae. *Korean J. Microbiol. Biotechnol.* **31(2)**: 95-102.

Olaizola, M. (2003). Commercial development of microalgal biotechnology: from the test tube to the marketplace. *Biomol. Eng.* **20(4-6)**: 459-466.

Pakdel, H. and Roy, C. (1991). Hydrocarbon content of liquid products and tar from pyrolysis and gasification of wood. *Energy Fuels* **5(3)**: 427-436.

Petrus, L. and Noordermeer, M.A. (2006). Biomass to biofuels, a chemical perspective. *Green Chem.* **8(10)**: 861-867.

Prakash, J., Pushparaj, B., Carlozzi, P., Torzillo, G., Montaini, E. and Materassi, R. (1997). Microalgal biomass drying by a simple solar device. *Int. J. Solar Energy* **18(4)**: 303-311.

Prins, M.J., Ptasinski, K.J. and Janssen, F.J.J.G. (2006). More efficient biomass gasification via torrefaction. *Energy*, **31(15)**: 3458-3470.

Pulz, O. (2001). Photobioreactors: Production systems for photo trophic microorganisms. *Appl. Microbiol. Biotechnol.* **57(3)**: 287-293.

Ran, C.Q., Chen, Z.A., Zhang, W., Yu, X.J. and Jin, M.F. (2006). Characterization of photobiological hydrogen production by several marine green algae. *Wuhan Ligong Daxue Xuebao* **28(Suppl.2)**: 258-263.

Richmond, A. (2000). Microalgal biotechnology at the turn of the millennium: A personal view. *J. Appl. Phycol.*, **12(3)**: 441-451.

Running, J.A., Severson, D.K. and Schneider, K.J. (2002). Extracellular production of L-ascorbic acid by *Chlorella protothecoides, Protothecaspecies,* and mutants of *P. moriformis* during aerobic culturing at low pH. *J. Ind. Microbiol. Biotechnol.* **29(2)**: 93-98.

Seefeldt, L.C. (2007). Utah group plans to make biodiesel from algae. *Ind. Bioprocess.* **29(3)**: 5-6.

Sheehan, J. *et al.* (1998). A look back at the US Department of Energy's aqueous species program: biodiesel from Algae; NREL/TP-580-24190; The National Renewable Energy Laboratory, U.S. Department of Energy.

Shi, X.M., Jiang, Y. and Chen, F. (2002). High-yield production of luteinby the green microalga *Chlorella protothecoides* in heterotrophic fed batch culture. *Biotechnol. Prog.* **18(4)**: 723-727.

Survase, S.A., Bajaj, I.B. and Singhal, R.S. (2006). Biotechnological production of vitamins. *Food Technol. Biotechnol.* **44(3)**: 381-396.

Tamer, E., Amin, M.A., Ossama, E.T., Bo, M. and Benoit, G. (2006). Biological treatment of industrial wastes in a photobioreactor. *Water Sci. Technol.* **53(11)**: 117-125.

Tornabene, T.G., Holzer, G., Lien, S. and Burris, N. (1983). Lipid composition of the nitrogen starved green alga *Neochloris oleoabundans*. *Enzyme Microb. Technol.* **5(6)**: 435-440.

UN World Energy Assessment Report (2003). Energy and the Challenge of Sustainability; United Nations Development Program; United Nations: New York.

Valencia, I., Ansorena, D. and Astiasaran, I. (2007). Development of dry fermented sausages rich in docosahexaenoic acid with oil from the micro algae *Schizochytrium* sp.: Influence on nutritional properties, sensorial quality and oxidation stability. *Food Chem.* **104(3)**: 1087-1096.

Vunjak-Novakovic, G., Kim, Y., Wu, X., Berzin, I. and Merchuk, J.C. (2005). Air-lift bioreactors for algal growth on flue gas: Mathematical modeling and pilot-plant studies. *Ind. Eng. Chem. Res.* **44(16)**: 6154-6163.

Wang, Z., Pan, Y., Dong, T., Zhu, X., Kan, T., Yuan, L., Torimoto, Y., Sadakata, M. and Li, Q. (2007). Production of hydrogen from catalytic steam reforming of bio-oil using C12A7-O-based catalysts. *Appl. Catal., A* **320**: 24-34.

Wen, Z.Y. and Chen, F. (2003). Heterotrophic production of eico sapentaenoic acid by microalgae. *Biotechnol. Adv.* **21(4)**: 273-294.

Wu, X. and Merchuk, J.C. (2004). Simulation of algae growth in a bench scale internal loop airlift reactor. *Chem. Eng. Sci.* **59(14)**: 2899-2912.

Xu, H., Miao, X. and Wu, Q. (2006). High quality biodiesel production from a microalga *Chlorella protothecoides* by heterotrophic growth in fermenters. *J. Biotechnol.* **126(4)**: 499-507.

Zwart, R.W.R., Boerrigter, H. and van der Drift, A. (2006). The impact of biomass pretreatment on the feasibility of overseas biomass conversion to Fischer-Tropsch products. *Energy Fuels* **20(5)**: 2192-2197.

3

BIODIESEL FROM MICROALGAE: FUEL OF 21ST CENTURY

Jitendera, A.S. Ahluwalia[1*] and Anjuli Sood[2]

Abstract

Energy demands are increasing for the subsistence of civilizations. Increasing demands of this resource and depleting (and hence escalating price) have compelled the scientists to invent alternative and environment friendly sources. Microalgae has been receiving maximum attention among various plant sources due to their better yield, short life cycle, simple nutritional requirements wider distribution etc. These tiny organisms do not leave any waste by product to be disposed of during their processing. In addition there is better yield and quality of oil produced from these photosynthetic organisms which do consume CO_2 from air efficiently to aid improvement in our climate stability. Microalgae therefore seem to be the major favoured system to be exploited for biodiesel production to provide sustainable energy source for future generations.

Key words: Microalgae, Biodiesel, Biofuel, Transesterification

Introduction

Subsistence of human civilization and further evolution is critically dependent on the availability of continuous supply of fuel. The discovery and utilization of coal as fuel in 18th century and petroleum in 19th century has played a vital role in increased speed of industrialization and modernization. It takes millions of years to convert biomass into

1 Department of Botany, Panjab University, Chandigarh - 160 014, India
2 Department of Botany, University of Delhi, Delhi - 110 007, India
* *Corresponding author*: E-mail: phykos@pu.ac.in

fossil fuels that is not renewable within a time scale mankind can use. These fossil fuels have taken around 500 million years to accumulate but will vanish in less than one millionth of time, if consumed at the present pace. At present each and every country is power hungry and wants to ensure continuous supply of fuel, whose available reserves are projected to persist only for 50-100 years. The prices of this scarce resource are automatically soaring with increased demand. In a developing country like India which satisfies its 70% of energy requirements through imports, the import bill has direct bearing on overall finance budget. This leaves only a little of the total budget to be spent on social sector. The prices of crude oil has touched its all time high US $ 147 per barrel in July 2008. Also the burning of fossil fuels is tremendously increasing the concentration of Green House Gases in the atmosphere, leading to rapid climate change (Schenk *et al.,* 2008), which results in global warming. Depleting coal and petroleum reserves and sky rocketing prices have compelled us to switch over to renewable sources of energy. There is dire need to switch over to energy sources like Solar Energy, Wind Energy, Hydroelectric Energy, Nuclear Energy etc. But all these so called renewable sources have a lot of problems associated with them in terms of their adoption, commercial exploitation and regional restrictions (Fischer and Schrattenholzer, 2001). In the wake of all the above discussed problems, bio-fuels such as bioethanol, biomethane, biohydrogen and biodiesel are seen as amicable and reliable solutions (Hu *et al.*, 2008; Hankamer *et al.*, 2007). Biological material can be utilized in production of energy by burning of biomass (wood and gasses) (Fischer and Schrattenholzer, 2001), in production of biogas (methane), generation of hydrogen (Jo *et al.*, 2006), production of ethanol by fermentation (Boyle, 1996) and use of plant derived oils (Mushrush *et al.*, 2006). Though the production and use of most of these bio-fuels has yet to come up to an appreciable level, biodiesel has taken the lead both at research labs and industrial level.

Biodiesel and Its Advantages

Chemically, biodiesel refers to a diesel equivalent processed fuel consisting of short chain alkyl (methyl or ethyl) esters of plant based fats and oils, which can be used alone or blended with conventional diesel (Vasudevan and Briggs, 2008). Commercially, the lipid and fatty acid components of microalgae are converted to biodiesel via a transesterification reaction with methanol or ethanol in the presence of a catalyst, to yield mono alkyl esters, which is generically known as biodiesel (Fukuda *et al.*, 2001; Meher *et al.*, 2006). Biodiesel is a liquid, which varies in color between golden and dark brown depending on the type of feedstock used. It has viscosities similar to conventional diesel, is practically immiscible in water and has low vapour pressure but high boiling point. Although biodiesel cannot entirely replace petroleum-based diesel fuel, there are several reasons that justify its production: [WINDOWS-1252?]• It provides a market for excess production of vegetable oils and animal fats. [WINDOWS-1252?]• Biodiesel is one of the better sources of energy. Large-scale introduction of biodiesel could contribute to sustainable development on several fronts, environmentally, socially and economically. It decreases, although will not eliminate, the country's dependence on imported petroleum. Biodiesel can prove one of the most important components to mitigate greenhouse gas emissions and substitute of fossil fuels. The exhaust emissions of carbon monoxide, unburnt hydrocarbons, and particulate emissions from biodiesel are lower than with conventional diesel fuel. Unfortunately, most emission tests have shown a slight increase in oxides of nitrogen (NOx). When added to regular diesel fuel in an amount equal to 1–2%, it can convert fuel with poor lubricating properties, such as modern ultra-low-sulfur diesel fuel, into an acceptable fuel. The burning of an enormous amount of fossil fuel has increased the CO_2 level in the atmosphere, causing global warming. Biodiesel is a renewable resource and it fixes CO_2 in the atmosphere through photosynthesis. If biomass is grown in a sustained way, its combustion has no impact on the CO_2 balance in the atmosphere, because the CO_2 emitted by the burning of biomass is offset by the CO_2 fixed by photosynthesis. The major hurdle in the use of vegetable oil as biodiesel is the type of TAGs (triacylglycerols) produced by these plants which cannot be directly used as diesel fuel substitute mainly because of viscosity problems which is of the order of 10-20 times that of conventional diesel and hence requires prior processing before use (Aresta *et al.*, 2005).

Sources of Biodiesel

Biodiesel can be obtained either from Hydrocarbons (compounds containing Carbon and Hydrogen) produced by some plants such as members of family Euphorbiaceae, milk weed (*Asclepias* sps) or leguminous tropical trees like *Copiafera multijuga*; or it can also be extracted from lipids accumulated in seeds of many plants (Sood and Singh, 2007). These lipids have high-energy value and can be used as diesel engine fuels. Numerous crops have been proposed or being used for the commercial production of biodiesel in various countries like

Europe, USA, Spain, India etc. The focus has been mainly on oils like soyabean, rapeseed, sunflower, safflower and coconut, which are essentially edible in nature (Schmidt, 2007). A number of non-edible tree born oil seeds like Karanja (*Pongamia glabra*), Jatropha (*Jatropha curcus*) and Mahua (*Madhuca indica*) have great potential for making biodiesel. Few attempts have also been made for producing biodiesel from non-edible sources like frying oil, grease, crude tall oil and tallow (Zhang *et al.*, 2003a, b; Altiparmak *et al.,* 2007). The choice of feedstock is mainly based on its local availability, cost of production, government support and performance as a fuel. Although the biodiesel derived from oil crops is a potential renewable alternative to the petroleum fuel but unfortunately, biodiesel from oil crops, waste cooking oil and animal fat can not realistically satisfy even a small fraction of the existing demand for transport fuel. The major limitation impeding a more widespread use of biodiesel is the extensive acreage of land required for production of biodiesel. Production is also dependent upon the availability of feedstock. Therefore, industrial scale biodiesel production will remain geographically and seasonally restricted particularly in cases where vegetable oil producing crops are used as feedstock. In addition to this, the use of these feed stocks is leading to the global food crisis since a major part of the food crops is being diverted for its use in biodiesel production. In a country like India, where there is already a very high population pressure on available land resources. Government cannot even think of either using arable land for energy crops or diverting food crops for biodiesel production. Recent global food crisis and high inflation in year 2008 has its origin in diverting food crops for biodiesel > production, which was also acknowledged by FAO summit 2008. Sugar prices in Brazil, Maize prices in U.S.A. and Mexico, soybean prices in Europe and China has witnessed quantum jump in recent years. More over these energy crops require extensive acreages of land and at least a few months to years to produce required amount of biomass to provide continuous supply of biodiesel, which makes them unsuitable candidate as a source of biodiesel. Role of Algae as Savior An overview of all the above discussed limitations points out an urgent need for the discovery of such alternatives that could practically address the energy security of the country and meet the problems of climate change. In this regard, algal biodiesel is seen as a panacea as it appears to be the only source of biodiesel that has the potential to displace the conventional diesel with minimum limitations (harvesting and supply of CO_2 for high efficiency). Algae produce vegetable oil, which can be refined to biodiesel. Algae can grow anywhere and thrives on sunshine, which is plentiful and free. (Shay, 1993) reported that algae were one of the best sources of biodiesel. In fact algae are the highest yielding feedstock for biodiesel. Producing biodiesel from algae may be the only way to produce enough automotive fuel to replace current gasoline usage. The best algae for biodiesel would be microalgae. Microalgae are organism, less than 2 mm in diameter capable of photosynthesis. Macroalgae, like seaweed, is not as widely used in the production of biodiesel mainly because of their complex structure and longer life cycle unlike microalgae. Microalgae have much more oil than macroalgae and it is much faster and easier to grow. Although a number of algal species belonging to Xanthophyta, Phaeophyta and Chrysophyta are known to accumulate the storage products in the form of fats and lipids but still the microalgae particularly belonging to Chlorophyta and Cyanophyta (such as *Botryococcus, Dunaliella, Chlorella, Scenedesmus*) are being commercially exploited mainly because they are able to double their biomass within 24 hrs (Song *et al*, 2008). Microalgae can provide different types of renewable biofuels which are produced either by anaerobic digestion of algal biomass or derived from microalgal oil (Chisti, 2007, 2008). These biofuels include anaerobically produced methane, photobiologically produced biohydrogen and biodiesel (Ghiradi, 2006). Advantages of using Microalgae as source of biodiesel Because of their higher photosynthetic efficiency, higher biomass production and fast growth as compared to other energy crops microalgae appear as promising candidate for biodiesel production (Ginzburg, 1993; Dote *et al.*, 1994; Minowa *et al.*, 1995; Ratledge, 2004). It has many other advantages:

Oil content in microalgae can exceed up to 80% by weight of dry biomass (Metting, 1996; Spolare *et al.*, 2006). Oil levels of 20-40% are common among algae. It indicates oil content varies with different species of algae (Li *et al.*, 2002; Rezanka *et al.*, 2003) and oil productivity varies with growth rate, age of culture, factors indirectly affecting biomass production (Ginzburg, 1993). The yield of oil from microalgae are orders of magnitute much higher than traditional crop oils, if per hector soya crop results in 400 liters of oil, the same amount of algae results in 100,000 liters of oil. The yield is much superior and productivity per unit area is 2-5 folds higher as compared to other terrestrial crops. The productivity rates of microalgae can be up to 12.5 kg/m2/yr. Production of microalgal biomass can be carried out either in raceways ponds (open ponds) (Richmond, 1992) or in much sophisticated photobioreactors (Molina Grima *et al.*, 1999; Ugwa *et al.*, 2007). The major limitations with outdoor systems have been the problem of contaminants and biomass production varies with outdoor environmental conditions (Bisoyi and Singh, 1988; Richmond, 1992). Closed photobioreactors designs vary from tubular reactor (of wide range of configurations); plate reactor, helical reactor to hanging bags of various dimensions (Richmond, 2004). Although photobioreactors provide much higher yield of oil per hectare because of higher biomass productivity as compared to open raceways but they also increase overall cost of production of microalgae. Another system of indoor production of microalgae under polyhouse/glasshouse conditions has been successfully utilized at Centre for Conservation and Utilization of Blue-Green algae, Indian Agricultural research Institute, New Delhi, India for biomass production of microalgae, is economically less expensive as compared to photobioreactors. The production of microalgae under such polyhouse condition is faster as there are less chances of contamination than that of open systems and allows throughout the year production of microalgae (Singh and Pabbi, 1998). Algae like higher plants use photosynthesis to convert the sun's energy into chemical energy. One million tons of algal biomass can consume 1.6 million tons of CO_2 and release 1.2 million tons of O_2. Therefore, algae can solve two major problems: air pollution resulting from CO_2 evolution, hence can reduce global warming and future cries due to shortage of energy sources. Microalgae are basically aquatic organisms, their cultivation in liquid environment allows them to have better access to resources: water, CO_2 and nutrients. It is for this reason that microalgae are capable of synthesizing 30 times more oil per hectare than other terrestrial plants.

Microalgae system consumes less water than traditional oil seed crops and also land is hardly a limitation in their production. An interesting factor with microalgal production is the possibility of using human sewage or other waste products such as polluted streams from fertilizer run off to feed the microalga with nutrients and convert it into a useful feed crop for biodiesel production. This will not only answer our problem of land space for production but also helps in bioremediation. Microalgae complete their life cycle in within few days as compared to any other crop and can be harvested daily from regrown biomass. The microalgal oil can be extracted and purified using conventional technologies without the need for any expensive units dealing only with microalgal oil. Also, the extracted oil can be processed for biodiesel production in the same manner as the oil extracted from oil seeds. Therefore microalgae appear to be the best option for biodiesel production for self-reliance and sustainability of a country like India.

Economic Feasibility of Biodiesel Production

The economic feasibility of biodiesel production depends upon the fact that the feedstock biomass should be producible with low-input (*i.e.* less fertilizer, pesticides and energy) on land with low agricultural value and requires low-input energy to convert feedstock's to biodiesel (Haas *et al.*, 2006; Hill *et al.*, 2006). They also reported a direct linear relationship between the cost of feedstock and biodiesel production cost. Ramesh *et al.* (2007) calculated a net cost of production of Jatropha biodiesel is Rs 28/L, with raw material (*Jatropha* oil) costs Rs 22/L and processing costs Rs 9/L. It is important to note in all the cases cited above, feedstock cost comprise a very substantial proportion of overall biodiesel cot. This highlights the need to explore alternate feedstock with main aim to minimize cost raw material. In the light of above mentioned observations it can be safely concluded that using microalgae as source of biomass for biodiesel production will be much cheaper than other plant crops; because many unicellular algae, like bacteria, multiply their biomass with in no time. Thus use of microalgae for biodiesel production poses no problem at least at economic level.

Current Status

The idea of using microalgae for production of biodiesel is not new but is now being taken up seriously to cope up with the dual challenge of obtaining renewable energy sources and maintaining clean environment. So it has become more imperative for organizations and countries to invest more time and efforts into research on suitable feedstock such as algae. The concept of producing biodiesel from microalgae has started gaining momentum worldwide. Attempts are being made constantly by the scientists all over the world in order to increase the lipid content of microalgae from 20% to 80% using a number of advanced techniques like genetic engineering. This issue was also discussed at the 1st International Algal Biomass Summit in San Francisco in November 2007. Both developed and developing countries are in race to harness this potential renewable source of energy. For sure, research is being conducted by brilliant and devoted scientists all over in this direction. The algae-for-fuel' concept has been explored in the USA and some other developing countries. In India also experiments are 'on' in different universities and premier research institutes including Indian Institutes of Technology (IITs) to master this new technology, but still more efforts are required to bring this technology from lab to the industry.

References

Altiparmak, D., Keskin, A., Koca, A. and Guru, M. (2007). Alternative fuel properties of tall oil fatty acids methyl ester-diesel fuel blends. *Bioresource Technology* **98**: 241-246.

Aresta, M., Dibenedetto, A., Carone, M., Colonna, T. and Fragale, C. (2005). Production of biodiesel from macroalgae by supercritical CO_2 extraction and thermochemical liquefaction. *Environmental Chemical Letter* **3**: 136-139.

Bisoyi, R.N. and Singh, P.K. (1988). Effect of seasonal changes on cyanobacterial production and nitrogen yield. *Microbiology Ecology* **16**: 149-154.

Boyle, G. (1996). Renewable Energy. Oxford University Press, Oxford.

Chisti, Y. (2007). Biodiesel from microalgae. *Biotechnology Advances* **25**: 294-306.

Chisti, Y. (2008). Biodiesel from microalgae beats bioethanol. *Trends in Biotechnology* **26(3)**: 126-131.

Dote, Y., Sawayama, S., Inoue, S., Minowa, T. and Yokoyama, S. (1994). Recovery of liquid fuel from hydrocarbon-rich microalgae by thermochemical liquification. *Fuel* **73**: 1855-1857.

Fischer, G. and Schrattenholzer, L. (2001). Global boienergy potentials through 2050. *Biomass and Bioenergy* **20**: 151-159.

Fukuda, H., Kondo, A. and Noda, H. (2001). Biodiesel fuel production by transesterification of oils. *Bioscience and Bioenergy* 92: 405-416.

Ghiradi, M.L. (2006). Hydrogen production by photosynthetic green algae. *Journal of Biochemistry* **43**: 201-206.

Ginzburg, B.Z. (1993). Liquid fuel (oil) from halophylic algae: a renewable source of non-polluting energy. *Renew Energy* **3**: 249-252.

Hankamer, B., Lehr, F., Ruppercht, J., Mussgnug, J.H., Posten, C. and Kruse, O. (2007). Photosynthetic biomass and H_2 production by green algae: from bioengineering to bioreactor scale up. *Plant Physiology* **131**: 10-21.

Hass, M.J., McAloon, A.J., Yee, W.C. and Foglia, T.A. (2006). A process to estimate biodiesel production costs. *Bioresource Technology* **97**: 671-678.

Hill, J., Nelson, E., Tilman, D., Polasky, S. and Tiffany, D. (2006). Environmental, economic, and energetic costs and benefits of biodiesel and ethanol biofuels. *PNAS (Proceedings of the National Academy of Sciences)* **103**: 11206-11210.

Hu, Q., Sommerfeld, M., Jarvis, E., Ghirardi, M., Posewitz, M., Seibert, M. and Darzins, A. (2008). Microalgal triacylglycerols as feedstocks for biofuel production: Perspectives and advances. *The Plant* **54**: 621-639.

Jo, J.H., Lee, D.S. and Park, J.M. (2006). Modelling and optimization of photosynthetic hydrogen gas production by green alga *Chlamydomonas renhardtii* in sulfur-dervied circumstances. *Biotechnology Progress* **22**: 431-437.

Li, X., Fan, X., Han, L. and Lou, Q. (2002). Fatty acids of same algae from Bohai Sea. *Phytochemistry* **5**: 157-161.

Meher, L.C., Vidya Sagar, D. and Naik, S.N. (2006). Technical aspects of biodiesel production by transesterification a review. *Renewable and Sustainable Energy Reviews* **10**: 248-268.

Meher, L.C., Kulkarni, M.G., Dalai, A.K. and Naik, S.N. (2006). Transesterification of Karanja *(Pongamia pinnata)* oil by solid basic catalysts. *European Journal of Lipid Science and Technology* **108(5)**: 389-397.

Metting, F.B. (1996). Biodiversity and applications of microalgae. *Journal of Indian Microbiology* **17**: 477-489.

Minowa, T., Yokoyama, S.Y., Kishimato, M. and Okakurat, T. (1995). Oil production from algal cells of *Dunaliella tertiolecta* by direct thermochemical liquefcation. *Fuel* **74**: 1735-1738.

Molina Grima, E., Acien Fernandez, F.G., Garcia Camacho, F. and Chisti, Y. (1999). Photobiorectors: light regime, mass transfer, and scaleup. *Journal of Biotechnology* **70**: 231-247.

Mushrush, G.W., Wynne, J.H., Willauer, H.D. and Lloyd, C.J. (2006). Soyabean-derived biofuels and home heating fuels. *Journal of Environmental Science and Health, Part A: Toxic/Hazardous Substances and Environmental Engineering* **41**: 2495-2502.

Ramesh, D., Samaoathrajan, A. and Venkatachalam, P. (2007). Production of biodiesel from *Jatropha curcus* oil by using pilot biodiesel plant. www.jatropha.de/journal/ pilot_plant_biodiesl_leaflet

Ratledge, C. (2004). Fatty acid biosynthesis in microorganisms being used for single cell oil production. *Biochimie* **86**: 807-815.

Rezanka, T., Viden, I., Go, J.V. and Dembitsky, V.M. (2003). Polar lipids and fatty acids of three wild cyanobacterial strains of genus *Chlorococcidiopsis*. *Folia Microbiology* **48**: 781-786.

Richmond, A. (1992). Open systems for mass production of photoautotrophic microalgae outdoors: physiological principles. *Journal of Applied Phycology* **4**: 281-286.

Richmond, A. (2004). Biological principles of mass cultivation: *In*: Richmond, A. (*Eds.*) Handbook of Microalgal Culture: Biotechnology and Applied Phycology, Blackwell, pp. 125-177.

Schenk, P.M., Themas- Hall, S.R., Stephens, E. *et al.* (2008). Second generation biofuels: High efficiency microalgae for biodiesel production. *Bioenergy Resources* pp. 20-43.

Schmidt, C.W. (2007). Biodiesel: cultivating alternative fuels. *Environmental Health Perspectives* **115(2)**: A86-A91.

Shay, E.G. (1993). Diesel fuel from vegetable oils: status and opportunities. *Biomass and Bioenergy* **4**: 227-242.

Singh, P.K. and Pabbi, S. (1998). Mass Production of Blue-green Algal Biofertilizer, NCCUBGA, IARI, New Delhi, India, p. 67.

Song, D., Hou, L. and Shi, D. (2008). Exploitation and utilization of rich lipids microalgae as new feedstock's for biodiesel production: a review.

Sood, A. and Singh, P.K. (2007). Microalgae as a source of Biodiesel. *Indian Science Abstracts* **43**: 14-17.

Spolare, P., Joannis-Cassan, C., Duran, E. and Isambert, A. (2006). Commercial applications of microalgae. *Journal of Bioscience and Bioengineering* **101**: 87-96.

Ugwa, C.U., Aoyagi, H. and Uchiyama, H. (2007). Photobiorectors for mass cultivation of algae. *Bioresource Technology* p. 20.

Vasudevan, P.T. and Briggs, M. (2008). Biodiesel production: current state of the art and challenges. *Journal of Industrial Microbial Biotechnology* **35(5)**: 421-430.

Zhang, Y., Dube, M.A., Mchean, D.D. and Kates, M. (2003a). Biodiesel production from waste cooking oil: Process design and Technological assessment. *Bioresource Technology* **89(1)**: 1-16.

Zhang, Y., Dube, M.A., McLean, D.D. and Kates, M. (2003b). Biodiesel production from waste cooking oil. II. Economic assessment and sensitivity analysis. *Bioresource Technology* **90**: 22

4

PROSPECTS OF MICROALGAE OF NORTH EAST INDIA FOR BIODIESEL PRODUCTION

Jayanta Talukdar[1*] and M.C. Kalita[2]

Abstract

The triad of environmental, energy and economic security are the key issues for the sustainability of a real sense developmental process, which are now being seriously jeopardized with our impetuous activities. Various options of generating renewable energy alternatives are being explored due to the growing concern of diminishing fossil fuel reserves and their contributions towards the enrichment of air with major Greenhouse Gas CO_2*, which induce serious environmental consequences. An elevated level of* CO_2 *in the atmosphere is chiefly responsible for the warming of global climate. Adoptions of effective mitigation strategies with clean energy development are therefore need of the hour for the sustainability of the developmental processes. Alternative fuel productions from renewable photosynthetic biological sources have the potential to provide an environment friendly sustainable energy system for the society besides mitigating* CO_2*. The use of biomass for energy is largely motivated from the stand point of global energy crisis and global environmental issues. Alternative fuels that derived from biomass would reduce GHGs concentration potentially and could replace fossil fuels currently being used. Biodiesel is a clean burning renewable liquid biofuel, currently gaining worldwide acceptance as an able alternative to petroleum fuel. Biodiesel fuel production from microalgae has attracted much attention in recent times as a promising bio-feedstock for alternative liquid fuel*

1 Microalgae Biotechnology Division, Environment Biotechnology Laboratory, Gauhati University, Guwahati – 781 014, Assam

2 Department of Biotechnology, Gauhati University, Guwahati – 781 014, Assam

* *Corresponding author*: E-mail: jayanta_talukdar@yahoo.com

production over other non-edible oil seed crops owing to their higher oil yielding efficiency, faster growth rate, higher CO_2 *fixation and wide range of tolerance to varying environmental conditions. Large scale cultivation of microalgae for liquid biofuel production with an integrated and innovative approach for feeding* CO_2 *generated by industrial process using sewage / waste water might be an optimal solution to global warming, rising fuel crisis and saving our precious fresh water resources. The efficiency of microalgal liquid biofuel production largely depends on the selection of potent strains and appropriate technology. Exploration of the existing microalgal biodiversity has therefore a key role in finding an appropriate strain for sustainable production of biodiesel. India is facing formidable challenges in meeting its energy demands and in providing adequate energy of desired quality in various forms in a sustainable manner and at competitive prices. India produces only 22% of the mineral fuel requirement of a year and the rest is imported at a huge cost of foreign money. With the escalating crude oil prices in the international market and to ward off Global Warming that threatens our very existence, it is pertinent that India's need for biofuel development is the order of the day. India has vast expanse of waste and marginal land, tropical climate and vast manpower for developing sustainable biodiesel production towards its national energy security.The North Eastern region of India, more particularly Assam has an ambient of harboring rich diversity of freshwater microalgae. There was no earlier record regarding the oleaginous microalgae of the region. Accordingly a study was conducted and 22 microalgal species were isolated and screened for oil content. Among the isolated microalgae, Botryococcus braunii, Ankistrodesmus sp., Scenedesmus sp., Euglena sp., Navicula sp. and Nitzchia sp. are a few oleaginous microalgae noteworthy for biodiesel production. Hydrocarbon and lipids were quantitatively evaluated and measured gravimetrically in cultured B. braunii (21.9–60.7%), Ankistrodesmus sp. (55%), Chlorococcum sp (11.3–12.2%) and Scenedesmus sp (17.3%) following standard procedure.*

Key words: *Ankistrodesmus*; *Botryococcus braunii,* Biodiesel, Biodiversity, *Chlorococcum,* Global Climate Change, Global Energy Crisis, Microalgae, *Scenedesmus,* Sustainable

Introduction

Global Climate Change and Energy Crisis are the two most debated problems of the present era – a veracity the entire human civilization is facing. Global Warming is primarily attributed to the elevated level of CO_2 in the atmosphere (Wang *et al.,* 2008; Kondili and Kadellis, 2007; Roman - Leshkov *et al.*, 2007), which has been increasing sharply since the dawn of the Industrial Revolution. The global atmospheric CO_2 concentration has increased from nearly 280 ppm to over 380 ppm over the last 1000 years (Fig. 1) of which the largest changes have been observed in the last 200 years. Increasing concentration of CO_2 and other heat trapping green house gases (GHGs) in the atmosphere, primarily due to the combustion of fossil fuels in addition to other anthropogenic activities are responsible for rapid warming of the planet (IPCC 4AR, GW1, 2007). Our continued reliance on fossil fuels has sufficiently enriched the atmosphere with CO_2 and about 70% of the global warming problem is indeed caused by the combustion of fossil fuels for energy production and use, which contributes about 40% of energy-related global CO_2 emissions. The most rapidly growing world's transport sector accounts for about 25% of the total annual global anthropogenic Green House Gases emissions (Burk *et al.*, 2008), which has emerged as the prime sector for increasing global primary fuel demand and is predicted to grow by 1.3% per year up to 2030, envisaged to reach an overwhelming 116 million barrels of fossil fuel per day (The Royal Society, 2008). The rapidly growing economies of the world, particularly the developing countries including China and India adding

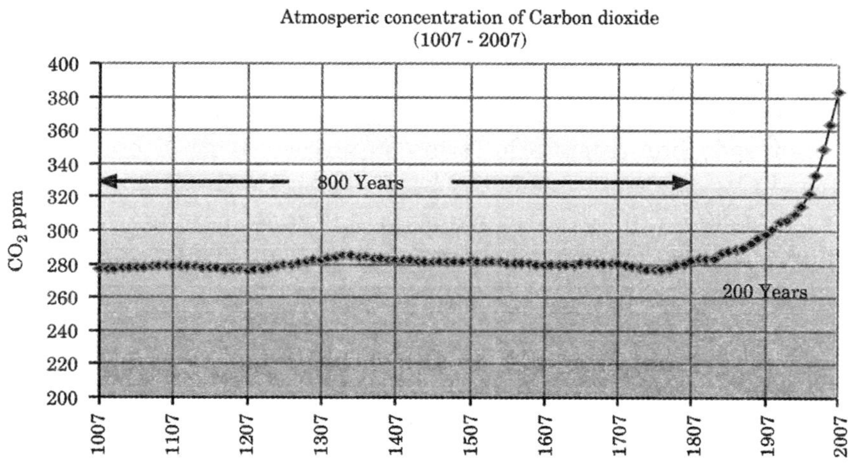

Fig. 1: Global atmospheric CO_2 concentration (1007-2007).
Source: Data from Mauna Loa by Scripps Institute of Oceanography, CDIAC, and NOAA/ESRL at www.esrl.noaa.gov/gmd/ccgg/trends, updated January 2008, with long term historical data compiled by Worldwatch Institute from Scripps, ORNL, and IPCC. http://www.earth-policy.org/Indicators/CO_2/2008_data3.htm

with increasing population growth have also mounted increasing demand of fossil fuels, which ascertain an uprising trend in fuel prices with further increase in the emissions of GHGs.

India, the second most popular country in the world, is one of the fastest growing economies with an average of 8.7% of exponential GDP growth drawing about 3.6% of the global primary energy demand (The Planning Commission, Govt. of India 2007; IEA, 2007). India's share towards the global CO_2 emissions currently levels at about 4.23% (1.1 Gt), ranking 5th among the top five largest global CO_2 emitters leading by the USA. 21.44% (5.8 Gt) followed by China 18.80% (5.2 Gt), Russia 5.69% (1.5 Gt) and Japan 4.47% (1.2 Gt) (Fig. 2, IEA, 2007). With strong economic growth and heavy reliance on fossil fuels, world CO_2 emissions are expected to be continued steadily from 28.1 billion metric tons in 2005 to 34.3 billion metric tons in 2015 and 42.3 billion metric tons in 2030 (EIA, 2008).

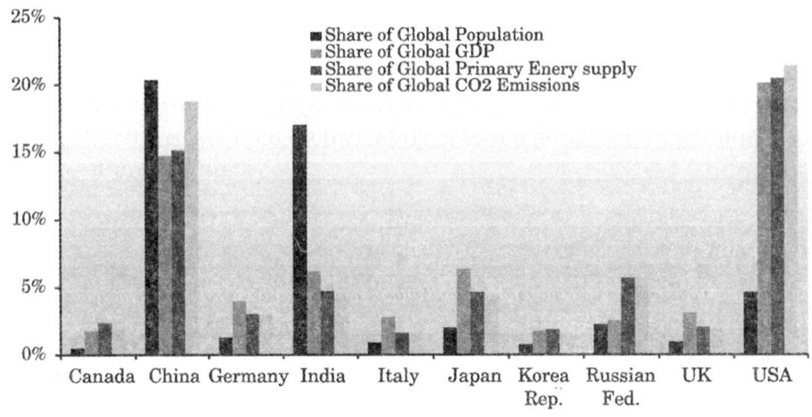

Fig.2: Global share of Population, GDP, Primary Energy Supply and CO_2 Emissions of the top ten largest CO_2 emitter countries (Data source: IEA, 2007).

The entire world is facing a formidable fuel crisis due to the declining non-renewable energy resources popularly known as 'Peak Oil' while the energy demand is exploding (Campbell, 2008). World's oil production is estimated to decline in between one and ten decades (Crookes, 2006) with the result of an impeding energy crisis. Driven by such dwindling supply, the cost of petroleum fuel has gone up sky high in the recent times, which has ominously jeopardized the economic progress of the nations.

The Global economy literally runs on energy. The inevitable role of energy in economic development of a country can hardly be avoided that dictates the economic progress of a nation. However, the current trend in soaring fuel prices poised by diminishing fossil fuel reserves is emerging as a potential threat to the global economic security. Despite the uprising fuel prices, energy demand is also increasing considerably with the increasing population growth and rapid economic progress around the globe. Global energy demand as driven by strong economic growth adding with world's growing population has poised a potential threat towards environmental security. Given the projected uprising trend in world population growth and energy demands, the most challenging issue facing the international community today is how to attain a sustainable economic development *vis-à-vis* environmental security.

Realizing the consequences of global warming resulting from the elevated CO_2 emissions propelled by our ravenous appetite of fossil fuels, issues like minimizing the emission level by declining the use of fossil fuels and CO_2 mitigation strategies have become a major focus of all environmental agenda. An alternative energy policy that improves the environmental security and also eases our dependence on fossil fuels with the development of clean energy is the order of the day to secure national energy security and sustainable economic development. Development of renewable sources of alternative energy has, hence become crucial in the national and international effort towards maximum self-reliance – the corner stone of our energy security strategy. The use of biomass as a renewable source of alternative energy such as biofuels is largely motivated from the stand point of global energy crisis and global environmental issues. However, biofuels derived from grains and seed oils of land based higher plants are not climate neutral as they release large amounts of N_2O, initiate land clearing, potentially compete with net food production and release about 10-100 times more CO_2 as is saved (Moomaw, 2008; Searchinger *et al.*, 2008; Scharfemann and Larance, 2008; Fargione *et al.*, 2008; Crutzen *et al.*, 2007). On the other hand alternative fuels derived from algal biomass have significant benefits as a technology to CO_2 Capture and Storage (CCS), which would not only mitigate CO_2 besides reducing other GHGs concentration in the atmosphere but also could replace fossil fuels currently being used (IPCC, 2007; Moomaw, 2008). Among the different mitigation strategies of CO_2, bio-mitigation by using microalgae is one of the options for low cost CO_2 mitigation strategy with multiple advantages (Borowitzka, 1999; Chisti, 2007; Li *et al.*, 2008; Wang *et al.*, 2008).

In this present pretext we have reviewed and summarized issues confronting India's national energy cum economic security. The advantages of microalgae technology in Indian perspectives have also been discussed. Possible corroborate production of biofuels is possible with the utilization of indigenous potent strain, which has been over sighted in the light of our original research findings. The ultimate goal of our research was to explore the existing oleaginous microalgal diversity of India's North Eastern Region for possible corroborate development of biofuel.

Petroleum Fuel Scenario of India

India ranks 7th as the world's largest energy producer accounting for about 2.49% of the world's total annual energy production and world's 5th largest energy consumer with about 3.6% of the global primary energy demand (The Planning Commission, Govt. of India 2007; IEA, 2007). Despite, being amongst the largest energy producer, India is a net importer of energy largely due to huge imbalance between energy consumptions and productions (Fig. 3). About 30% of India's total primary energy needs are being met by petroleum oil and about 76% of which are imported. With an appreciable economic growth of nearly 8.7% presently, India is expected to swell over in the coming years bound to develop an insatiable appetite for fuel. The indigenous energy resources of India will not be sufficient in the long run to sustain the envisaged growth of economies at 9% of GDP (The Planning Commission,

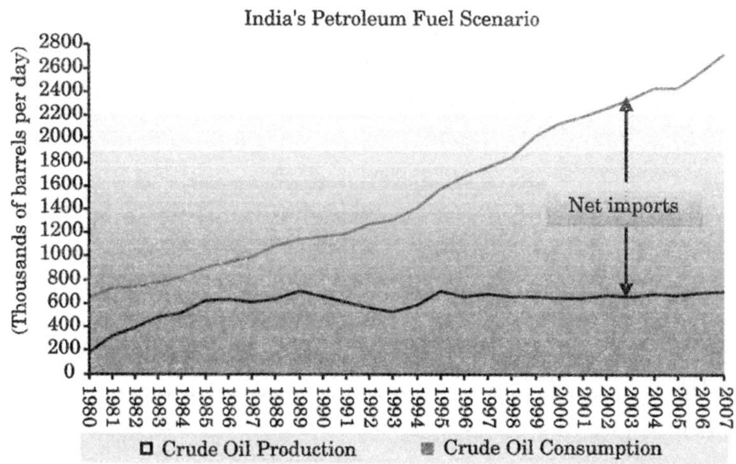

Fig. 3. India's Petroleum Fuel Production and Consumption (1980–2007). Data *Source*: Planning Commission, GoI 2007-2006; EIA, 2007.

Govt. of India, 2007). The energy supply system of India continues to be dependent on fossil fuels, which are contemplated to exhaust. India had 5.6 billion barrels of proven oil reserves, the second-largest (behind China) among the Asia-Pacific region (EIA, 2006), which can sustain the current level production for the next 35 years only (The Planning Commission, Govt. of India, 2007). Most of these crude oil reserves are located offshore near Mumbai and onshore in Assam state, although substantial undeveloped reserves are located in the offshore Bay of Bengal and in Rajasthan state (EIA, Govt. of India, 2006). The exponential growth of rising oil consumption and fairly stable production levels placed India increasingly dependent on foreign oil imports to meet its consumption demands. In 2006, the country produced an average of 0.846 million barrels per day of total oil liquids, of which 0.648 million barrels was crude oil, while in the same year, India consumed about 2.63 million bbl/d of oil. On the price front, international crude oil prices have increased by about six times from USD 23.32 per barrel in March 2002 to USD 147.02 per barrel in July 2008, poised India to spend huge amount of foreign reserves on its crude oil imports. The oil import bill for the year 2007-08 was nearly USD 77.02 billion, which is 40% more than that of 2006-07 (Economic Advisory Council to the Prime Minister, Govt. of India, 2008). However, at the present level of crude oil prices reaching all time high over USD 147 per barrel in the month of June-July, 2008 and likely to be persist over USD 100 per barrel for the rest of the year coupled with depreciation in value of rupee against dollar, the oil import bill is expected to swell over USD 100 billion. The impact of crude oil price at close to USD100 per barrel may have a dampening impact on economic activity with larger inflationary burden. This rising oil import bill has been the focus of serious concern due to the pressure it placed on national economy, scarce foreign exchange resources, inflation and also, because it is largely responsible for national security reason (Economic Advisory Council to the Prime Minister, Govt. of India, 2008). This rising oil import bill has been the focus of serious concern due to the pressure it placed on national economy, scarce foreign exchange resources, inflation and also, because it is largely responsible for national security reason.

The Rational of Renewable Alternative Fuel Develop-ment in India: Current Feedstock and Constrains

India is facing formidable challenges in meeting its energy demands and in providing adequate energy of desired quality in various forms in a sustainable manner and at competitive prices. India needs to sustain an 8-10% economic growth for the next 25 years to eradicate poverty and meet its human development goals (The Planning Commission, Govt. of India, 2007). The energy demand of the country is expected to grow at an annual rate of about 4.8%. Most of the energy requirements are currently satisfied by fossil fuels. Petroleum oil provides energy for about 95% of transportation of the country and the demand for transport fuel continues to rise. However, domestic production of crude oil can only fulfill between 25-30% of the national consumptions and the rest have to be met by import. Our dependence on imports of foreign oil will continue to increase in the foreseeable future. The projected crude oil demand of India is in between 350-486 Mtoe by 2030 of which only about 35 Mtoe could be met by domestic production (The Planning Commission, Govt. of India, 2007). Availability and easy access to energy are considered as catalysts for sustainable economic development. Meeting the energy access challenges and ensuring lifeline supply of clean energy to all is essential if growth is to be inclusive (The Planning Commission, Govt. of India, 2007). Undoubtedly, biofuels are going to play an immensely important role in meeting India's surging energy demands and an environmentally sustainable developmental process.

Biodiesel – a clean burning alternative liquid biofuel is currently gaining worldwide acceptance as an environmentally friendly solution to the growing global energy crisis. It has emerged as the most able alternative to petroleum diesel fuel owing to its environmentally friendly characteristics, biodegradability and renew-ability. Presently biodiesel is produced from vegetable oils of plant and animal fats, offers promising renewable source of alternative transport fuel with environmental, economic and national energy security benefits. It burns in conventional diesel engines with or without any modifications while reducing pollutions (100% less sulfur dioxide, 37% less unburned hydrocarbons, 46% less carbon monoxide, and 84% less particulate matter) in comparison to conventional diesel fuel. Biodiesel is prominently an acceptable option in Indian perspectives for energy security, reduction in oil imports, rural employment generation, and improving agriculture economy.

The Indian subcontinent with an area of 329 million hectares has a tropical advantage of growing several species of energy crops including sugarcane, wheat, castor, jatropha, pongamia etc for biofuel production. India's current feedstock for bioethanol production (as a substitute of petrol) is largely concentrated on sugarcane, which is a water intensive crop. Among the other feedstock *Jatropha curcas* has gained much attention for biodiesel production in India. Jatropha plant bears fruits from second year of its plantation and the economic yield stabilizes from $4^{th}/5^{th}$ year onwards having an average life with effective yield up to 50 years. It can produce around 2 kg of seeds/plant and the economic yield is considered to be 4-6 tones per hectare per year depending on agro-climatic conditions and agricultural practices. The yield of *Jatropha* on average soil is about 1.6 tones of oil per hectare per year. The Government of India has introduced a nationwide programme under the National Mission on Biofuel in 2003 with the envisage of achieving a target of meeting 10.14 million tones of biodiesel (@ 20% blending) from *J. curcas* by 2012 and to achieve the target about 27 billions of planting materials are required to be planted over 11.2 million hectares of land (Planning Commission, Govt. of India, 2003). On the other hand India will require about 111.9 million tones of 100% biodiesel to completely replace its petroleum diesel demand by 2020. India has about 63.9 mha of wasteland mainly because they are unsuitable for cultivation at its present stage (Table 1). However among the total wastelands only 27% or 17.3 mha (land amount included in categories 11, 12 and 13 in the Table 1) are considered potential for *Jatropha* plantation (Rajagopal, 2002), which implies that Jatropha alone cannot fulfill India's total biodiesel requirement to replace its petroleum diesel demand or otherwise India will have to sacrifice more and more

Table 1: Wasteland and its categories in India

S. no.	*Category of wasteland*	*Area (mh)*	*% of total wasteland*
1	Snow covered/Glacial	5.6	9%
2	Barren rocky/Sheet rock	6.5	10%
3	Sand-in-land/Coastal	5.0	8%
4	Land effected by salinity/alkalinity	2.0	3%
5	Gullied/ravenous land	2.1	3%
6	Upland with or without scrub	19.4	30%
7	Water logged & marshy	1.7	3%
8	Steep sloping area	0.8	1%
9	Shifting cultivation land	3.5	6%
10	Mining/industrial wastelands	0.1	0%
11	Degraded/pastures/grazing land	2.6	4%
12	Under utilized/degraded notified forest land	14.1	22%
13	Degraded land under plantation crop	0.6	1%
	Total	64.0	100%

Source: www.dolr.nic.in/wasteland.htm

productive land to completely replace the projected demands of petroleum diesel with *Jatropha* biodiesel. India with just 2.4% of global area supports more than hundred millions of the human population; it may not be possible to set aside its farmland for energy crops for biodiesel production due to the pressure of producing food grains. This is where India must need to find additional, reliable, cost-effective and sustainable feedstock for biodiesel production. In this context biodiesel from microalgae seems to be a suitable substitute for petroleum or diesel fuel in the long run.

Microalgae - A Viable Option for Biological CO_2 Mitigation and Potential Biofuel Production

The current consensus of the scientific community is that atmospheric CO_2 concentrations will continue to rise in direct proportion to fossil fuel use with significant consequences for global climate; alternatively, minimizing the use of fossil fuel will diminish the risk of many negative outcomes of GHGs induced climate change. Clearly, the optimal mitigation scenario for the future involves a transition to renewable source alternative energy (Huntley and Redalje, 2007). Renewable energy such as biomass energy has attracted much attention as a promising alternative way of Biological CO_2 mitigation. Bio-mitigation of CO_2 involves CO_2 fixation via photosynthesis leading to the production of biomass energy such as biofuels that offers the promises of improving environmental quality, reducing the carbon imprint of the transportation sector, and contributing towards improved national energy security by offsetting petroleum imports. The potential for increased CO_2 capture in agricultural crop based Biological CO_2 mitigation was estimated to contribute only 3-6% of fossil fuel emissions (Skjanes *et al.*, 2007) mostly due to slow growth rates of the conventional terrestrial plants. On the other hand microalgal biomass with a much faster growth rate have natural ability to sequester CO_2 while trapping solar energy with an efficiency of 10-50 times higher than that of terrestrial plants (Li *et al.*, 2008; Ushi and Ikenouchi, 1997). The concept of mitigating CO_2 using the microalgae *Chlorella* cultures has been described as a method to reduce CO_2 emissions (Jeong *et al.*, 2003).

Microalgae are capable of producing more the amount of oil per unit area of land in comparison to that of all other known oil producing crops (Campbell, 2008; Chisti, 2007; Schneider, 2006; Haag, 2007). The annual yields of microalgal oil per unit area are in order of magnitude greater than those of higher plants. The per hectare yield of microalgal oil has been estimated between 58,700–136,900 liters per year, which is 7–31 times greater than the next best crop *i.e.* palm oil (5950 L $ha^{-1}y^{-1}$) (Chisti, 2007) or *Jatropha* (1892 L $ha^{-1}y^{-1}$). According to an estimate algae can produce about 10 tones of biodiesel per acre per year (Mitchell, 2008) and only about 4.52 mha of land will be required to meet India's fossil diesel requirement by 2020. Realizing such high oil yielding potentialities with much faster growth rate and efficient CO_2 fixers, microalgae appear to be the best option as a renewable source of biofuel that has the potentiality to completely displace petroleum diesel fuel. To support the rapid growth of microalgae there needs an algal production unit with additional amounts of CO_2 that could be supplied from industrial sources like power plants, cement factories or ethanol factories which would have otherwise been a potential source of global warming. Thus a potential source (CO_2) of global warming could be converted to a source of clean burning biofuel. In addition, all other methods used for mitigation of CO_2 only sequester CO_2 but do not convert it into much necessitate fuel, implies microalgae a sensible strategy of choice to combat climate change.

Biodiesel produced from current feed stocks of oil bearing seed crops alone cannot substantially fulfill the existing demand. Moreover an increasing demand of biodiesel worldwide will force to convert more and more fertile land for energy cropping, which potentially compete with net food production and also, will pace up other environmental issues like deforestation, loss of biodiversity, water crisis etc. Knowing that majority of the countries including India is already thriving under the acute pressure of Global Food Crisis, the strategy of large scale energy cropping with land crops for biofuel production is virtually unsustainable.

Integrated Microalgal Biomass Technology for Sustainable Biofuel Production

Microalgal Biomass has gained increasing attention as the most promising renewable source of biofuel owing to its obvious contributions towards the sustainable development - environmentally, economically and socially. Research and Development on microalgae cultivation for Renewable Energy Production and CO_2 utilization has been carried out in the US and other countries for several decades. During 1970's researches in the US with the active support from the Department of Energy, Govt. of USA (DOE) emphasized the process combing wastewater with conversion of harvested microalgal biomass into methan (biogas) by anaerobic digestion, and now successfully implemented in the States (Benemann, 2003). The "Microalgal Biomass Technology" for alternative fuel production offers numerous advantages (Fig. 4). Firstly, microalgae have higher efficiency of CO_2 fixation and faster growth rates as compared to conventional oleaginous crop plants (Borowitzka, 1999; Chisti, 2007; Li *et al.*, 2008). Secondly, microalgae could completely recycle CO_2 by converting it into the chemical energy via photosynthesis that can be converted again into fuels using existing technologies (Demirbas, 2004). Thirdly and more importantly as aquatic plant, algae do not compete with arable land for cultivation. Algae can be grown in freshwater or saline, and salt concentrations up to twice that of seawater can be used effectively (Brown and Zeiler, 1993; Aresta *et al.*, 2005) that implies algae need not compete with other users for fresh water (Campbell, 2008). With an innovative and integrated approach the microalgal biofuel production technology can be made economically cost-effective and environmentally sustainable by combing it with other processes such as wastewater treatment (Wang *et al.*, 2008). The utilization of wastewater that is rich in N/P content may bring about remarkable advantages by providing N and P nutrients for growing microalgae which in turn will remove nitrogen and phosphorus

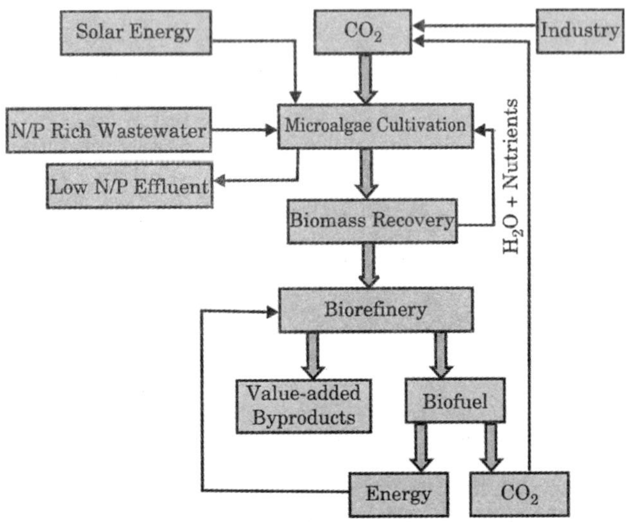

Fig.4: A conceptual microalgae cultivation system for combined biofuels production, CO_2 bio-mitigation & N/P removal from waste water. Input carbon source Co_2: Nitrogen and phosphorus source: N/P rich wastewater; Energy source: Solar energy. Outputs: low N/P effluent, value added by product and biofuels (Redrawn after Want *et al.* 2008 & Chisti 2008)

(Mallick, 2002) as well as metal ions. Such integrated efforts to cleanup industrial flue gas with microalgal culture by combining it with wastewater treatment will significantly enhance the environmental and economical benefits of the technology for biofuel production by minimizing the additional cost of nutrients and savings of precious freshwater resources. Fig. 4 depicts a conceptual flow-chart of the integrated microalgal technology for biofuel production by fixing CO_2 from industrial exhaust gases and solar energy with complete "recycling" of CO_2 and waste water treatment. The merits of CO_2 bio-mitigation locates primarily in the fact that the biomass produced in the process of CO_2 fixation can be converted efficiently into biofuel. Usage of fast growing oleaginous microalgae species could further enhance the feasibility of the technology.

Algae as Source of Biomass for Liquid Biofuel Production: Selection of Potent Strains

The prospects of microalgae utilization in the production of biofuel was extensively studied in the USA by the US DOE during 1970 to 1994 under the much acclaimed "Aquatic Species Program" (ASP). The algae that are considered for biofuel production are usually unicellular aquatic microalgae with faster growth rates and high oil yield (Table 2). Providing optimum conditions some microalgae can double its biomass in less than 24 hours (Chisti, 2007; Schneider, 2006). This high yield and high density biomass is ideal for intensive culture and an excellent renewable source for liquid biofuels. Microalgae produce many different kinds of lipids, hydrocarbons and other complex oils (Chisti, 2007; Banerjee *et al.*, 2002; Metzer and Lageru, 2005) and all of which are not satisfactory for making biodiesel. Therefore selection

Table 2. Oil content of some selected microalgae

Species	*Oil content (%dry weight)*
Botryococcus braunii	25 – 75
Chlorella sp	28 – 32
Cylindrotheca sp	16 – 37
Dunaliella primolecta	23
Nannochloris sp	20 – 35
Nannochloropsis sp	31 – 68
Nitzchia sp	45 – 47
Scenedesmus sp	16 – 40
Ankistrodesmus sp	31

of species with high oil content and potential for biodiesel production is an important criterion. The growth conditions in the culture technique, harvesting and oil extraction methodologies for optimum yields are equally important for sustainable production. Xu and Wu (2006) demonstrated the cultivation *Chlorella protothecoids* in a light deprived, heterotrophic environment with inexpensive hydrolyzed corn starch as the sole food source with increased population density of 15.5 g/L and increased oil content from 30% to 55.3%. *B. braunii*, on the other hand a green colonial microalga, with world wide distribution has been studied extensively due to its unusually high liquid hydrocarbon content (Metzger *et al.*, 1985; Sawayama *et al.*, 1992, 1994; Casadevall *et al.*, 1985; Villarreal-Rosales *et al.*, 1992; Qin, 2005; Banerjee *et al.*, 2002; Dayananda *et al.*, 2007). *B. braunii* is able to produce a high level of hydrocarbons in the range between 17-76% of its dry weight (Metzer *et al.*, 1985; Sawayama *et al.*, 1994). On hydrocraking the hydrocarbon rich oil of *B. braunii* yields 67% petrol, 15% aviation turbine fuel, 15% diesel fuel and 3% residual oil (Hillen *et al.*, 1982). The microalgae *Ankistrodesmus* in outdoor field experiments was reported to achieved a total lipid content of 24.5% of the total organic weight (Hill and Feinberg, 1984). More recently Hossain and Salleh (2008) have demonstrated that biodiesel also can be produced from most commonly encountered algal blooms caused by species like *Spirogyra* and *Oedogonium* sps.

The Oleaginous Microalgal Diversity of North East India: Prospects of Microalgal Biofuel Production

India has a rich repository of natural resources. The North East of India is acclaimed to be a genetic treasure land of plant, animal and microbial resources, which forms a distinctive part of the Indo-Burma Hotspot that ranks the 6th among the 34 biodiversity hotspots of the world. The region also falls in the bio-geographic tri-junction of the Indian, the Himalaya and the Mountains of Southwest China (Conservation International, 2007). This active centre of speciation is a centre of gene diversity of domesticated crops and a secondary centre for several economically important plants.

Assam is one of eight northeastern states of India located south of the eastern Himalayas (26.15°N 91.77°E) comprising of the Brahmaputra valley, the Barak valley, the Karbi Anglong and the North Cachar Hills. With an area of 78,438 sq kms, Assam is almost equivalent to the size of Ireland or Austria. Assam is surrounded by the rest of the

Seven Sister States: Arunachal Pradesh, Nagaland, Manipur, Mizoram, Tripura and Meghalaya, together comprising the North Eastern States of India including Sikkim also shares international borders with Bhutan and Bangladesh. The state also shares the cultures, peoples and climate with South-East Asia. The average annual rainfall in the plains of the Brahmaputra and Barak river valley is about 2,818 mm with per humid to humid temperate climate and annual temperature falls in the range between ± 10°C in winter to ± 32.2°C in summer. Besides having such diverse ago-eco-climatic condition, Assam is endowed with a wide range of wetlands covering an area of 1012.3 sq kms (Boruah, 1997) harboring a rich diversity of freshwater algae. These rich and diverse bio-resources of the region are yet to be explored for their potential biotechnological applications. Our study in the University of Gauhati was envisaged with the isolation and screening of freshwater microalgae of Assam in search of potential strains for sustainable production of renewable biofuel and the results have yielded commendable success. However, it was beyond the scope of this study to screen all the microalgal species for screening of lipid yield and therefore, a few species were selected on the basis of earlier literature and published works (Feinberg, 1984; Chisti, 2007).

Methodology Approached

Water samples were collected randomly from eight districts of the lower Brahmaputra valley of Assam during August to November 2007 in transparent plastic bottles of volume 100-250 ml. All together 98 nos. of water samples were collected within the period mostly from wetlands, road side water reservoirs, domestic ponds, fishery ponds and paddy fields.

Isolation of unicellular algae/colonies from natural collection were performed by repeated streaking of diluted samples (up to 10^{-3}) on BG11 medium solidified with 1% of agar. Single algal cells or colonies were picked up under microscope using micro needles or pasture pipette and grown in sterilized screw capped culture vials (3-5 ml). Isolated unialgal cells were cultured on 150 ml of BG11 growth media in 250 ml Erlenmeyer flask for all isolated strains. pH of the culture media were adjusted to 7.5 prior to autoclaving. Purities of the isolated samples were checked periodically by microscopic observation as well as growing in solidified agar plates. Pure unialgal cultures of the isolated species were grown at temperature 25 ± 2°C and photoperiod of 16 hrs light: 8 hrs dark cycle under cool white light of 1.2 ± 0.2 k lux intensity provided by fluorescent bulb. For quantitative analysis of lipid yield isolated monocultures of the selected species were grown in 250 ml of Erlenmeyer flasks taking 150 ml of BG11 medium. Three sets of cultures for each of the selected species were inoculated with 10 ml of six weeks old algal suspensions. All the experimenting culture sets were maintained in the same laboratory growth conditions as mentioned above.

Algal suspensions (20-30 ml) of 4-12 weeks old cultures of each of the experimenting samples were harvested by using pre combusted and pre weighted GF/C filter paper. Prior to harvesting the media free algal cells were washed 2-3 times with 2% NaCl solution, centrifuged and finally washed with distilled water. Harvested algal cells were combusted at 100°C for 3 hrs prior to measure its dry weight gravimetrically (Dayananda *et al.*, 2007). Total lipids were repeatedly extracted 20-30 ml of a solvent system of Chloroform: Methanol (2:1) from the homogenized dried algal biomass in a mortar and pestle in presence of glass powder. The chloroform layers were collected in pre weighted glass vials. The solvent was evaporated and dried under vacuum over Na_2SO_4. Total lipids were measured gravimetrically and expressed as dry weight percentage (Lee *et al.*, 1998; Bligh and Dyer, 1959). The hydrocarbon contents of 6-12 weeks old cultured *B. braunii* isolate (BbDHNOI) were extracted by using n-hexane and measured gravimetrically after evaporating the solvent to dryness. The yield was expressed as percentage dry weight (Sawayama *et al.*, 1992). Crude hydrocarbon extracts were purified by column chromatography on silica gel with n-hexane as eluent. Preliminary GC analysis was done on BP-5 capillary column in Oil India Ltd, Duliahjan. The various steps involved in isolation of microalgae and culture in the laboratory are showing in the Fig. A (Plate 1). Figs. B, C & D (Plate 1) showing solvent extraction of algal lipid and crude oil extracts.

Results and Discussion

In our study 22 freshwater microalgal species were isolated from randomly collected water samples of eight districts of Assam state. During the collection period large scale natural algal blooms were observed, which reveled favorable environmental conditions for large scale open pond culture of microalgae in the region (Figs. A-C, Plate 2). The isolated species were belonging to Cyanophyceae, Chlorophyceae and Bacillariophyceae. The present study has revealed a rich freshwater oleaginous microalgal diversity, which could be explored for the development of future renewable sources of alternative energy. *Ankistrodesmus* sp., *Botryococcus braunii*, *Chlorella* sp.,

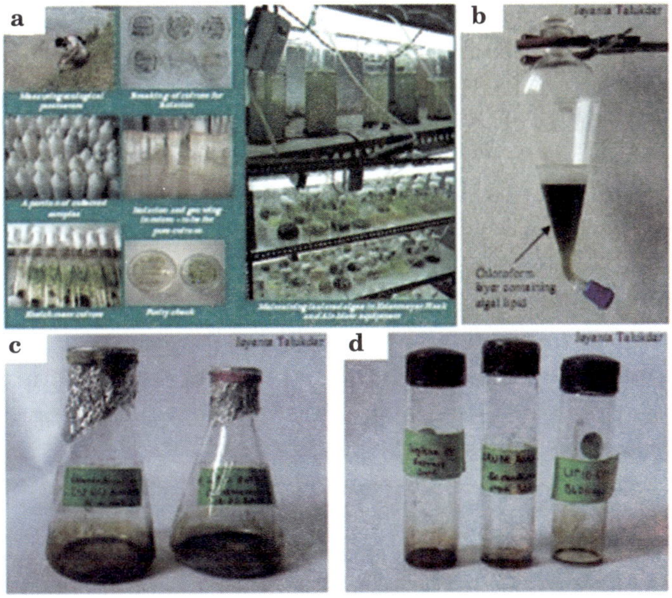

Plate 1. (a-d): Photographs showing methods of isolation and algal oil extraction A. Field collection, isolation and unialgal culture in the laboratory. B. Solvent extraction of algal oil (Total lipid).

Plate 2. a-c Photographs of natural algal blooms and some selected microalgae species of Assam. d-j Photographs some isolated microalgae (not to scale, magnification 400x) d. *Ankistrodesmus* sp, e. *Scenedesmus* sp, f. *Nevicula* sp, g. *Euglena saguinea* h. *Nitzchia* sp, i. *Hantzcchia* sp, j. *Chlorococcum* sp.

Chlamydomonas sp., *Scenedesmus* sp., *Hantzschia*, *Navicula* sp., *Nitzchia* sp, and *Cyclotella* sp. were some of the isolated microalgal species of the region noteworthy for biodiesel production (Feinberg, 1984), which have been studied extensively in different parts of the world including India. Figs. D-J (Plate 2) depicts some of the isolated microalgal species of the region. The liquid hydrocarbon producing green microalgae *B. braunii* isolates of the region (Figs. A-F, Plate 3) were of special interest owing to its worldwide recognition as a potent renewable energy source through large scale cultivation (Okada *et al.*, 1997) besides other microalgae, which could be utilized for potential production of biodiesel. *Ankistrodesmus* sp., *B. braunii*, *Chlorococcum* sp. and *Scenedesmus* sp. were selected for the preliminary screening of total lipid contents in this study. The results are shown in the Table 3. The yield of total lipids for *Ankistrodesmus, Chlorococcum* and *Scenedesmus* are found to be in the range between 21.7-42.0%, 11.3-12.2% and 17.3-19.1% respectively while the hydrocarbon content of the *B. braunii* (BbDHNOI) isolate was found to be in range between 21.9-60.7%. *B. braunii* isolates of the region are found to be prospective in the light of its high hydrocarbon content and wide distribution in natural condition that can be utilized potentially for sustainable biofuel production.

Table 3. Total lipids contents of a few selected freshwater microalgae of Assam

Species/strains	*Family*	*Total lipid (% dry weight)*		*Reference(s)*
		Found	*Reported*	
Ankistrodesmus sp.	Chlorophyceae	21.7 – 42.0	24.5 – 31.0	SERI, 1984
Botryococcus braunii (BbDHNOI)	Chlorophyceae	21.9 – 60.7*	25.0 – 75.0	Casadevall *et al.*,1985
Scenedesmus sp.	Chlorophyceae	17.3 – 19.1	16.0 – 40.0	Becker, 1994
Chlorococcum sp.	Chlorophyceae	11.3 – 12.2	_	

* *Represent crude hydrocarbon contents*

Concluding Remarks

The looming threats of climate change and the inevitable role of energy in economic development clearly resurgent the urgency the global transition to renewable alternative fuel. India's quest for a sustainable alternative fuel policy is of immensely important as it has been indicated from its fossil fuel scenario and commendable economic development. The current strategy of biodiesel production in India needs to be precisely rethought for sustainable development. Biodiesel production from microalgal feedstock could one of the better options. Microalgal biodiesel is technically feasible and seems to be the only renewable biofuel that can replace liquid fuels derived from petroleum. Our study has revealed the potentiality of microalgal resources of the North Eastern Region of India beyond any doubt. Majority of the species considered here have potential to become superior biomass feedstock. *B. braunii, Ankistrodesmus,* and *Scenedesmus* species have already been considered as potential microalgae for liquid biofuel production. The indigenous isolates of *B. braunii* with its high hydrocarbon production and *Ankistrodesmus* with high lipid content as shown by their respective cultures in the laboratory condition reconcile the fact that the microalgae of the region could be the potential bio-feedstock towards sustainable production of renewable alternative biofuel. Economics of producing liquid biofuels from microalgae could be substantially improved with the adoption of an integrated approach of biomass-production conversion system. The most exciting reason about microalgal oils is the potential of chemically converting it into petrol, diesel and jet fuel. Thus, microalgal biofuels are a sustainable energy resource. Biomass production via large scale microalgae cultivation with an integrated approach could be the optimal option for CO_2 biomitigation, greenhouse gas abatement to ward off Global warming and to secure our much needed energy security with the development of sustainable renewable alternative sources of biofuels. In this context the oleaginous microalgae of the North Eastern Region of India have tremendous prospects in the development of microalgal biofuel technology.

Acknowledgements

This study was supported in part by a Grant-in-Aid from the Defense Research Laboratory (DRDO), Tezpur, Assam, India. We are grateful to Dr. R.B. Srivastava, Director, DRL his kind assistance moral support during the study. We are also thankful to Dr. H.C. Das, Chief Research Scientist, Oil India Ltd, Duliahjan, Assam for providing technical support during the study.

References

Aresta, M., Dibenedetto, A., Carone, M., Colonna, T. and Fagale, C. (2005). Production of biodiesel from macroalgae by supercritical CO_2 extraction and thermochemical liquifaction. *Environ. Chem. Letters* **3**: 136-139.

Banerjee, A., Sharma, R., Chisti, Y. and Banerjee, U.C. (2002). *Botryococcus braunii*: a renewable source of hydrocarbons and other chemicals. *Crit Rev Biotechnol.* **22**: 245-79.

Boruah, P. (1997). Wetlands of Assam. Scientific Note: ARSAC/RSAG-SAC/WLM/PR/01/97. Assam Remote Sensing Application Centre, Guwahati and Space Application Centre, Ahmedabad, India. p. 165.

Becker, E.W. (1994). *In*: Microalgae: biotechnology and microbiology. *Ed.* Baddiley, J. *et al.*, Cambridge Univ. Press, Cambridge, New York. p. 178.

Benemann, Jhon R. (2003). Biofixation of CO_2 and Greenhouse Gas abatement using Microalgae – Technology Road Map. Final report to US DOE (under subcontract agreement 7010000926) p. 30.

Bligh, F.G. and Dyer, W.J. (1959) A rapid method of total lipid extraction and purification. *Can J Biochem Physiol* **37**: 911-7.

Borowitzka, M.A. (1999). Commercial production of microalgae: ponds, tanks, tubes and fermenters. *J Biotechnol* **70**: 313-321.

Brown, A.C., Knights, B.A. and Conway, E. (1969). Hydrocarbon content and its relationship to physiological state in the green alga *Botryococcus braunii*. *Phytochemistry* **8**: 543-547.

Brown, L. and Zeiler, K.G. (1993). Aquatic biomass and carbon dioxide trapping. *Energy Conservation Management* **34(9-11)**: 1005-1013.

Burk, J., Bals, C., Beck, M. and Rüthlein, E. (2008). *In*: Climate change performance index. Germanwatch, Berlin. http://www.germanwatch.org/klima/ccpi2008.pdf

Campbell (2008). Biodiesel: Algae as a renewable source for liquid fuel. *Guelph Engineering Journal*. **(1), 2 – 7**: 1916-1107.

Casadevall, E., Dif, D., Largeau, C., Gudin, C., Chaument, D. and Desantit, O. (1985). Studies on batch and continuous cultures of *Botryococcus braunii*: hydrocarbon production in relation to physiological state, cell ultra structure, and phosphate nutrition. *Biotechnology and Bioengineering* **27**: 286-295.

Chisti, Y. (2007). Biodiesel from microalgae. *Biotechnol Adv*. **25**: 294-306.

Chisti, Y. (2008). Biodiesel from microalgae beats bioethanol. *Trends in Biotechnol* **26(3)**: 126-131.

Conservation International (2007) Biodiversity Hot Spots. Last updated 2007.

http://www.biodiversityhotspots.org/xp/Hotspots/hotspotsScience/Pages/default.asp

Crookes, R.J. (2006). Comparative biofuel performance in internal combustion Engines. *Biomass and Energy* **30**: 461-468.

Crutzen, P.J., Moser, A.R., Smith, K.A. and Winiwater, W. (2007). N_2O release from agro-biofuel production negates global warming reduction by replacing fossil fuels. *Atmos Chem Phys Discuss* **7**: 11191-11205.

Dayananda, C., Sarada, R., Kumar, V. and Ravishankar, G.A. (2007). Isolation and characterization of hydrocarbon producing green alga *Botryococcus braunii* from Indian freshwater bodies. *Electronic Journal of Biotechnology* **10(1)**: 78-91.

Demirbas, A. (2004). Current technologies for the thermo-conversion of biomass into fuels and chemicals. *Energy Source* **26**: 715-730.

Economic Advisory Council to the Prime Minister, Govt of India (2008). *In*: Review of Economy 2007-08, New Delhi.

Energy Information Authority (EIA) (2006). World oil markets. *In*: International Energy Outlook 2006 p28 www.eia.doe.gov

Energy Information Authority (EIA) (2008). Highlights. International Energy Outlook pp. 1-5 www.eia.doe.gov

Fargione, J., Hill, J., Tilman, D., Polasky, S. and Hawthorne, P. (2008). Land clearance and biofuel carbon debt. Science Express pp 1-4, 10.1126 www.scienceexpress.org

Feinberg, Daniel A (1984). Fuel options from microalgae with representative chemical compositions. SERI/TR-388-4625.20 Golden Colorado: Solar Energy Research Institute.

Haag, A.L. (2007). Algae bloom again. *Nature* **447**: 520-521.

Hill, A.M. and Feinberg. D.A. (1984). Fuel from microalgae lipid products. SERI/ TR-231-2348 Golden Colorado: Solar Energy Research Institute.

Hillen, L.W., Pollard, G., Wake, L.V. and White, N. (1982). Hydrocracking of oils of *Botryococcus braunii* to transport fuels. *Biotechnol Bioeng* XXIV: 193-205.

Hossain, A.B.M. Sharif and Salleh, A. (2008). Biodiesel fuel production from algae as renewable energy. *American J Biochem and Biotechnology* **4(3)**: 250-254.

Huntley, M.E. and Redalje, D.G. (2007). CO_2 mitigation and renewable oil from photosynthetic microbes: a new appraisal. *Mitig Adapt Strategies Glob Chang* **12**: 573-608.

International Energy Agency (IEA) (2007). *In*: World Energy Outlook. Executive summary, China and India insight, p. 14.

http://www.iea.org/textbase/npsum/WEO2007SUM.pdf

IPCC 4AR WG I (2007). Intergovernmental Panel on Climate Change Fourth Assessment Report, Working Group I and Synthesis Report.

Jeong, M.L., Gills, M.J. and Hwang, J.Y. (2003). Carbon dioxide mitigation by microalgal photosynthesis. *Bull. Korean Chem. Soc* **24(12)**: 1763-1766.

Kondili, E.M. and Kadellis, J.K. (2007). Biofuel implementation in East Europe: current status and future prospects. *Renew Sustain Energy Rev* **11**: 2137-2151.

Lee, S.J., Yoon, B.D. and Oh, H.M. (1998). Rapid methods for the determination of lipid from the green *Botryococcus braunii. Biotechnology Techniques* **12(7)**: 553-556.

Li, Y., Horsman, M., Wu, N., Lan, C.Q. and Dubois-Calero, N. (2008). Biofuels from Microalgae. Biotech Prog (in press) ASAP Article, DOI 10.1021/ bp070371kS8756-7938(07)00371-2.

Mallick, N. (2002). Biotechnological potential of immobilized algae for wastewater N, P and metal Removal: a review. *Biometals* **15**: 377-390.

Metzger, P., Berkaloff, C., Couté, A. and Casadevall, E. (1985). Alkadiene and botryococcene-producing races of wild strains of *Botryococcus braunii. Phytochemistry* **24**: 2305-2312.

Metzer, P. and Lageru, C. (2005). *Botryococcus braunii*: a rich source of hydrocarbons and related ether lipids. *Appl Microbiol and Biotechnol* **66**: 486-496.

Mitchell, G. (2008). UCSD consortium to develop biofuels from microalgae: a concept; progress one project at a time (Presentation), University of California. San Diego. http://spg.ucsd.edu/algae/pdf/080129_SDCMA_biofuel_symp_bgm.pdf

Moomaw, W. (2008). Renewable Energy and Climate Change: An overview. IPCC scoping meeting on renewable energy sources (Proceedings). Lubeck, Germany 20-25 January 2008. pp. 3-11.

Okada, S., Murakami, M. and Yamaguchi, K. (1997). Characterization of Hydrocarbons from the Yayoi Strain of the Green Microalga *Botryococcus braunii. Phytochem Annalysis* **8**: 198-203.

Qin, J. (2005). Bio-hydrocarbons from algae. RIRDC Australian Govt Publication No.05/025. www.rirdc.gov.au/fullreports/index.html

Rajagopal, D. (2002) Rethinking Current Strategies for Biofuel Production in India http://www.iwmi.cgiar.org/EWMA/files/papers/rajagopal_biofuels_final_Mar02.pdf

Roman-Leskov, Y., Barret, C.J., Liu, Z.Y. and Dumesic, J.A. (2007). Production of Dimethylfuran for liquid fuels from biomass derived carbohydrates. *Nature* **447**: 982-985.

Sawayama, S., Minowa, T., Dote, Y. and Yokoyama, S. (1992). Growth of the hydrocarbon rich microalga *Botryococcus braunii* in secondarily treated sewage. *Appl Microb Biotech* **38(1)**: 135-138.

Sawayama, S., Inoue, S. and Yokoyama, S. (1994). Continuous culture of hydrocarbon-rich microalga *Botryococcus braunii* in secondarily treated sewage. *Applied Microbiology and Biotechnology* **41**: 729-731.

Schneider, D. (2006). Grow your own? : would the wide spread adoption of biomass-derived transportation fuels really help the environment. *American Scientist* **94**: 408-409.

Searchinger, T., Heimlich, R., Houghton, R.A., Dong, F., Elobeid, A., Fabiosa, J., Tokgoz, S., Hayes, D. and Yu, T.H. (2008). Use of US croplands for biofuels increases greenhouse gases through emissions from land use change. Science Express pp.1-6/101126.1151861 www.scienceexpress.org/7

Scharfemann, J.P.W. and Larance, W.F. (2008). How green are biofuels? *Science* **319**: 43-44.

Skjanes, K., Lindblad, P. and Muller, J. (2007). Bio CO_2 – a multidisciplinary, biological approach using solar energy to capture CO_2 while producing H_2 and high value products. *Biomol Eng.* **24**: 405-413.

The Planning Commission (2007). *In*: 11th 5 year plan, Energy sector. Govt of India.

The Royal Society (2008). Sustainable biofuels: prospects and challenges. *In*: Royal Society Policy Document, ISBN 978 0 85403 662 2 www.roalsociet.org

Ushi, N. and Ikenouchi, M. (1997). The biological CO_2 fixation and utilization project by RITE(1): Highly effective photobioreactor system. *Energy Covers Manag* **38(Suppl 1)**: S487-S492.

Villarreal-Rosales, E., Metzger, P. and Casadevall, E. (1992). Ether lipid production in relation to growth in *Botryococcus braunii*. *Phytochemistry* **31(9)**: 3021-3027 (Abstract).

Wang, B., Li, Y., Wu, N. and Lan, Q. (2008). CO_2 bio-mitigation using microalgae. *Appl Microbiol Biotechnol* **79**: 707-718.

Xu, H., Miao, X. and Wu, Q. (2006). High quality biodiesel production from a microalgae *Chlorella protothecoides* by heterotrophic growth in fermenters. *J Biotechnol* **126**: 499-507.

5

HETEROTROPHIC PRODUCTION OF LUTEIN AND BIOMASS BY *CHLORELLA VULGARIS* WITH DIFFERENT NITROGEN SOURCES

J. Mary Leema[1*], R. Kirubagaran, A.K. Abdul Nazar and S. Kathiroli[2]

Abstract

Microalgae has captured widespread attention in the recent times due its increased application in pharmaceutical, nutraceutical, fish feed and animal feed industries. Among the microalgae the Chlorophycean algae Chlorella sp. has gained popularity as potential candidate for "food of the future" due to its nutritive value and potential for high value pigment production. The problems frequently confronted in phototrophic mass culture systems like low biomass production and high production cost have resulted in alternative mode of microalgal production viz. heterotrophic culture. Heterotrophic culture circumvents the requirement for light and facilitates large scale production in short period. During this study, the heterotrophic production potential of lutein by Chlorella vulgaris in seawater was investigated. The influence of various nitrogen sources (nitrate as sodium and potassium nitrate; ammonium as ammonium chloride and urea) on the heterotrophic growth of C. vulgaris were investigated using flask cultures. An alkali extraction method was used for the extraction of lutein from C. vulgaris cells. Lutein was separated by a HPLC system. No inhibitory effect on the algal growth was observed over a nitrogen concentration range of 0.06 to 0.12 M.

1 Andaman and Nicobar centre for Ocean Science and Technology, NIOT, R & D Complex, Port Blair, A&N Islands, India - 744 103

2 National Institute of Ocean Technology, Palikaranai, Chennai, Tamil Nadu, India - 601 302

* *Corresponding author*: E-mail: leemaroy2002@yahoo.com

Increase in initial nitrogen concentration from 0.06 M to 0.12 M showed a concomitant increase in biomass production, specific growth rate and lutein production. There was a significant difference in biomass production and specific growth rate among the cultures grown with different nitrogen sources. Highest biomass concentration was observed in the cultures with nitrate (Sodium nitrate: 11.18 ± 0.013 g dry cells / L Potassium nitrate: 11.48 ± 0.027 g dry cells / L) at an initial nitrogen concentration of 0.12M. The maximum lutein content in dry cells ranged from 0.49 mg lutein / g dry cells to 1.60 mg lutein / g dry cells. The heterotrophic culture of C. vulgaris was scaled up successfully in fermentor in which modified Basal medium containing 45 g / L of glucose as carbon source and 0.12M Potassium nitrate as nitrogen source was used. Maximum biomass concentrations of 22.23 g dry cells / L and specific growth rate of 0.91 / day were achieved in 96 hrs. The heterotrophic lutein producing ability of C. vulgaris suggested that the alga can be potentially developed for commercial production of lutein on a large scale in fermentors.

Key words: *Chlorella vulgaris*, Heterotrophic culture, Nitrogen source, Lutein production

Introduction

Lutein [(3R, 3'R, 6'R) - â-carotene-3, 3'-diol], has attracted wide attention in pharmaceutical and biotechnology industry due to its possible role in prevention of age related macular degeneration (Beatty *et al.,* 2004), cancer (Astorg, 1997); as feed additive in aquaculture and poultry industry (Lorenz and Cysewski, 2000). The most attractive property of xanthophyll carotenoids such as lutein is their ability to quench singlet oxygen and free radicals and act as superior antioxidants (Edge *et al.,* 1997). Health related benefits and antioxidant properties of xanthophyll carotenoids like lutein have created a multi million dollar market for these carotenoids. Currently industrial scale production of xanthophyll carotenoids is from synthetic sources. But recently there is a surge in the demand for natural sources of lutein and other carotenoids. At present marigold flowers (*Tagetes erecta*) is the most copious natural source of commercial lutein (Khachik, 1995). However, microalgae have an added advantage over higher plants due to their shorter generation time and possibility of producing large scale continuously in bioreactors (Shi *et al.,* 1997). Business communications report, 2005 (http://www.bccresearch.com) has predicted the market for lutein to increase at the rate of 6.1% per year and expected to reach US $187 million in 2009.

Recently lutein production has been reported from few chlorophycean microalgae like *Chlorella zofingiensis* (Del Campo *et al.*, 2004), heterotrophically grown *Chlorella protothecoides* (Shi *et al.*, 2002) and phototrophically grown *Muriellopsis* sp. (Del Campo *et al.*, 2001). Chlorella being a fast grower and chief feed supplement has overtaken other microalgae as prospective candidate species for lutein production (Del Campo *et al.*, 2004). Hence during this study an attempt was made to harness the advantages of heterotrophic growth for the production of biomass and lutein from another Chlorella species *Chlorella vulgaris* in seawater.

Carbon and nitrogen sources are two most important nutritional elements controlling microalgal growth. Three different groups of nitrogen source (nitrate, ammonium and urea) are commonly used in microalgal culture. The effect of different nitrogen sources and their concentrations on the biomass and lutein production of *Chlorella protothecoides* under heterotrophic condition has been reported by Shi *et al.* 2000. But microalgal species differ in their growth response to different nitrogen sources. Orûs MI and Martinez (1991) have previously described the heterotrophic potential of Spanish strain of *Chlorella vulgaris* (UAM 101) isolated from the sugar refineries waste water. However there are no studies dealing with the effect of different nitrogen source and concentrations on the growth and biomass production of Indian strain of *Chlorella vulgaris* in seawater. The objective of the present study was to investigate the influence of different nitrogen sources and their concentrations on the biomass and lutein production of *Chlorella vulgaris* in heterotrophic culture.

Materials and Methods

Microalgal strain and culture condition

The microalgal strain used in this study, *Chlorella vulgaris* was obtained from the culture collections of Central Marine Fisheries Research Institute, Tuticorin, India. The culture was maintained aseptically in modified Basal medium (Shi *et al.*, 1997) in 250 ml Erlenmeyer flasks. Axenic batch cultures were grown in 500 mL Erlenmeyer flasks containing 250 mL of modified Basal medium at 24° ± 1°C under 16:8 light/dark regime and continuous shaking (130 rpm) in temperature controlled orbital shakers with light (Orbitek LT, Scigenics, Chennai, India). For heterotrophic batch cultures the Erlenmeyer flasks (500 mL) containing 250 mL medium (pH adjusted to 6.1 prior to autoclaving at 121µ C for 20 min.) substituted with 45 g/l glucose (0.22 µ filter sterilized) was wrapped in aluminium foil and kept in temperature controlled orbital shakers in darkness at 24° ± 1°C and continuous shaking (130 rpm). An inoculum of 5% (v/v; average cell concentration of 0.25 gL^{-1} dry weight) was inoculated into 500 ml Erlenmeyer flasks each containing 250 ml medium under dim light and aseptic conditions. The cultures were then incubated at 24° ± 1°C with orbital shaking at 130 rpm under darkness. Four different nitrogen sources *viz.* potassium nitrate, sodium nitrate, ammonium chloride and urea were used at three different nitrogen concentrations 0.06 M, 0.09 M and 0.12 M. The experiment was repeated thrice with duplicates for each treatment during each trial.

The scale up experimentwas performed using 10 L fermentor (BioFlo 110, New Brunswick, New Jersey, USA – Plate 1). The media used for fermentor cultivation the modified Basal medium (Shi *et al.,* 1997) was supplemented with 45 g/L of glucose and 0.12 M potassium nitrate and inoculated with 5% exponentially growing cells (with cell concentration of 0.25 gL^{-1} dry weight). The cultivation conditions in the fermentor was computer controlled as follows: pH-6.1 ± 0.1, temperature - 26°C, agitation-200 rpm with aeration by sterilized air at a flow rate of 1vvm, D.O., 50% saturation. All operations were carried out aseptically.

Plate 1. Heterotrophic culture of *C. vulgaris* in 10 L fermentor

Determination of cell density and specific growth rate

The optical density of the culture fluid was measured daily at 540 nm (Becker, 1994) with spectrophotometer (Unicam UV 300, UV-VIS spectrophotometer, USA). Dense cultures were suitably diluted with distilled water to make the O.D below 0.6. From the O.D values daily biomass concentration was derived by using previously prepared standard calibration curves for optical density Vs biomass for each treatment. Specific growth rate (µ) at exponential phase was calculated using the following equation, $\mu = (\ln X_2 - \ln X_1)/(t_2 - t_1)$, where X_2 and X_1 represent dry cell weight (biomass) concentrations (g/l) at culture time t_2 and t_1 (Chen and Johns, 1996). Determination of biomass (cell dry weight) concentration Dry cell weight concentration was done according to Chen *et al.* (1996).

Determination of pigments

Lutein in the algal cells was extracted using alkali digestion method and its concentration determined according to Shi *et al.* 1997 and Shi and Chen 1997. The whole process was carried out in darkness.

For determination of pigment contents by HPLC, the chemical standards of astaxanthin, â–carotene and lutein, were purchased from Sigma Chemical Co. (St. Louis, MO, USA). HPLC-grade acetonitrile, dichloromethane and methanol were obtained from Merck (Germany). The extracted pigments (20 µl aliquots) were separated and analysed on an HPLC system (Shimadzu, Japan). The HPLC system used in this system comprised of a pump (Shimadzu SCL-10AD VP), a system controller (Shimadzu SCL-10A VP), injector (Rheodyne, 20 µl loop volume), a reverse phased C_{18} column and a UV detector (Shimadzu SPD-10AV VP). The mobile phase was methanol/dichloromethane/ acetonitrile/water (67.5:22.5:9.5:0.5, v/v) and the flow rate of mobile phase was set at 1 ml/min. The column was kept in room temperature (22-25 µC). The samples and standard were filtered through a 0.22 µm syringe filter (acrodisc) prior to injection. The samples/standards were injected in 20 µl volume and the elution was detected at 450 nm. The lutein concentration in the microalga was calculated by comparing the peak area with that of standard lutein (Kimura and Rodriguez-Amaya, 2002).

Statistical Analysis

The results were subjected to Analysis of Variance (ANOVA) at the 99% confidence interval ($p<0.01$).

Results and Discussion

Effects of different nitrogen sources and initial concentrations on the algal growth. The green microalgae, *Chlorella vulgaris* was grown in flasks heterotropically (in the dark) to investigate the effect of four different nitrogen sources (*viz.* Sodium nitrate, Potassium nitrate, Ammonium Chloride and Urea) on the growth and pigment production. The influence of different initial nitrogen concentrations (0.06 M, 0.09 M and 0.12 M) on the biomass production and lutein production was also analysed. As shown in Table 1 the highest maximum biomass concentration was observed in the cultures containing nitrate (Sodium nitrate: 11.18 ± 0.013 gL^{-1}; Potassium nitrate: 11.48 ± 0.027gL^{-1}) at an initial concentration of 0.12 M nitrogen. The maximum biomass concentration of the cultures containing urea and ammonium chloride were significantly lower than the nitrate groups ($P<0.01$). Similar trend was observed in all the three initial nitrogen concentrations. Muenscher 1923 has reported that at equal concentrations nitrate as nitrogen source gives a better growth (in darkness and light) than ammonium in *C. vulgaris*. The cultures with ammonium chloride as nitrogen source had the lowest maximum biomass concentration at all the three concentrations. The pH in the flask cultures was not controlled, hence a severe drop in pH (final pH -3.2) was observed in the culture grown with ammonium chloride as nitrogen source. Wen and Chen 2001 have suggested assimilation of ammonium ions by the algae as the most probable reason for the drop in pH. Similar observations of reduction in pH and concomitant reduction in biomass concentration while using ammonium as nitrogen were also reported in *Chlorella protothecoides* (Shi *et al.*, 2000); *Nitzschia laevis* (Wen and Chen, 2001) and *Phaeodactylum tricornatum* (Yongmanitchai and Ward, 1991). However Chlorella *vulgaris* being a species capable of tolerating higher ammonium chloride concentration displayed a higher maximum biomass concentration (0.06 M: 6.25 ± 0.0126 gL^{-1} Vs 4.8 gL^{-1}; 0.09 M: 6.99 ±

Table 1. Heterotrophic growth of *C. vulgaris* with various nitrogen sources and initial concentrations in flask cultures. Values are the means ± S.D of three independent experiments with duplicate cultures. Different superscript for significant differences at 1% confidence level

Nitrogen concentration in medium (M)	***Maximum biomass concentration on dry weight basis (gL^{-1})***			
	Nitrate (Sodium)	***Nitrate (Potassium)***	***Urea***	***Ammonium***
0.06	7.81 ± $0.017^{a,ab,ac}$	8.07 ± $0.020^{a,ab,ac}$	7.06 ± $0.33^{a,ab,ac}$	6.25 ± $0.022^{a,ab,ac}$
0.09	10.07 ± $0.019^{b,ab,ac}$	10.31 ± $0.016^{b,ab,ac}$	9.08 ± $0.017^{b,ab,ac}$	6.99 ± $0.032^{b,ab,ac}$
0.12	11.18 ± $0.023^{c,ac,ab}$	11.48 ± $0.047^{c,ac,ab}$	9.63 ± $0.031^{c,\ ac,ab}$	8.63 ± $0.029^{c,ac,ab}$

0.0183 gL^{-1} Vs 4.7 gL^{-1}; 0.12 M: 8.63 ± 0.01695 gL^{-1} Vs 4.9 gL^{-1}) when compared with that of *C. protothecoides* at similar concentrations (Shi *et al.*, 2000). Similar increase in maximum biomass concentrations with increase in initial nitrogen concentration has been reported in *Chlorella protothecoides* by Shi *et al.* 2000. In *C. pyrenoidosa* increase in initial nitrate concentration from 0.5 gL^{-1} to 10 gL^{-1} increased the biomass concentration from 7.0 g/l to 16.6 g/l (Wu *et al.*, 2007). Table 2 shows the biomass and lutein production in 10 L fermentor. The maximum biomass concentration of 22.23gL^{-1} was obtained at 96 hrs and the specific growth was 0.91d^{-1}. Both the values were much higher than the flask culture indicating the better performance in fermentor.

Table 2. Heterotrophic growth and lutein prodction *C. vulgaris* in fermentor

Maximum biomass concentration on dry weight basis (gL^{-1})	*Time appeared*	*Specific growth rate (d^{-1})*	*Maximal lutein content in dry cells (mgg^{-1})*
22.23	96 h	0.91	2.47

Nitrate is prevalently used as nitrogen source in many media used for *Chlorella* culture (Sorokin and Krauss, 1958). This preference comes from its suitability to be sterilized in an autoclave. However, few authors (Goldman, 1977; Stengel and Soeder, 1975; Syreth and Morris, 1963) have reported ammonium as an excellent nitrogen source for *Chlorella* due to the less expenditure needed for the absorption of ammonium than nitrate (Fig. 1). Ammonium has not been popular as a nitrogen source due to its inconvenience for sterilization in autoclaves and the lethality accompanying severe drop in pH.

Effect of different nitrogen sources and initial concentration on lutein production

Lutein production by the microalga *Chlorella vulgaris* was observed at all the three initial concentration (0.06 M, 0.09 M, 0.12 M) and four different nitrogen sources used. Table 3 depicts the lutein production by *Chlorella vulgaris* under heterotrophic condition with different nitrogen sources and various initial nitrogen concentrations. There was statistically significant difference in the maximum lutein production using the different sources ($P<0.01$). There was also a significant difference ($P<0.01$) in the maximum lutein production at the different initial nitrogen concentrations used. Regardless of the type of nitrogen source used the lutein production increased with the increase in nitrogen

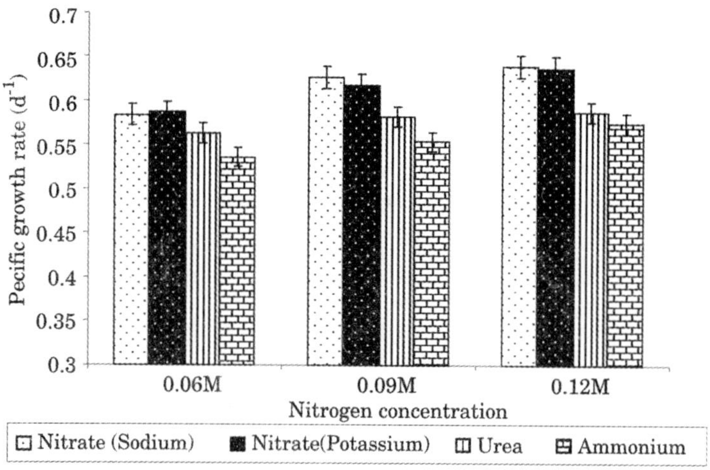

Fig. 1. Influence of different nitrogen sources and concentration on the mean specific growth rate of *C. vulgaris* under heterotrophic condition. Values are the means ± S.D of three independent experiments with duplicate cultures

Table 3. Heterotrophic production of lutein from *C. vulgaris* with various nitrogen sources and initial concentrations in flask cultures. Values are the means ± S.D of three independent experiments with duplicate cultures

Nitrogen concentration in medium (M)	*Maximum lutein content in dry cells (mgg^{-1})*			
	Nitrate (Sodium)	*Nitrate (Potassium)*	*Urea*	*Ammonium*
0.06	0.796 ± 0.002	0.86 ± 0.003	0.488 ± 0.035	0.527 ± 0.003
0.09	1.225 ± 0.035	1.0523 ± 0.002	0.922 ± 0.010	0.625 ± 0.005
0.12	1.599 ± 0.037	1.491 ± 0.006	1.412 ± 0.014	0.986 ± 0.017

concentration and the highest lutein content was recorded at highest nitrogen (0.12 M) concentration. Lutein production on dry weight basis declined significantly (1.599 mgg^1 to 0.527 mgg^{-1}) when the initial nitrogen level was lowered from 0.12M to 0.06M. This seems logical, since cell growth (both specific growth rate and the maximum biomass concentration) were significantly ($p<0.01$) lower at lower nitrogen concentration (0.06 M, 0.09 M) than at higher concentration (0.12 M). Shi *et al.* 2000 have also reported an increase in initial nitrogen resulting in consequent elevation in cellular lutein concentration. These authors have also suggested a 25–46% reduction in fermentation cycle when the initial nitrogen concentration was enhanced. Similar trend was observed in *Chlorella zofingiensis* by Del Campo *et al.* 2004. The highest maximum lutein content was noticed in cultures using nitrates (Sodium nitrate: 1.599 mgg^{1} and Potassium nitrate: 1.490 mgg^{1}) at an initial nitrogen concentration of 0.12 M. Lutein production in 10 L fermentor at 0.12 M initial concentration was higher than that of flask culture (2.470 mgg^{-1} Vs 1.599 mgg^{1}).

Nitrogen is an important component governing the formation of carotenoids in microalgae. Although nitrogen is not directly involved in the biosynthetic pathway of carotenoids it can alter the normal cellular metabolism such as protein synthesis and hence it can indirectly affect the pigment formation in microalgae (Ip *et al.,* 2005). Boussiba and Vonshak (1991) have emphasized the need for sufficient quantity of nitrogen for continuous synthesis of proteins responsible for supporting pigment formation. Lutein is a primary carotenoid essential for the appropriate structure and functioning of the photosystems (Demming –Adams and Adams III 2002). Hence there are ample reasons to believe that initial nitrogen concentrations have effect on the lutein production. For instance Delcompo *et al.* 2004 have reported decreased lutein production in chlorophycean microalgae *Muriellopsis* sp. at low initial nitrogen concentration and the lutein concentration doubled when the nitrogen source in the medium was increased.

Conclusions

Hence, the fast growth exhibited by this strain of *Chlorella*, its amenability to grow under heterotrophic conditions, its ability to grow with any type of nitrogen source, its ability to grow in sugar refinery effluent (Martinez *et al.,* 1987) can compensate other drawbacks and can be chosen as a prospective candidate species for mass production of lutein through adequate management of growth conditions, culture containers and genetic manipulations.

References

Astorg, P. (1997). Food carotenoids and cancer prevention: an overview of current research. *Trends Food Sci Technol* **8**: 406-413.

Beatty, S., Nolan, J., Kavanagh, H. and O'Donovan, O. (2004). Macular pigment optical density and its relationship with serum and dietary levels of lutein and zeaxanthin. *Arch Biochem Biophys* **430**: 70-76.

Becker, (1994). Measurement of algal growth. *In*: Microalgae Biotechnology& Microbiology. Cambridge University Press, pp. 56-62.

Boussiba, S. and Vonshak, A. (1991). Astaxanthin accumulation in the green alga *Haematococcus pluvialis*. *Plant Cell Physiol* **32**: 1077-1082.

Demming-Adams, B. and Adams, W.W. III (2002). Antioxidants in photosynthesis and human nutrition. *Science* **298**: 2149-2153.

Del campo, J.A., Rodríguez, H., Moreno, J., Vargas, M.A., Rivas, J. and Guerrero, M.G. (2001). Lutein production by *Muriellopsis* sp. in an outdoor tubular photobioreactor. *J. Biotechnol.* **85**: 289-295.

Del Campo, J.A., Rodríguez, H., Moreno, J., Vargas, M.A., Rivas, J. and Guerrero, M.G. (2004). Accumulation of astaxanthin and lutein in *Chlorella zofingiensis* (Chlorophyta). *Appl Microbiol Biotechnol* **64(6)**: 848-854.

Edge, R., McGarvey, D.J. and Truscott, T.G. (1997). The carotenoids as antioxidants—a review. *J Photochem Photobiol B* **41**: 189-200.

Goldman, J.C. (1977). Biomass production in mass cultures of marine phytoplankton at varying temperatures. *J Exp Mar Biol Ecol* **27**: 161-169.

Ip, P.F. and Chen, F. (2005). Production of astaxanthin by the green microalga *Chlorella zofingiensis* in the dark. *Process Biochem* **40**: 733-738.

Khachik, F. (1995). Process for isolation, purification, and recrystallization of lutein from saponified marigolds oleoresin and uses thereof. U.S. Patent 5,382,714.

Kimura M, Rodriguez – Amaya (2002). A scheme for obtaining standards and HPLC quantification of leafy vegetable carotenoids. *Food Chemistry* **78**: 389-398.

Lorenz, R.T. and Cysewski, G.R. (2000). Commercial potential for Haematococcus microalgae as a natural source of astaxanthin. *Trends Biotechnol* **18**: 160-167.

Martinez, F., Avendafio, C., Marco, E. and Orçs, M.I. (1987). Algal population and auxotrophic adaptation in a sugar refinery wastewater environment. *J Gen Appl Microbiol* **33**: 331-41.

Muenscher, W.C. (1923). Protein synthesis in *Chlorella*. *Botanical Gazette* **75(3)**: 249-267.

Olaizola, M. (2000). Commercial production of astaxanthin from *Haematococcus pluvialis* using 25,000-liter outdoor photobioreactors. *J Appl Phycol* **12**: 499–506

Orçs, M.I., Marco, E. and Martinez, F. (1991). Suitability of *Chlorella vulgaris* UAM 101 for heterotrophic biomass production. *Bioresource Technology* **38**: 179-184.

Shi, X.M. and Chen, F. (1997). Stability of lutein under various storage conditions. *Nahrung/Food* **41**: 38-41.

Shi, X.M., Chen, F., Yuan, J.P. and Chen, H. (1997). Heterotrophic production of lutein by selected *Chlorella* strains. *J Appl Phycol* **9**: 445-450.

Shi, X.M., Zhang, X.W. and Chen, F. (2000). Heterotrophic production of biomass and lutein by *Chlorella protothecoides* on various nitrogen sources. *Enzyme Microb Technol* **27**: 312-318.

Shi, X.M., Jiang, Y. and Chen, F. (2002). High-yield production of lutein by the green microalga *Chlorella protothecoides* in heterotrophic fed-batch culture. *Biotechnol Prog* **18**: 723-727.

Sorokin, C. and Krauss, R.W. (1958). The effect of light intensity on the growth rates of green algae. *Plant Physiol* **33**: 109-113.

Stengel, E. and Soeder, C.J. (1975). Control of photosynthetic production in aquatic ecosystems. *In*: Cooper, J.P. *editor*. Photosynthesis and productivity in different environments. Cambridge, UK: Cambridge University Press. pp. 645-660.

Syreth, P.J. and Morris, I. (1963). The inhibition of nitrate assimilation by ammonium in *Chlorella*. *Biochem Biophys Acta* **67**: 566-575.

Wen, Z.Y. and Chen, F. (2001). Optimization of nitrogen sources for heterotrophic production of eicosapentaenoic acid by the diatom *Nitzschia laevis*. *Enzyme Microb Technol* **29**: 341-347.

Wu, Z.Y., Shi, C.L. and Shi, X.M. (2007). Modeling of lutein production by heterotrophic *Chlorella* in batch and fed-batch culture. *World J Microbiol Biotechnol* **23**: 1233-1238.

Yongmanitchai, W. and Ward, O.P. (1991). Growth of and omega-3 fatty acid production by *Phaeodactylum tricornutum* under different culture conditions. *Appl Environ Microbiol* 57: 419-425.

6

WETLANDS OF ASSAM: BOURNE FOR EXPLORATION OF INDIGENOUS BIO-DIESEL POTENTIAL MICROALGAE

Simrat Kaur, H.K. Gogoi, R.B. Srivastava[1] and M.C. Kalita[2]

Abstract

Microalgae, the autotrophic organisms that are less than 1 mm in diameter with less complex structure, fast growth and high oil content are globally being targeted as an efficient source of biodiesel. A realistic estimate has been offered by University of North Hampshire for the replacement of vehicular fuel with biodiesel by utilizing algae with more than 55% oil content. China has also reported high quality diesel from microalgae Chlorella protothecoides by heterotrophic growth in fermenters. In Indian scenario, there exists every need for extensive R & D efforts to probe into a suitable indigenous strain, required for sustainable algal-oil production to meet our domestic demand of biodiesel. The state of Assam in North Eastern region of India, apart from being the mega bio-diversity hotspot in the world, has bestowed upon with vivid wetland resources. Four different types of wetlands namely pond, ox-bow lake, waterlogged wetland and swamp or marsh, total 3388 in number, are found in 23 various districts of Assam. The largest river island in the world, Majuli is also found here in one of the main district Jorhat, which is a natural heritage. Apart from natural wetlands, there are also several man made reservoirs and tanks. These water resources provide habitat and a conducive environment to support a rich diversity of fresh water population of phytoplanktons, which have hitherto remains unexplored. The prospects and potentials of algal bioresources from this fabulous biodiversity region is being explored in

[1] Defence Research Laboratory, Tezpur, Assam - 784 001, India

[2] Department of Biotechnology, Gauhati University, Guwahati - 781 014, Assam, India

the present study for algal-oil production and its conversion to biodiesel. The work on collection and screening for selecting highly oleaginous microalgal strains has been initiated at Defence Research Laboratory at Tezpur in Assam. Sampling has been done so far from four different districts viz. Dhemaji Lakhimpur, Nogaon and Sonitpur. Isolation and purification has been carried out to obtain the pure cultures of microalgae, which will be further screened for desired lipid content. An attempt will also be made for augmentation of lipid content under suitable and controlled laboratory conditions.

Key words: Assam, Algal Biodiversity, Algal Lipids, Indigenous Microalgae, Biodiesel

Introduction

Algae is a non taxonomic term that constitutes a heterogenous group with a broad spectrum of phenotypes that are categorized into six eukaryotic and one prokaryotic phylum. Each of these phyla is defined by the assemblage of algal species which show similar photosynthetic pigment units, reserve food storage organelles and structural complexity. The eukaryotic phyla include Chlorophyta (green algae), Euglenophyta, Dinoflagellates, Bacillariophyta (diatoms), Rhodophyta (red algae) and Phaeophyta (brown algae); Cyanophyta (blue green algae) is the only prokaryotic phylum.

Algae show a hierarchy in structure, size and thallus organization that transcends from Chlorophyceae to Phaeophyceae, which shows correlation and adaptation to niches in their varied habitats. The size of eukaryotic algae ranges from less than 1 micron, the diameter of smallest alga *Ostreococcus tauri*, to 60 meters, the length of giant kelp *Macrocystis pyrifera*. Although, most freshwater algae are microscopic and occur in large numbers so as to form the pond scums. Algae not only occupy aquatic ecosystems, freshwater or marine but also occur in almost all other habitats. *Dunaliella salina* is capable of growing in waters nearly saturated with NaCl. *Chlamydomonas nivalis*, cryophilic green algae, can survive in conditions such as low temperature, poor nutrition, permanent freeze-thaw cycles and high irradiation. Algae can grow on the bark of trees, rocks, soil and other natural or manmade surfaces in the form of biofilms (Hallmann, 2007).

Freshwater Habitats of Algae in Assam

The Indian state of Assam has the maximum number and area of water masses under floodplain wetlands, mainly associated with the river

Brahmaputra. The state also receives one of the highest rainfall in the world ranging between 178 and 305 cm, in some places exceeding 400 cm. The river Brahmaputra and its tributaries very often off change its course, leading to formation of various types of floodplain wetlands or beels.

Apart from various other water bodies such as waterlogged wetland, swamp and river island, Assam has around 1,392 beels and total area associated with beels is estimated at 92,000 ha. The floodplain wetlands or beels include low lying water bodies of diverse origin, size, depth, shape and ecological characteristics. These have been classified on the basis of morphometry, water retentivity, depth, size and shape. There are seasonal beels which periodically get inundated by monsoon rains and floods but completely dry up during summer months; the perennial beels which are deeper and permanently retain water round the year.

Beels are generally considered as highly eutrophicated system with high rate of primary productivity. The microalgae are ecologically known as phytoplanktons. The various phytoplankton groups associated with the beels are Chlorophyceae, Bacillariophyceae, Cyanophyceae, Euglenophyceae, Dinophyceae and Chrysophyceae. In beels of upper Assam zone, the phytoplanktons are represented by four classes in the order of Chlorophyceae> Bacillariophyceae> Cyanophyceae> Dinophyceae.

The order of dominance of various classes of phytoplanktons in central Assam zone is as follows: Chlorophyceae>Dinophyceae> Bacillariophyceae> Cyanophyceae> Euglenophyceae. The lower Assam zone of Brahmaputra valley show richer phytoplankton diversity. Low altitude, high temperature and gradual slopes provide conducive conditions for rich plankton growth. The order of domination of phytoplankton in lower Assam beels is reported as Chlorophyceae> Cyanophyceae> Dinophyceae> Euglenophyceae> Bacillariophyceae> Chrysophyceae. The notable feature in the beels of lower Assam is the high rate abundance of Euglenophyceae. The main genera of microalgae which have been reported from these phytoplankton groups are *Ankistrodesmus*, *Mougeotia*, *Closterium*, *Botryococcus*, *Spirogyra*, *Microsterium*, *Cosmarium*, *Microspora*, *Pleodorina*, *Eudorina*, *Pandorina*, *Volvox*, *Gyrosigma*, *Zygnema* (Chlorophyceae); *Nitzschia*, *Pinnularia*, *Melosira*, *Navicula*, *Surirella*, *Diaptoma*, *Synedra* (Bacillariophyceae); *Nostoc*, *Spirullina*, *Oscillatoria*, *Anaebena*, *Nodularia* (Cyanophyceae); *Euglena*, *Phacus* (Euglenophyceae); and *Dinobryon* (Chrysophyceae) (Sugunan and Bhattacharya, 2000).

The accounted diversity of phytoplankton has so far been studied and exploited mainly as primary producers and a source of aquaculture feed.

Exploration and Utilization of Algae - An optimistic foresight for biodiesel

Keeping in view, the great diversity of water bodies and favorable environmental conditions of Assam, there is an ample scope to discover many potential oleaginous microalgal species which can provide promising feedstock for biodiesel. Thus to begin with, exploration, culturing and screening of suitable strains of indigenous microalgae is the need of hour which would be necessary to sustain for our domestic need for biofuel.

Culturing Microalgae

Microalgae may be obtained in great numbers from water bodies or moist places. Once obtained from natural source, there are several steps involved in obtaining the desired strain of microalgae in pure form. Certain species of microalgae being fast growing and dominant can be easily isolated and cultured; others may vanish soon from the collected water samples. Thus time is an important factor while isolating algae. Other factors which affect isolation are mimicking of the nutrient status and other physiochemical parameters of the habitat; taxonomic knowledge to know the particular nutrient requirements of target species; and elimination of contaminants (bacteria, fungi and zooplankton) and those undesired algae that outcompete the target species. To obtain desired alga, the sample is enriched with culture medium, soil-water extract or particular nutrients and the detailed techniques for culturing algae have been followed (Andersen, 2005).

Screening of Oleaginous Microalgal Strains

The conventional method of lipid estimation involves solvent extraction which is tedious and also requires large quantities of biomass. Therefore, to screen the desired strain for large scale production, a sensitive and rapid method is to be selected. A fluorescent method for *in situ* lipid estimation is advantageous since it requires few micro liters of culture broth and is sensitive to detect the lipid concentration in the range of 2-5000 µg/ml (Kimura *et al.*, 2004). Nile red, a dye that fluoresces at defined wavelength has been used to determine the neutral lipid content of Microalgal cells (Cooksey *et al.*, 1987; Elsey *et al.*, 2007).

Lipid Composition of Microalgae

Average lipid content of microalgae is of the order of 20–40% of dry weight which can be enhanced by manipulating the cultural conditions. Among the different classes of microalgae, Chlorophyceae and Chrysophyceae contain maximum content of total lipids. Microalgal lipids are triglycerides and fatty acids with a chain length of C_{14} to C_{22}, which may either be saturated or unsaturated. Saturated and monounsaturated ones are the predominant forms of lipid found in eukaryotic algae. The most common storage forms are neutral lipids or triglycerides which may also constitute up to 80% of total lipid fraction. Among the different algal taxas, some variations are found in types of lipids. The neutral triglyceride, linolenic (18:3) is common in green algae, whereas other similar form linoleic (18:2) is only a minor constituent in diatoms (Borowitzka and Borowitzka, 1988).

Augmentation in Algal Lipids

CO_2 concentration, light and nutrient stress that affect growth rate and cell yield in algal cultures also affect the lipid and fatty acid composition of algae. CO_2 concentration during growth gives varying effects on microalgae, including change in fatty acid composition. Remarkable changes in linoleic and linolenic acid associated with CO_2 concentration has been reported in *Chlorella vulgaris*. The relative content of linoleic acid was higher in cells grown in high CO_2, while that of linolenic acid was much lower in cells grown in low CO_2 (Tsuzuki *et al.*, 1990). The species *Chlorella minutissima* has a fast growth rate with high PUFA content and it is easily cultivated in different stress conditions (Seto *et al.*, 1984).

The best studied factor has been the N_2 limitation. In all the algal species, there is an increase in chlorophyll as well as in biomass and protein contents with an increase in N_2 levels. At low N_2 levels, the green algae are reported to have a high lipid content of 35% to 58% of the dry weight of which 70% are neutral lipids such as triglycerols, mainly 16:0 and 18:1 fatty acids (Borowitzka and Borowitzka, 1988). At low concentration of nitrogen, *Chlorella* and *Euglena* synthesized saturated (16:0) and monounsaturated (18:1) fatty acids, whereas at high N_2 concentration, 16:2, 16:3, 16:4 and 18:2 fatty acids predominated (Amotz and Tornabene, 1985). Formation of chloroplastic and extraplastidial lipid bodies containing both triacylglycerols and carotenoids under stress conditions such as high irradiance and N_2 starvation has been described in various green microalgal species such as *Dunaliella bardawil*, *Chlorella zofingiensis*, *Scenedesmus* sp. (Thompson, 1996) and *Haematococcus pluviallis* (Boussiba, 2000).

Conversion of Algal Lipids to Biodiesel

The liquid fuel or biodiesel is obtained by the conversion of algal triglycerides to methyl esters by a process called transesterification. It is accomplished by adding excess alcohol plus an alkaline catalyst. The conversion of triglyceride to ester fuel occurs at about 98% efficiency (Borowitzka and Borowitzka, 1988). The process for obtaining biodiesel from *Chlorella protothecoides* has been optimized using 100% catalyst quantity (based on oil weight) with 56:1 molar ratio of methanol to oil at temperature of 30°C. This reduces the product specific gravity from an initial value of 0.912 to a final value of 0.864 in about 4 hours reaction time (Xu *et al.*, 2006).

Economic Feasibility of Biodiesel Production

To make algal diesel cost competitive with petrodiesel, a reasonable target price of $ 0.48/L has been estimated for microalgal oil. The estimated cost of oil recovery from the lower-cost biomass produced in photobioreactor is $ 2.80/L and this cost get reduced to $ 0.72, if the oil content of algal biomass is enhanced to 70% by weight (Chisti, 2007). It is also arguable that cost reduction can rightly be achieved at the level of biomass production. Economical and sustainable biomass production can be obtained by using cheap alternatives of nutrients or by directly growing algae in the eutrophicated sources of water such as industry effluents or sewage water. Moreover, the prevailing environmental conditions of Assam are highly conducive for growth of microalgae. Thus, it has been foresighted that hot and humid climate of the state will favor the growth of microalgae in open pond cultivation system.

Conclusions

Looking at the present scenario of global production of biodiesel from microalgae, the R & D organizations in India should also venture into a project which can explore the biowealth from its major biodiversity hotspot region. This is also necessary to sustain our future needs for energy as well as to enable us to earn the carbon credit by adopting the green-clean technology for biofuel production. Exploration for alternate energy resources has hitherto remained unexplored from the richest treasure trove of biodiversity hot spot of North Eastern India. Assam possesses tremendous freshwater bodies such as ponds and beels, which are enriched with diverse microalgae species. Looking at the tremendous biodiversity of this region, Defence Research Laboratory is putting all efforts to tap the potentials of indigenous strains for biodiesel production.

References

Amotz, A.B. and Tornabene, T.G. (1985). Chemical profile of selected species of microalgae with emphasis on lipids. *J. Phycol.* **21**: 72-81.

Anderson, R.A. (2005). Algal Culturing Techniques. Elsevier Academic Press, USA.

Borowitzka, M.A. and Borowitzka, L.J. (1988). Microalgal Biotechnology. Cambridge Univ. Press, Cambridge, New York.

Boussiba (2000). Carotenogenesis in the green alga *Haematococcus pluvalis*. Cellular physiology and stress response. *Physiol Plant* **108**: 111-117.

Chisti, Y. (2007). Biodiesel from microalgae. *Biotech. Adv.* **25**: 294-306.

Cooksey, K.E., Guckert, J.B., Williams, S. and Callis, P.R. (1987). Fluorometric determination of the neutral lipid content of Microalgal cells using Nile Red. *J. Microb. Methods* **6**: 333-345.

Elsey, D., Jameson, D., Raleigh, B. and Cooney, M.J. (2007). Fluorescent measurement of Microalgal neutral lipids. *J. Microb. Methods* **68**: 639-642.

Hallmann, A. (2007). Algal transgenics and biotechnology. *Transgenic Plant J.* **1(1)**: 81-98.

Kimura, K., Yamaoka, M. and Kamisaka, Y. (2004). Rapid estimation of lipids in oleaginous fungi and yeasts using Nile red fluorescence. *J. Microb. Methods* **56**: 331-338.

Seto, A., Wang, H.L. and Heseltine, C.W. (1984). Culture conditions affect eicosapentaenoic acid content of *Chlorella minutissima*. *J. American Oil Chemists Society* **61(5)**: 892-894.

Sugunan, V.V. and Bhattacharjya, B.K. (2000). Ecology and Fisheries of Beels in Assam. Bull. No. 104. Central Inland Capture Fisheries Research Institute.

Thomson, G.A. (1996). Lipids and membrane function in green algae. *Biochem. Biophys. Acta* **1302**: 17-45.

Tsuzuki, M., Ohnuma, E., Sato, N., Takaku, T. and Kawaguchi, A. (1990). Effects of CO_2 concentration during growth on fatty acid composition in microalgae. *Plant Physiol.* **93**: 851-856.

Xu, H., Miao, X. and Wu, Q. (2006). High quality biodiesel production from a microalga *Chlorella protothecoides* by heterotrophic growth in fermenters. *J. Biotech.* **126(1)**: 499-507.

7

POTENTIAL OF A RED ALGA, *LEMANEA* FOR USE AS A SOURCE OF BIOFUEL

Jayashree Rout[1*], B.B. Chaugule[2] and Rahul Bhosale[3]

Abstract

Algae are the highest yielding feedstock for biodiesel which produce multiple times more oil then the other leading feedstock. The advantages of deriving biodiesel from algae are the rapid growth rate, incredible space efficiency, high per acre yield, sulphur free fuel, non toxicity and high biodegradability. The present paper is an attempt to highlight the potential of a fresh water red alga, Lemanea fluviatilis as a source of biofuel. This is the only fresh water red alga with known use as food from this part of country. This specific alga has the potentials for biodiesel perspective. The microalgae contain lipids and fatty acids as membrane components, storage products metabolites and sources of energy. While exploring for biofuel, an interesting array of fatty acid profile has been identified by Gas chromatography. The data present 53.10% total saturated fatty acid and 46.8% total unsaturated fatty acid.

Key words: Rhodophyceae, Red algae, *Lemanea*, Biofuel

There has been a renewed interest on algae owing to their significance as bio-fertilizers, supplements to nutritious food and feed, bioactive compounds, fine chemicals, pharmaceuticals products and even

1 Department of Ecology & Environmental Science, Assam University, Silchar - 788 011, Assam, India

2 Department of Botany, University of Pune, Pune - 411 007, Maharashtra, India

3 Department of Biotechnology & Bioinformatics, University of Pune, Pune -411 007, Maharashtra, India

* *Corresponding author*: E-mail: rouyjaya@rediffmail.com

biofuel. The prime requirement for launching various programmes involving algae is the availability and identity of the organism itself. The information on some economic benefits of algae such as human food, fish food, insectivorous plant food, utility in ornamental fish aqua culture programme etc. have been widely reported. These tiny organisms are so efficient that they can double their weight several times a day, producing oil in the process-30 times more oil per acre than soyabeans (Inhabitat, 2008). Being in rich biodiversity hotspot area, our group working on algae in this University are keen to develop through collaborative tie-ups with industry to explore viable algal species for renewable fuel production.

Biofuels are really a form of solar energy. By the process of photosynthesis the solar energy is converted into chemical energy, then in form of oils leading to biofuels. More efficient a particular plant is at converting solar energy into chemical energy; the better it is from biofuel perspective. Algae are a clear winner in this regard. Sklar (2006) highlighted algae hold promise for ethanol, butanol, biofuel and synthetic oils. The San Diego based company also boasted that unlike other biofuel, growing algae does not need vast areas of agricultural land and therefore there is no "food versus fuel" trade off (Salter, 2008). Algae, the highest yielding feedstock for biodiesel, produces on an average 24 times more oil per acre, then the next leading feedstock-palm oil (at 635 gallons/acre/year) (Hodge, 2007). Some types of algae are about 50% oil, suitable for biodiesels (CNN, 2008). The advantages of deriving biodiesel from algae are manifold: (i) rapid growth rates (ii) incredible space efficiency, a high per acre yield (iii) sulphur free fuel (iv) Non toxic (v) high biodegradability.

Feeding the algae with extra CO_2 and organic material could boost the oil yield to as much as 40% by weight. A published scientific report on this aspect highlights on a plan to greatly increase algae oil yields by feeding the algae with extra CO_2 (the main green house gas from power plants) and organic material like sewage and waste water, meaning the algae could simultaneously produce biofuel and cleanup environmental problems (Colosi, 2008).

The present paper is an attempt to highlight the potential of a red alga, *Lemanea fluviatilis* an endemic form from this part of India for use as a source of biofuel. This is a fresh water Rodophyta of bluish green colour found in Indian Sub-continent (Pandey and Trivedi, 1995). This alga grows in rivers of Manipur state in North-East part of India. The filaments have beautiful plant bodies with feathery branches lacier and delicate and usually attached to rocks or solid substratum in water by holdfast and poorly adapted to turbulent habitats (Fig. 1a, b). The indigenous technology for field harvest and easy handling is reported. This is the only fresh water red alga with known use as food from this part of country. This unique alga is a delicious item in the diet, relished by the local people. While exploring the neutraceutical aspect, this specific alga, we got interested to assess it's potentials for biodiesel perspective.

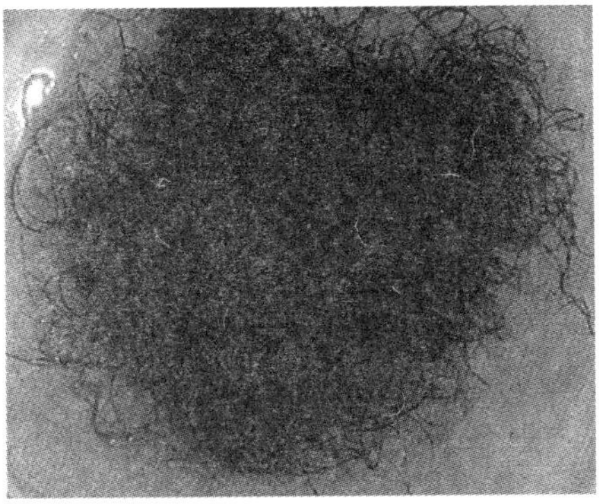

Fig. 1a. The *Lemanea* plant with feathery branches

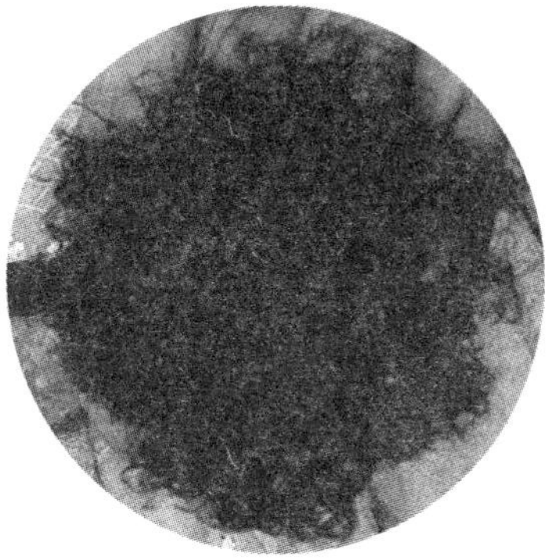

Fig. 1b. The dry form of *Lemanea* plant used commercially

Biodiesel algae grow in winter as envisaged by AutoblogGreen (2008). Green Star products grow algae in outdoor environments in low temperature even in snowfall period. In the present paper, we have identified the juvenile stage of *Lemanea* maturing during winter, fruiting strands reaching their mature size in the spring and spores becoming evident in late spring. This algal bloom during winter period can be harnessed for possible source with biofuel potential. The alga shows various nutritional constituent as per analysis (air dry sample): protein-20%, carbohydrate-50% and lipid-10%. In addition to a number of trace elements, a rare trace element like silver is present in this alga. This is not reported to occur in any other herbal sources earlier.

Microalgae contain lipids and fatty acids as membrane components, storage products metabolites and sources of energy. In our pilot-scale exploration for biofuel, an interesting array of fatty acid profile has been identified by Gas chromatography. The data (Table 1) present different fatty acid components with varying relative percentage: total saturated fatty acid-53.10%, total unsaturated fatty acid-46.8% [Total Mono Unsaturated Fatty Acids-22.30% and Total Poly Unsaturated Fatty Acids (PUFA content)-24.5%] and the least contributor is Trans

Table 1. Relative percentage of Fatty acids detected in *Lemanea fluviatilis*

S. No.	***Name of the test***	***Result***
A	***Saturated fatty acids***	
1	Caproic acid methyl ester (C4:0)	2.70%
2	Caprylic acid methyl ester (C6:0)	1.60%
3	Tridecanoic acid methyl ester (C13:0)	0.40%
4	Myristic acid methyl ester (C14:0)	5.10%
5	Pentadecanoic acid methyl ester (C15:0)	0.30%
6	Palmitic acid methyl ester (C16:0)	39.40%
7	Stearic acid methyl ester (C18:0)	1.00%
8	Heneicosanoic acid methyl ester (C21:0)	0.20%
9	Behenic acid methyl ester (C22:0)	0.30%
10	Tricosanoic acid methyl ester (C23:0)	1.40%
11	Lignoseric acid methyl ester (C24:0)	0.70%
	Total saturated fatty acids	**53.10%**
B	***Mono unsaturated fatty acids***	
12	Palmitoleic acid methyl ester (C16:1)	1.30%
13	Oleic acid methyl ester (c18:1n9c)	9.40%
14	cis-11-Eicosenoic acid methyl ester	0.10%
15	Erucic acid methyl ester (C22:1n9)	11.50%
	Total mono unsaturated fatty acids	**22.30%**
C	***Poly unsaturated fatty acids***	
16	Linoleic acid methyl ester (C18:2n6c)	2.90%
17	Linolenic acid methyl esters (C18:2n6c)	0.10%
18	cis-11,14-Eicosadienoic acid methyl	0.60%
19	cis-11,14,17-Eicosatrienoic acid methyl ester (C20:3n3)	0.20%
20	cis-13,16-Docosadienoic acid methyl	19.60%
21	cis-, 5,8,11,14,17-Eicosapenaenoic acid	1.10%
	Total poly unsaturated fatty acids	**24.50%**
D	***Trans fatty acids***	
22	Elaidic acid methyl ester (C18:1n9t)	0.20%

Fatty Acid-0.20%. Among the long chain carbon, Palmitic acid methyl ester (C16) (39.4%) shared highest in saturated Fatty Acid followed by Myristic acid methyl ester (C14) (5.1%). Among the Mono Unsaturated Fatty Acids, Eucic acid methyl ester (C22) (11.5%) contributed highest followed by Oleic acid methyl ester (C18) (9.4%). Out of the total PUFA content cis-13.16 Docosadienoic acid methyl shared the highest percentage (19.6%) followed by Linoleic acid methyl ester (C18) (2.9%). Other components reflect their meagre share in the fatty acid profile. The oil content analysis was not been able to be performed due to technical difficulties.

We are in a process to identify various algal species and make them robust and ensure to bring out a scale of production that is meaningful and economically viable. The algae are planned to be grown in a bioreactor or in the fish tank of pan shape 1800 mm x 900 mm x 600 mm size fashioned parallel with interconnections, aeration facility and fed with sunlight and water. The algae are then harvested and turned into biodiesel fuel. Green fuel Technologies announced they had began construction of their commercial scale algae plant while Petrosun announced they would be taking their pilot algae commercial farm (Science Daily 2008). Most of the work on algae on fuel production in 1980's and 1990's focussed almost entirely on the open pond approach. Now working in this field with collaboration most groups have shifted on the use of proprietary photobioreactors. On encountering any problem, vulnerability most importantly temperature fluctuations, high evaporation loss etc by using tanks (as proposed) enclosed photobioreactors can immediately be taken over for mass production of algae for biofuel development.

The alga *Lemanea* is comprised of about 50% carbohydrate and 10% lipid. A two-in-one approach both food and fuel seems to be meaningful and economic. The starch product can be reclaimed for ethanol fermentation and the oil for biodiesel production - Green Fuel Plan.

References

Autoblog Green, (2008). (http://www.autobloggreen.comFeb.2/2008/02/green-star-products-grows-biodiesel-algae-in-cold-montana-winter/).

CNN, (2008). (http://www.cnn.com/2008/TECH/Science/04/01/algae.oil/index.html#cnnSTCvideO).

Colosi, L. (2008). Biodiesel from algae. Mongabay.com.Aug.15.

Hodge, N. (2007). Algal biofuel: Biodiesel Bliss-The second coming.

Inhabitat, (2008). (http://www.inhabitat.com/2008/04/15/oilgae-the-next-biofuel-from algae/).

Pandey, S.N. and Trivedi, P.S. (1995). A textbook of Algae. Vikas Publishing House Pvt. Ltd. p. 342.

Salter, J. (2008). Green pond algae could be made into biofuel.Telegraph.Co.UK.

Science Daily (2008). In the world of alternative fuels, there may be nothing greener than pond scum.

Sklar, S. (2006). A consumer Guide to Solar Energy.

8

EXPLORATION AND UTILIZATION OF MICROALGAE FOR BIODIESEL PRODUCTION: HINDRANCE TOWARDS TRANSGENIC APPROACH

J. Purkayastha*, H.K. Gogoi and R.B. Srivastava

Abstract

Continued use of fossil based fuels has been widely recognized as unsustainable because of very fast depleting source and their major contribution towards environmental pollution. As a potential renewable and carbon neutral energy source biodiesel fuel is becoming essential for the global fuel market. As demonstrated in this communication, microalgae with higher photosynthetic efficiency, larger biomass and faster growth rate compared to that of oil crops appear to be an appropriate source of renewable biodiesel that is capable of meeting the global demand for transport fuels. Oil content of many microalgae is usually 80% of its dry weight, greatly exceeding than that of the oil productivity of the best oil producing crops. Algae as third generation biodiesel need transgenic manipulation to deal with "weeds", light penetration, photoinhibition, carbon assimilation, etc. Transgenic microalgae with high oil-bearing productivity through genetic manipulations are capable of making microalgal derived biodiesel economically competitive with petrodiesel. In this correspondence, the potential of microalgal biodiesel and the advantage of microalgae as the raw material for biodiesel has been analyzed. Moreover, key enzymes in microalgal fatty acid biosynthesis and problems in manipulation of lipid production via genetic engineering, especially the approaches for exploration and utilization of high oil yielding transgenic microalgae for biodiesel production have also been highlighted.

Defence Research Laboratory Tezpur, Post Bag No. -2, Tezpur -784 001, Assam, India

* *Corresponding author*: E-mail: jubilee_p@rediffmail.com

Key words: Microalgae, Biodiesel, Oil contents, Transgenic microalgae, Fatty acid biosynthesis

Introduction

Concern over energy, the environment and natural resources is increasing across the globe. The use of microalgae is gaining popularity as a potential high volume feedstock for biodiesel production. Reducing accumulation of CO_2 in the environment is now a major priority and there is worldwide effort to replace petroleum sourced fuels with a sustainable source of bioenergy. The large amount of land needed to fulfil the requirement for crop based biofuel makes this an unrealistic route forward, and it is becoming increasingly apparent that microalgae have many advantages over using land crops for the production of biofuel (Chisti, 2007). Like plants, microalgae use sunlight to produce oils but they do so more efficiently than crop plants. Microalgae can provide feedstock for several different types of renewable biofuels. Biodiesel from oil has recently become the focus of much activity.

When planning for biodiesel use and its potential as an ongoing resource, it's important to review the current status of utilization of biodiesel in national and international bio-fuel market, analyze the feasibility, progress and hindrances in manipulation of lipid production in microalgae via genetic engineering etc. Moreover, the review studies offer the scopes for handling the present problems associated with adoption of microalgal biodiesel along with the suggested solutions.

Advantages of Biodiesel

Because plants (including microalgae) produce oils from sunlight and air, and can do so year after year on cropland, these oils are renewable. Animal fats are produced when the animal consumes plant oils and other fats, and they too are renewable. Used cooking oils are mostly made from vegetable oils and animal fats, and also are both recycled and renewable. The biodiesel is made by chemically combining any natural oil or fat with an alcohol such as methanol or ethanol. These chemicals are called long chain mono alkyl esters, or also referred to as long chain fatty acid methyl esters.

Biodiesel is a cleaner fuel than petroleum diesel. It is virtually free of sulfur, thereby eliminating the production of sulfur oxides during combustion. Emissions of hydrocarbons, carbon monoxide, and particulates during combustion are also significantly reduced in comparison to emissions from petroleum diesel. These properties make biodiesel useful in facilitating compliance with the Clean Air Act Amendments of 1990 (CAAA). An additional benefit of biodiesel is that it is thought to be a nontoxic, biodegradable fuel that can be used in standard, unmodified diesel engines. Biodiesel provides essentially the same energy content and power output as petroleum-based diesel fuel while reducing emissions. Biodiesel also contains 11% oxygen by weight, as well as a slightly higher cetane number, which provides for more complete combustion and a reduction in most emissions. HFRR (microns) 300–600 <300 Ethanol fuel made from fermentable starchy and lignocelluloses plants can also reduce tailpipe emissions, including air toxics. However, the heating value of ethanol fuel is only 2/3 of gasoline and more less than petrodiesel. The combustibility of ethanol fuel may descend due to its any blend ratio with water (Yuan *et al.*, 2005). Because biodiesel has its fine characteristic advantage over other biomass fuel, many governments worldwide successively developed the research programs and the industrial production of biodiesel raw material. The aim is to displace fossil fuel which reserves reduces and seriously pollutes the environment day by day.

Potential of Microalgae as Biodiesel Resource

Microalgae are microscopic aquatic plants with the potential to produce large quantities of lipids (plant oils). The use of microalgae is gaining popularity as a potential high volume feedstock for biodiesel production. Microalgae, (including cyanobacteria, diatoms, dinoflagellates) combine the benefits of (1) plants in sequestering atmospheric CO_2 and using energy from the sun and (2) microbes in high productivities and controllability.

Comparison of various sources of biodiesel shows microalgae as the most potent source of biodiesel (Table 1). Microalgae are of interest because of their high growth rates and tolerzihce to varying environmental conditions, and also because the lipids they produce. Unlike other oil crops, microalgae grow extremely rapidly and many are exceedingly rich in oil. Microalgae commonly double their biomass within 24 h. Biomass doubling times during exponential growth are commonly as short as 3.5 h. Oil content in microalgae can exceed 80% by weight of dry biomass (Metting, 1996; Spolaore *et al.*, 2006). Oil levels of 20–50% are quite common (Table 2). Oil productivity, that is the mass of oil produced per unit volume of the microalgal broth per day,

Table 1. Comparison of some sources of biodiesel

Crop	*Crop oil yield (L/ha)*
Corn	172
Soybean	446
Canola	1190
Jatropha	1892
Coconut	2689
Oil palm	5950
Microalgae [b]	136,900
Microalgae [c]	58,700

[a] 70% oil (by wt) in biomass; [b] 30% oil (by wt) in biomass.

Table 2. Oil content of some microalgae

Oil content of some microalgae	*Oil content (% dry wt)*
Botryococcus braunii	25–75
Chlorella sp.	28–32
Crypthecodinium cohnii	20
Cylindrotheca sp.	16–37
Dunaliella primolecta	23
Isochrysis sp.	25–33
Monallanthus salina	20
Nannochloris sp.	20–35
Nannochloropsis sp.	31–68
Neochloris oleoabundans	35–54
Nitzschia sp.	45–47
Phaeodactylum tricornutum	20–30
Schizochytrium sp.	50–77
Tetraselmis sueica	15–23

depends on the algal growth rate and the oil content of the biomass. Microalgae with high oil productivities are desired for producing biodiesel. Microalgal biodiesel would come on line later to fulfill larger market demands from the diesel market, and enhance the long-term potential of biodiesel as a renewable energy source. Depending on species, microalgae produce many different kinds of lipids, hydrocarbons and other complex oils (Banerjee *et al.,* 2002; Metzger and Largeau, 2005; Guschina and Harwood, 2006). Not all algal oils are satisfactory for making biodiesel, but suitable oils occur commonly.

Using microalgae to produce biodiesel will not compromise production of food, fodder and other products derived from crops. Potentially, instead of microalgae, oil producing heterotrophic microorganisms (Ratledge, 1993; Ratledge and Wynn, 2002) grown on a natural organic carbon source such as sugar, can be used to make biodiesel; however, heterotrophic production is not as efficient as using photosynthetic microalgae. This is because the renewable organic carbon sources required for growing heterotrophic microorganisms are produced ultimately by photosynthesis, usually in crop plants. Production of algal oils requires an ability to inexpensively produce large quantities of oil-rich microalgal biomass.

Genetic Engineering: Improving Microalgal Strains with High Lipid Content

To genetically engineer an organism, such as a microalgal, to carry out a particular function (*e.g.*, lipid accumulation) better, the DNA composition of the organism must be altered by manipulating the genes that encode proteins of regulatory significance. A gene can be thought of as a stretch of DNA that gives an organism the ability to perform a specific function. For most genes, the DNA is "transcribed into messenger RNA, which is then "translated" into a protein. In many cases, the protein product is an enzyme that catalyzes a particular chemical reaction.

Improving strains for biodiesel production has been challenging due to the lack of information on algal pathways and metabolic control information. This is changing and new opportunities are opening for algal transgenics. Unlike bacteria which can not perform post-translational modifications, mammalian cells which are expensive and unstable on a large scale, or transgenic plants which lead to concerns over crop containment and chemicals, microalgae can offer a platform for recombinant protein, lipid and chemical production which is sustainable with high yields.

The challenge for microalgae has been in producing stable transformants. In the past five years, however, new techniques and protocols specific to various prokaryotic and eukaryotic microalgae have been developed resulting in the mapping of complete genomes for an increasing number of algal species. This rapid increase in algae genetic information is giving scientists the ability to relate form and function and to determine applicable microbial transformation tools. With dynamic modeling of particular regulatory processes using a systems biology approach, new studies can be developed to improve transgenic expression and further lipid production.

Manipulation of Fatty Acid Biosynthesis Enzymes

Acetyl-coenzyme a carboxylase (ACCase)

ACCase catalyses the rate-limiting reaction for fatty acid synthesis by irreversible carboxylation of acetyl-coenzyme A (acetyl- CoA) to produce malonyl-CoA through its two catalytic activities, biotin carboxylase (BC) and carboxyltransferase (CT) (Fig. 1) (Davis *et al.,* 2000). ACCase is a multi-subunit enzyme in most prokaryotes, whereas it is a large, multi-domain enzyme in most eukaryotes. Recent studies has suggested that soybean seeds in the beginning of development show an increase in expression of *ACC* gene as high oil content specie of soybean, from the beginning to mid-period of development, exhibits about a 2-fold increase in ACCase activity than that of low oil content specie of soybeans (Somers *et al.,* 2001), It has also been found that transformation of B. *napus* with *Brassica napus* napin storage-protein gene promoters fused to the *Arabidopsis ACC* gene resulted in 10- to 20-fold higher ACCase activity in mature seeds in which the total oil content was increased approximately by 5% (Ohlrogge *et al.,* 1999) suggesting that ACCase overexpression altered seed fatty acid composition with the largest effect being an increase in oil content. Antisense expression of ACCase in *B. napus* resulted in lower lipid content (Sellwood *et al.,* 2000). In prokaryotes like *Escherichia coli*, overexpression of four ACCase subunits resulted in a 6-fold increase in the rate of fatty acid synthesis (Davis *et al.,* 2000). All of the above evidences are in the line that overexpression of *ACC* gene is able to increase ACCase activity and the total oil content. National Renewable Energy Laboratory (NREL), which was supported by the United States Department of Energy's Office of Fuel Development during 1991–1996 (Sheehan *et al.,* 1998), conducted extensive research on the manipulation of algae lipid mechanism for high oil content. The first report of cloning a full-length ACCase sequence and its genetic transformation from any photosynthetic organism was demonstrated by Scientists in NREL during the early 1990s. They developed genetic transformation methods for microalgae and accomplished successfully *ACC* gene transformation of *Cyclotella cryptica* strains with demonstrated potential for biodiesel production in 1993–1995 (Sheehan *et al.,* 1998; Dunahay *et al.*, 1995; Roessler and Ohlrogge, 1993). However further continued studies had to be stopped due to scarcity of funds and was closed eventually with no budget in 1996 (Sheehan *et al.,* 1998). Currently very little is known about the molecular biology of oleaginous microalgae and the genetic regulation of lipid biosynthesis pathways.

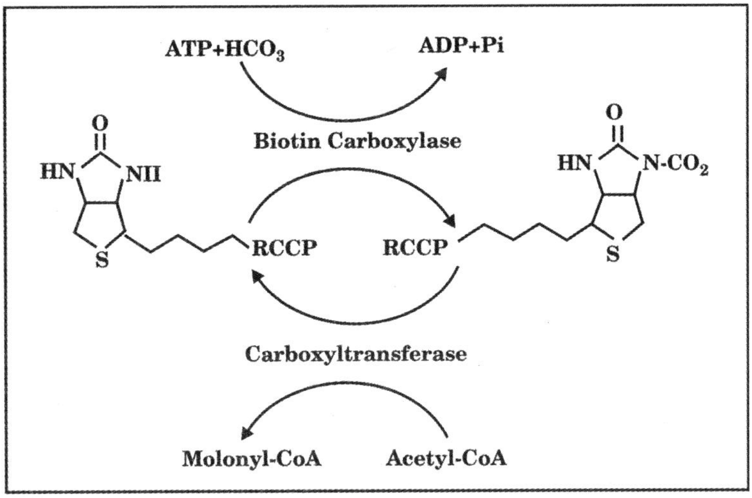

Fig. 1. Reaction mechanism of *E. coli* ACCase

Phosphoenolpyruvate carboxylase (PEPC)

Carbon fixation in photosynthesis for "C4" plants and photoautotrophic microalgae like cyanobacteria is catalyzed by the enzyme PEPC, which catalyzes the addition of CO_2 to phosphoenolpyruvate (PEP) to form the four-carbon compound oxaloacetate. The fatty acids biosynthesis in oil crop depended on the relative activity of ACCase and PEPC (Chen *et al.,* 2002). The biosyntheses of either protein or lipid take PEP as the common substrate. It is converted into oxaloacetic acid by PEPC and then directly participates in the anabolism of protein. On the other hand, PEP can be converted into pyruvate by the pyruvate kinase. Pyruvate is then transformed to acetyl-CoA by pyruvate dehydrogenase, yielding malonyl-CoA by the ACCase, finally participated in the anabolism of fatty acid in this form (Chen *et al.,* 2002). Overall, the relative activity of PEPC and ACCase determines the flow direction of PEP and the ration of proteins/lipids produced in oilseeds crops (Fig. 2). In the early 1950s, it was found that the contents of proteins and lipids were inversely correlated to each other in plants, as if the oil content would negatively relate to the activity of the PEPC (Chen *et al.,* 2002). Antisense expression of *B. napus* PEPC gene resulted in PEPC transgenic rapeseeds having 6.4%–18%, and even to 50% higher oil content, than control seeds (Chen *et al.,* 2002; Chen *et al.*, 1999; Chen *et al.*, 1999). Similarly, Antisense expression of soybean PEPC gene increased oil content in transgenic plant seeds and yielded a new strain of high oil content transgenic soybean (Zhao *et al.*, 2005). Therefore, the antisense

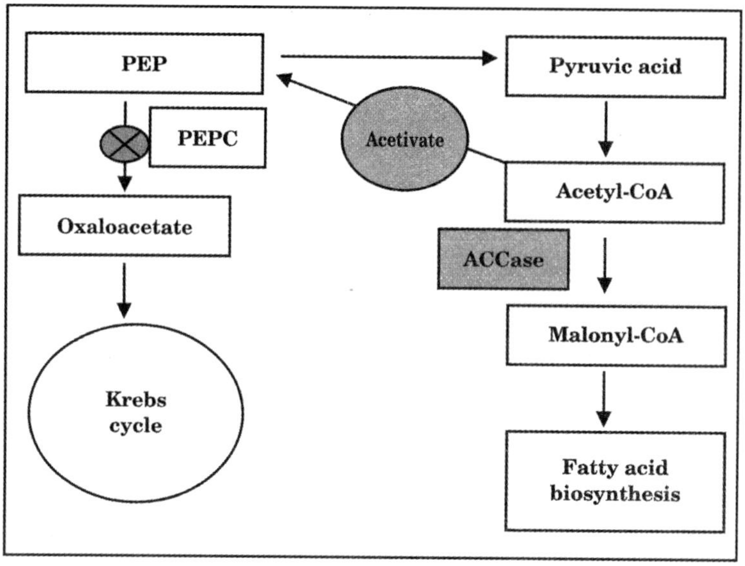

Fig. 2. Possible role of PEPC in biochemical processes of fatty acid biosynthesis pathway

expression of *PEPC* gene for the modulation of fatty acids biosynthesis in oil crops is an effective method. Recently, the DNA sequence of *PEPC* in many microorganisms, including *Anabaena* sp. PCC 7120 (Luinenburg and Coleman, 1992), *Anacystis nidulans* (Katagiri *et al.,* 1985), *E. coli* (Sabe *et al.,* 1984), *Microcystis aeruginosa* NIES-843 (Kaneko *et al.,* 2007), *Synechococcus vulcanus* (Chen *et al.,* 2002) and *Synechocystis* sp. PCC 6803 (Sato *et al.,* 2007) have been sequenced with the detailed functional notes (http://bacteria.kazusa.or. jp/cyanobase/). However, no studies on expression of PEPC for the modulation of fatty acids biosynthesis have been reported; also, there were no reports about the role of PEPC in regulating oil content of microalgae.

Current Problems

Even though the importance of algal biotechnology and genetic engineering has quickly increased, there are at least two major problems in the technological level that need to be solved in microalgae genetic engineering; one is the growth rate of microalgae, and the other is the expression efficiency of exogenous genes in microalgae.

Growth rate of microalgae

Algae is accepted as a promising source for expression products and other compounds, however, the commercial application of wild-type algae is still limited (Borowitzka, 1999) and that of transgenic algae is in its basics.

The major limiting factor for algal growth is light. Although, micro algae grow up to high densities in photo-bioreactors or even open pond systems, however, the light becomes limiting in dense cultures beyond the first few centimeters slowing cell growth. One suggestion solving such problems is the development of special bioreactors (Morita *et al.*, 2002), the use of stirred tanks, or shallow ponds.

An alternative solution that, besides the sunlight and the external light source, the minimum transmission route and the maximized surface area of lighting for illumination could be adopted, and the toughened fiber-glass biophotoreactor installed with a built-in light source could also be applied to increase the growth rate of microalgae. Another approach is the use of heterotrophic algae and addition of the required organic substrate. Recently, genetic engineering approach has transformed photoautotrophic algae into heterotrophic algae by introducing a gene for a sugar transporter into their genome *viz.*, in *Volvox carteri* (Hallmann and Sumper, 1996) and *Phaeodactylum tricornutum* (Zaslavskaia *et al.*, 2001). The diatom *Phaeodactylum tricornutum* grew heterotrophically even in the dark with glucose as the only carbon source (Zaslavskaia *et al.*, 2001).

The quantity of CO_2 venting should also be considered in the mass production of microalgae. The problem of the carbon sourcc could bc solved by utilizing CO_2 from coastal power plants. In addition, mixed culture of photoautotrophic and heterotrophic stage of microalgal cells under the condition of light or dark is also to be possible for the purpose of high cell density culture. In recent years, the method of mixed culture in a 100 liter biophotoreactor with CO_2 venting has been applied to improve the increasing rate and high cell density culture of microalgae in the group (Zhu *et al.,* 2006).

Growth in bioreactors provide limited culture volume and sterilization is quite expensive. Large scale production is done in open pond systems requires large starter cultures grown in closed photo-bioreactors in order to avoid overrun of the species of interest by other species in the unsterile open pond system (Walker *et al.*, 2005). Open pond systems also require the development of strategies to reduce the probability of gene flow between transgenic algae and their wild relatives. Moreover, more work is needed to optimize algal biotechnology for extensive commercial use in order to deal with the problems like harvesting of algae from an open pond, extracting and purifying bioproducts etc.

Expression efficiency of exogenous genes in microalgae

Although all elements required for transcription and translation had been included and the construct was integrated into the genome, gene construct was not expressed as desired in transgenic algae. The expression efficiency of exogenous genes is mostly 0.1%–0.9% of soluble proteins in host cells of microalgae. This gene silencing occurs *e.g.* through methylation, and is caused by positional effects and epigenetic mechanisms. Often, screening of a larger number of transformants for a transformant with high expression solves this problem. When DNA constructs with a heterologous origin are used, common codons of the DNA donor species are rarely found in the genes of the target organism, hence the corresponding tRNA abundance is low leading to low rate of translation and, as a consequence, less expression rate. One strategy to solve this problem is to look for heterologous genes that have codon usage similar to genes of the target organism. This approach was used to identify *Streptoalloteichus hindustanus ble* gene as a useful selectable marker for transformation of *Chlamydomonas reinhardtii* (Stevens *et al.*, 1996) and *Volvox carteri* (Hallmann and Rappel, 1999). Another way is to adhere the heterologous gene to the codon usage of the target species. This way *Aequorea victoria GFP* gene and the *Renilla reniformis luc* gene have been converted to ideal reporter genes for expression in *Chlamydomonas reinhardtii* (Fuhrmann *et al.*, 1999, 2004). Heterologous genes should not contain their own introns as they will not be spliced correctly; cDNAs should be used. However, genes without introns are often poorly expressed. This problem can be solved by introducing homologous introns into the heterologous coding region. For example, in *Chlamydomonas reinhardtii* (Lumbreras *et al.*, 1998) and *Volvox carteri* (Hallmann and Rappel, 1999), usefulness of such chimeric genes has been demonstrated.

Problems have been reported concerning insufficient DNA delivery, failure to integrate into the genome, or improper transport into the chloroplast or through the plasma membrane into the extracellular compartment.

There is a need to improve plasmid vectors, promoters, enhancers, terminators and SD sequences of microalgae for increasing expression efficiency of exogenous genes in transgenic algae, and the endeavor resulted in the amount of transgenic expressed in soluble proteins increasing from the initial 0.01% to now 3%–5% which is the hi ghest expression efficiency in the world (Shi *et al.*, 2004; Song *et al.*, 2006; Shi *et al.*, 2005; Ren, *et al.*, 1998; Liu *et al.*, 1999; Luo *et al.*, 2000; Dai *et al.*, 2001). Clearly more work is needed to uplift algal transgenics, especially in those species that are model organisms in basic research and in those that are particularly interesting for biofuel and other industries.

Meeting the objective: biodiesel production from microalgae

Survey programs on resources of oil-bearing microalgae should be mapped out and key scientific and technological problems in the program should be tackled by cooperation amongst relative academics, algae breeding farms and oil refineries.

Only development of transgenic technology of microalgae will not serve the purpose but also development of new large-scale bioreactors with the most suitable condition for the culture of microalgae is essential. Expressions of exogenous genes in cyanobacteria have relative matured technologies but the technology for gene transformation in eukaryotic algae is still at infancy and needs to be strengthened further. Expression efficiency of exogenous genes in high oil content microalgae can be increased greatly by the application of the existing knowledge and technology of genetic engineering, such as cloning of genes related to lipids metabolism, analysis of new regulation sequences and construction of new transformation system. Total lipids content in the cells of transgenic microalgae should also be increased by effective methods, such as the regulation of C/N ratio in culture medium and extension cell growing period.

There is an urgent need for long-term supply of purified microalgae on a large scale, the harvesting of mass production and lipids extraction from mass production for biodiesel.

(1) Continuous production of algal biomass is a prerequisite for biodiesel production. Hence, coordination between microalgal breeding farms and oil refineries is essential for regular supply of oil bearing microalgae.

(2) The suitability of transgenic microalgae for biodiesel production should be evaluated under different environmental conditions. For example, for large scale cultivation and for aseptic pure culture, transgenic algal photobioreactors can be built near some industrial plant from which large volume of CO_2 exhaust could be used as a carbon source.

(3) One of the crucial step in commercial biodiesel production from microalgae is the harvesting of algal biomass. Suitable technology should be developed for *in situ* collection of algal biomass and extraction of fats and lipids.

(4) Recent technology of enzyme immobilization will effectively eliminate the negative impact of methanol and glycerine byproduct on on the enzymatic activity and stability (Deng *et al.*, 2003).

Once biodiesel is generally accepted and popularized in the nation, the specialized nationwide sales networks for biodiesel would be established quickly.

Concluding Remarks

Algae can meet several requirements that other systems cannot sufficiently accomplish for which algal systems gain more and more influence in the production of substances of economic, industrial, and pharmaceutical importance.

Much more basic research that needs to be performed before algal transgenics and algal biotechnology can be commercially adopted. Genetic modifications that enhance the physiological properties of algal strains and optimization of algal production systems will further improve the potential of this innovative technology in the future.

References

Banerjee, A., Sharma, R., Chisti, Y. and Banerjee, U.C. (2002). *Botryococcus braunii*: a renewable source of hydrocarbons and other chemicals. *Crit Rev. Biotechnol* **22**: 245-79.

Borowitzka, M.A. (1999). Commercial production of microalgae: ponds, tanks, tubes and fermenters. *Journal of Biotechnology* **70**: 313-321.

Chen, J., Lang, C. and Huang, R. (2002). A method for producing seeds having altered protein/fatty acid composition. European Patent EP1229120.

Chen, J.Q., Huang, Y.Z. and Lang, C.X. (1999a). Molecular cloning and sequencing of the PEP gene from *Brassia napus* and the construction of the antisense PEP gene. *J Zhejiang Univ (Agr & Life Sci)* **25**: 365-367.

Chen, J.Q., Lang, C.X. and Hu, Z.H. (1999b). Antisense PEP gene regulates t ratio of protein and lipid content in *Brassica napus* seeds. *Chin J Agr. Biotechnol.* **7**: 316-320.

Chen, L., Omiya, T. and Hata, S. (2002). Molecular characterization of a phosphoenolpyruvate carboxylase from a thermophilic cyanobacterium, *Synechococcus vulcanus* with unusual allosteric properties. *Plant Cell Physiol.* **43**: 159-169.

Chisti, Y. (2007). Biodiesel from microalgae. *Biotechnol Advan.* **25**: 294-306.

Dai, W., Shi, D.J. and Zhang, H. (2001). Expression of human epidermal growth factor gene in cyanobacteria. *Acta Bota Sini.* **43**: 1260-1264.

Davis, M.S., Solbiati, J. and Cronan, J.E. (2000). Overproduction of Acetyl-CoA carboxylase activity increases the rate of fatty acid biosynthesis in *Escherichia coli*. *J Biol Chem.* **275**: 28593-28598.

Deng, L., Tan, T.W. and Wang, F. (2003). Studies of enzymatic synthesis of biodiesel. *Chin J Biotech.* **19**: 97-101.

Dunahay, T.G., Jarvis, E.E. and Roessler, P.G. (1995). Genetic Dunahay, T.G., Jarvis, E.E. and Roessler, P.G. Genetic *Navicula saprophila*. *J Phycol.* **31**: 1004-1012.

Fuhrmann, M., Oertel, W. and Hegemann, P. (1999). A synthetic gene coding for the green fluorescent protein (GFP) is a versatile reporter in *Chlamydomonas reinhardtii*. *Plant Journal* **19**: 353-361.

Fuhrmann, M., Hausherr, A., Ferbitz, L., Schödl, T., Heitzer, M. and Hegemann, P. (2004). Monitoring dynamic expression of nuclear genes in *Chlamydomonas reinhardtii* by using a synthetic luciferase reporter gene. *Plant Molecular Biology* **55**: 869-881.

Guschina, I.A. and Harwood, J.L. (2006). Lipids and lipid metabolism in eukaryotic algae. *Prog Lipid Res.* **45**: 160-86.

Hallmann, A. and Rappel, A. (1999) Genetic engineering of the multicellular green alga *Volvox*: a modified and multiplied bacterial antibiotic resistance gene as a dominant selectable marker. *Plant Journal* **17**: 99-109.

Hallmann, A. and Sumper, M. (1996). The *Chlorella* hexose/H^+ symporter is a useful selectable marker and biochemical reagent when expressed in *Volvox*. *Proceedings of the National Academy of Sciences USA* **93**: 669-673.

Kaneko, T., Nakajima, N. and Okamoto, S. (2007). Complete genomic structure of the bloom-forming toxic cyanobacterium *Microcystis aeruginosa* NIES-843. *DNA Res.* **14**: 247-256.

Katagiri, F., Kidaki, T. and Fujita, N. (1985). Nucleotide sequence of the phosphoenolpyruvate carboxylase gene of the cyanobacterium *Anacystis nidulans*. *Gene* **38**: 265-269.

Liu, F.L., Zhang, H.B. and Shi, D.J. (1999). Construction of shuttle vector expressing human tumor necrosis factor-á gene and expression of the gene in *Anabaena* sp. PCC 7120. Sci in China (Series C) **29**: 217-224.

Luinenburg, I. and Coleman, J.R. (1992). Identification, characterization and sequence analysis of the gene encoding phosphoenolpyruvate carboxylase in *Anabaena* sp. PCC 7120. *J General Microbiol.* **138**: 685-691.

Lumbreras, V., Stevens, D.L. and Purton, S. (1998). Efficient foreign gene expression in *Chlamydomonas reinhardtii* mediated by an endogenous intron. *Plant Journal* **14**: 441-447.

Luo, N., Ning, Y. and Shi, D.J. (2000). Cloning and expression of human pro-urokinase gene in the cyanobacterium *Synechococcus* sp. PCC 7002. *Acta Bota Sini.* **42**: 931-935.

Metting, F.B. (1996). Biodiversity and application of microalgae. *J Ind Microbiol* **17**: 477-89.

Metzger, P. and Largeau, C. (2005). *Botryococcus braunii*: a rich source for hydrocarbons and related ether lipids. *Appl Microbiol Biotechnol.* **66**: 486-96.

Morita, M., Watanabe, Y. and Saiki, H. (2002). Photosynthetic productivity of conical helical tubular photobioreactor incorporating *Chlorella sorokiniana* under field conditions. *Biotechnology and Bioengineering* **77**: 155-162.

Ohlrogge, J.B., Roesler, K.R. and Shorrosh, B.S. (1999). Methods of increasing oil content of seeds. United States Patent 5925805.

Ratledge, C. (1993). Single cell oils — have they a biotechnological future? *Trends Biotechnol.* **11**: 278-84.

Ratledge, C. and Wynn, J.P. (2002). The biochemistry and molecular biology of lipid accumulation in oleaginous microorganisms. *Adv Appl Microbiol.* **51**: 1-51.

Ren, L., Shi, D.J. and Dai, J.X. (1998). Expression of the mouse metallothionein-I gene conferring cadmium rersistance in a transgenic cyanobacterium. *FEMS Microbiol Lett.* **158**: 127-132.

Roessler, P.G. and Ohlrogge, J.B. (1993). Cloning and characterization of the gene that encodes acetyl-coenzyme A carboxylase in the alga *Cyclotella cryptica. J Biol Chem.* **268**: 19254-19259.

Sabe, H., Kondo, S. and Shmuzev, A. (1984). Properties of human.interleukin-2 receptors expressed on non-lymphoid cells by cDNA transfection. *Mol Biol Medi*, **2**: 379-396.

Sato, S., Shimoda, Y. and Muraki, A. (2007). A large-scale protein-protein interaction analysis in *Synechocystis* sp. PCC6803. *DNA Research* **14**: 207-216.

Sellwood, C., Slabas, A.R. and Rawsthorne, S. (2000). Effects of manipulating expression of acetyl-CoA carboxylase I in *Brassica napus* L. embryos. *Biochem Soci Trans.* **28**: 598-600.

Sheehan, J., Dunahay, T. and Benemann, J. (1998). A look back at the U.S. Department of Energy's Aquatic Species Program: biodiesel from algae. Close-out report. NREL/TP-580-24190.

Shi, D.J., Deng, Y.G. and Zhao, X.G. (2004). Cyanobacterial genetic engineering technology for recombinant pharmaceutical products. Proceedings of 2004 National Medicine Bioengineering Seminar of Chinese Society of Biotechnology. p. 54.

Shi, D.J., Duan, R. and Song, D.H. (2005). Cyanobacterial gene engineering: promising biotechnological application. *In*: The 10th International Conference on Applied Phycology, Qingdao.

Shi, D.J., Ran, L. and Li, Y. (2004). The efficient expression box and its vector of filament cyanobacterium. Chinese Patent, ZL00132268.0–10–13.

Somers, D.A., Wyse, D.L. and Gronwald, J.W. (2001). Transgenic plants expressing maize acetyl CoA carboxylase gene and method of altering oil content. United States Patent 6222099.

Song, D.H., Liu, J. and Shi, D.J. (2006). The expression of granulocyte colony-stimulating factor (G-CSF) in *Synechococcus* sp. PCC 7002. Proceedings of the Symposium for National Marine Biotechnology and Marine Pharmaceuticals. pp. 200-203.

Spolaore, P., Joannis-Cassan, C., Duran, E. and Isambert, A. (2006). Commercial applications of microalgae. *J Biosci Bioeng* **101**: 87-96.

Stevens, D.R., Rochaix, J.D. and Purton, S. (1996). The bacterial phleomycin resistance gene *ble* as a dominant selectable marker in *Chlamydomonas. Molecular and General Genetics* **251**: 23-30.

Walker, T.L., Purton, S., Becker, D.K. and Collet, C. (2005). Microalgae as bioreactors. *Plant Cell Reports* **24**: 629-641.

Yuan, Z.H., Wu, C.Z. and Ma, L.L. (2005). The principle and technology of biomass utilization. Beijing: Chemical Industry Press.

Zaslavskaia, L.A., Lippmeier, J.C., Shih, C., Ehrhardt, D., Grossman, A.R. and Apt, K.E. (2001). Trophic conversion of an obligate photoautotrophic organism through metabolic engineering. *Science* **292**: 2073-2075.

Zhao, G.L., Chen, J.Q. and Yin, A.P. (2005). Transgenic soybean lines harbouring anti-PEP gene express super-high oil content. *Mol Plant Bre*. **3**: 792-796.

Zhu, X.Y., Li, Q. and Liu, D.L. (2006). Studies on the culture of the transgeneic TNF-á *Anabaena* IB02 in 100L photobioreactor. Xiamen: Proceedings of the Symposium for National Marine Biotechnology and Marine Pharmaceuticals Academic Meeting. pp. 220-222.

9

MACROALGAE FOR SECOND GENERATION BIOFUELS: MYTH OR REALITY? A EUROPEAN PERSPECTIVE

Stefan Kraan*

Abstract

The global demand for bio-fuels continues unabated. Driven by skyrocketing prices, rising concerns over environmental pollution and global warming, and the movement to alternate fuels, the world ethanol market is projected to reach 27.7 billion gallons per annum by the year 2012. However, the demand for ethanol needs to be balanced by the problems associated with the supply of the raw material feedstock. Corn and sugar cane are the major feed-stocks for the bio-ethanol industry, and the continued use of these crops will drive the food versus fuels debate even more as demand for ethanol increases. Not only does the large-scale production of corn and sugar damage the environment by the use of harmful pesticides, it uses two other valuable resources: arable land and enormous quantities of water. The challenge is to find a feed-stock which is abundant and carbohydrate-rich. This crop needs to be sustainable, use less or no agricultural inputs (pesticides, fertiliser, land, water), and not be part of the human or animal food chain. Seaweed can be that feed stock. Current cultivation technologies available will be examined and the economic aspects of using brown seaweeds for bio-ethanol production will be presented together with some personal views on the feasibility of biofuel production from seaweed.

Key words: Macroalgae, Seaweed biorefinery, Biofuel, European algae, *Saccorhiza polyschides*, *Alaria esculenta*, *Saccharina latissima*

Scientific Director and Co-founder Ocean Harvest Technology, N17 Business Park, Milltown, Tuam, Co. Galway, Ireland

* *Corresponding author*: E-mail: stefan@oceanharvest.ie

Background

Over the last century global economic development has been driven by the availability of cheap resources and human materialism. The cornerstone for this industrial and economic revolution was the abundance of a cheap and easily modified energy source, oil. However, the current rate of consumption should result in the depletion of oil by the middle of this century (Peak Oil, 2009). Moreover, the general consensus is that fossil oil usage has been responsible for another global impact, *i.e.*, global warming and the effects on climate through green house gas emission (GHG) by burning fossil oil (Global Warming, 2009). It has become urgent to identify alternatives to oil and to manage the effects of releasing carbon by-products into the atmosphere (Copenhagen Climate Conference, 2009). Alternative fuel sources are emerging in the form of renewable (solar, wind, wave and biofuels) and nuclear options. While nuclear could contribute to our energy needs, it too is finite and creates another environmental legacy. The renewable options are part of the answer but in their current form will only supply a fraction of the energy requirement. The solution lies in science and the exploitation of nature's energy cycle, photosynthesis. Photosynthesis is the biochemical process and biomass is the biochemical agent that can provide our energy, chemicals and materials in sufficient quantities for a sustainable future. Society has to make a transition from a hydrocarbon to a carbohydrate economy. Plants and thus macroalgae are efficient solar energy converters. The chemical reactions create sugars, lipids and protein, the basis for all current energy generated from oil in use today.

The road transport sector in the European Union (EU) is the main consumer of petroleum based products (47% in 2006; Annon, 2008a) and accounts for more than 30% of the total energy consumption in the EU (Annon, 2006a). Transportation fuels are thus promising targets for a reduction in GHG-emissions. The aim of replacing 10% of transport fuels by 2020 with biofuels is motivated not only by improving the CO_2-balance, but also reducing the dependency on imported oil products and securing the energy supply (Annon, 2006b). The only commercially available biofuels today are first generation biofuels, mainly bioethanol and biodiesel, produced from *e.g.* sugar cane and corn, and rapeseed, respectively. However, continued use of these crops will drive the food versus fuels debate even more as demand for ethanol increases. Not only does the large-scale production of corn and sugar cane damage the environment by the use of pesticides, it uses two other valuable resources: arable land and enormous quantities of water. For instance, the production of corn in the USA uses over 3 trillion liters a year in 2007 (Chiu *et al.*, 2009).

Nevertheless, the global demand for bio-fuels continues unabated. Driven by high oil prices, concerns over environmental pollution and global warming, and the movement to alternate fuels, the world ethanol market is projected to reach 80-100 billion litres per annum by the year 2012 (Licht, 2009). Stringent legislation on emission standards and governmental intervention by way of subsidies and tax incentives are expected to foster significant market growth in the medium to long term. Increased demand and the competition with food production has called for the development of second generation biofuels, based on utilization of ligno-cellulosic biomass, such as wood and agricultural waste. Second generation biofuels do not compete with food as a feedstock, but they compete for land and fresh water resources.

Therefore the challenge is to find a feedstock which is abundant and carbohydrate-rich. This crop must be sustainable, use no agricultural inputs (pesticides, fertiliser, land, water), and must not be part of the human or animal food chain. Such a feedstock and an alternative to terrestrial biomass are marine macroalgae or seaweeds.

Macroalgae and aquatic biomass are emerging as one of the most promising potential sources for biofuels production. The cultivation, transformation and final use of aquatic biomass are attracting the interest of the scientific community, the industry and of an increasing number of stakeholders at an international level. This is first of all related to the ability to further improve the sustainability of the production of biofuels. One of the most critical aspects of biofuel sustainability stands in the way in which biofuels raw materials are produced, especially with respect to their CO_2 balance and to the competition with food agriculture areas that they create. Macroalgae and aquatic biomass can represent a major progress in this sense: higher carbohydrate levels and biomass yields, their widespread availability, the absence of a competition with food or agricultural surfaces, the high quality of the by-products, their use as a mean to capture CO_2 and their suitability for integrating in wastewater treatments to reduce pollution make them one of the most attractive renewable sources for a sustainable energy strategy.

Species and Production

Several species of macroalgae accumulate high levels of carbohydrates, which are suitable as substrate for microbial conversion processes, *e.g.*

for production of bioethanol or other desirable chemicals. Five European kelp species are suitable to fulfil that role, *i.e.*, *Saccorhiza polyschides* (Lightfoot) Batters; *Alaria esculenta* (L.) Greville; *Laminaria hyperborea* (Gunnerus) Foslie; *Laminaria digitata* (Hudson) J.V. Lamouroux; and *Saccharina latissima* (Linnaeus) C.E. Lane, C. Mayes, Druehl & G.W. Saunders (Werner & Kraan, 2004). They differ in various aspects, such as morphology, ecophysiology and longevity. *Laminaria digitata* and *Laminaria hyperborea* are the only species that form extended monospecific kelp beds. The vertical distribution of these five species on the shore depends on factors such as light penetration, tolerance to desiccation, interspecific competitiveness and adaptation to rigorous wave impact (Werner and Kraan, 2004). Kelp inhabit the continuously submersed sublittoral zone with the upper sublittoral occasionally emerging at extreme low water levels. The lower limit for algal growth in the sublittoral is naturally determined by light levels. In coastal waters rich in particles, the depth limit for kelp growth may be 10-15 meters below mean low water, whereas in clearer waters of the open Atlantic coast kelps are found in depths down to 30-40 meters (Lüning, 1990).

The biomass productivity of macroalgae ranges from 3.3 to 11.3 kg wet weight m^{-2} $year^{-1}$ or converted to carbon about 1 to 3.4 kg carbon m^{-2} $year^{-1}$ (Gao and McKinley, 1994; Mann, 1982; Mohammed and Fredriksen, 2004). Seaweed com-munities of the European coasts have an annual productivity of approxi-mately 2 kg C per m^2, which is far higher than, for example, temperate tree plantations or grass-lands with a productivity of generally less than 1 kg C m^{-2}, $year^{-1}$ (Mann, 1982; Chapman, 1987; Thomas, 2002; Lûning & Pang, 2003; Mohammed and Fredrikson, 2004), and 2.8 times higher than for sugar cane (Gao and McKinley, 1994). Macroalgae can be cultivated in the open sea (Zhang *et al.*, 2008; Bartsch *et al.*, 2008). Ocean farming of sea-weed does not depend on fresh water and does not occupy land areas (Yarish and Pereira, 2008). Sustainable utilization of algal biomass - a largely unexplored feed stock resource can be a complement to terrestrial biomass for the future European energy and carbon security and thereby also strengthen the European maritime economy.

Ocean farming of seaweed has the potential to produce in the order of 40 ton dry weight biomass per hectare per annum (Table 1). An area of 2500 km^2, the size of Luxembourg, would be able to provide 10 million ton dry biomass, representing 5.6-5.8 million ton carbohydrates. With current 90% enzymatic conversion into ethanol (McGaughren, 2008) this would yield close to 2 billion litres of bio-ethanol. This is about 2-

Table 1. Production of some kelp species in the wild and cultivated dry weight per hectare per year

Species	*Kelp forest ton dry per hectare y^{-1}*	*Wet Kg m^{-1} of rope*	*Ton dry hectare y^{-1}*	*References*
Kelp species	33–113			Gao and McKineley, 1994
Saccharina latissima		3.75	20	Buck and Buchholz, 2004
		3–28		Druehl *et al.*, 1988
Alaria esculenta			1.5–2	Kain and Dawes (1987)
		7–45		Kraan and Guiry, 2001
Laminaria digitata		3–15		Arzel (1998); Werner and Kraan, 2004
Macrocystis pyrifera		38–62		Chenoweth, 2002
Saccharina japonica		15–60		Tseng, 1987
			54–115	Chenoweth, 2002
		–	131	Brinkhuis *et al.*, 1987

3% of the world's global bioethanol pro-duction (Licht, 2009). The total production of bioethanol in the EU member states was 2.9 billion litres in 2007 (Annon, 2008b). Besides bioethanol, algal biomass can be used for sustainable production of a great variety of products, including fuels, chemicals and materials. The use of algal biomass has the potential to not only replace fossil resources, and thereby mitigate climate change, but also aid in the recycling potential of nitrates and phosphates in near and inshore waters. Seaweed biomass represents an abundant and carbon neutral renewable resource with potential to reduce green house gas (GHG) emissions and the man-made impact on climate change.

Cultivation of Carbohydrate-Rich Macroalgae

Among marine macroalgae, species of the brown algal order Laminariales (so-called kelp species) belong to the fastest growing plants in the world. These fast growing, large size brown algae prefer the growth conditions of the temperate zone, which in the case of Europe stretches from northern Portugal to Northern Norway. In the America's in the northern hemisphere kelp grows from Mexico to Alaska and in the southern hemisphere from Peru down to the tip of Chile (Guiry and Guiry, 2009).

Cultivation of seaweed has a long tradition in East-Asia (Carmona, 2006). The annual production of 14 million t wet weight represents the largest aquaculture production on a global basis and 93% of the worldwide commercial harvest of seaweeds in 2004 (Bartsch *et al.*, 2008). Seaweed cultivation in Europe is still in its infancy with a few commercial attempts, notably in France, Germany and Ireland (Buck and Buchholz 2004; Kraan and Guiry, 2006; Kaas, 2006). Total production of these farms combined is less than 50 ton wet weight and all used in the food industry.

The main cultivated species in order of importance are: *Saccharina japonica* (J.E. Areschoug) C.E. Lane, C. Mayes, Druehl & G.W. Saunders, previously called *Laminaria japonica, Undaria pinnatifida* (Harvey) Suringar, *Porphyra yezoensis* Ueda, *P. Tenera* Kjellman, *Kappaphycus alvarezii* (Doty) Doty ex P.C. Silva and *Gracilaria* spp. Greville (Carmona *et al.*, 2006). There are two main approaches to cultivate seaweeds; one is based on the sexual reproduction of the various species and the other on the ability of some species to propagate vegetatively (Bartsch *et al.*, 2008; Hernández-González *et al.*, 2007; Titlyanov *et al.*, 2006). New approaches using micro-propagation proved to be efficient but still require improvement (Reddy *et al.*, 2008). During the last twenty years various trials and experiments have been carried out on species interesting for their hydrocolloid content like *Gelidium sesquipedale* (Clemente) Thuret their value as edible seaweeds (*Undaria pinnatifida, Alaria esculenta, Palmaria palmata* (Linnaeus) Kuntze (Kraan and Guiry, 2006), or their ability to extract nutrients from effluent (*Ulva* spp. Linnaeus) while having a potential value like *Gracilaria* spp. (Buschmann *et al.*, 2001). The integration of seaweeds for the treatment of shellfish and finfish aquaculture effluents or urban sewage is of global interest (Carmona *et al.*, 2006; Chopin *et al.*, 2004). Not only can they remove large amounts of inorganic nutrients but they have also the capacity to accumulate heavy metals and other contaminants. Knowing that shellfish and finfish aquaculture represents a yearly production of about 30 million tons and producing large quantities of ammonia, integrated systems including shellfish and/or finfish and macroalgae have been tested with success in order to remove this excess of nitrogen in land based systems as well as in open sea cultures (Neori *et al.*, 2000; Schuenhoff *et al.*, 2003; Troell *et al.*, 2009).

Among European kelp species *Saccharina latissima* (previously known as *Laminaria saccharina*) is the fastest-growing macroalgal species. This species is similar to *S. japonica* of which 3 million ton fresh weight is harvested annually from aquaculture in northern China, and 1 million ton fresh weight additionally in Japan (Wu and Pang, 2006; Ohno and Largo, 2006). Other relevant brown algal species for cultivation as a possible energy crop in Europe are *Alaria esculenta*, *Laminaria* spp. and *Saccorhiza polyschides*.

Kelp as a rule contains approximately 60% carbohydrates of the dry weight. Laminaran (a glucose polymer) and mannitol are energy storage compounds, resembling starch in land plants, while alginates are structural compounds and correspond to cellulose and lignin in land plants. The seasonal variations of the carbohydrate composition are considerable (Honya *et al.*, 1993; ISC, 2009). The content of storage carbohydrates has a maximum in autumn. During the winter season, with nutrients in excess, the stored carbohydrates are utilized as energy source for protein synthesis reproduction and growth. Growth conditions and harvesting time may therefore strongly affect the quality of the seaweed biomass as substrate for microbial production processes.

Cultivation Systems

Over the last couple of decades different cultivation systems for seaweeds have been developed and improved ranging from intertidal fixed and floating bottom farms for Eucheuma/Kappaphycus and *Gracilaria* (*e.g.*, Philippines, Vietnam and Thailand) to elaborate floating net structures for *Porphyra* and long-line systems for kelp in China, Korea and Japan (Carmona *et al.*, 2006). Modifications of long-line systems have been tested at small scale in Europe amongst them a novel ring system from Germany (Buck and Buchholz, 2004). These cultivation systems show that there is potential to develop large-scale ocean cultivation of seaweeds. However, existing cultivation and harvesting technology is labour intensive and need to be optimised to reduce costs and energy demand (Troell *et al.*, 2009).

A major bottle-neck for mass cultivation of kelp biomass in the sea is the production of juvenile sporo-phytes on ropes from zoospores via female and male gametophytes (Pérez *et al.*, 1984, 1992; Kaas, 2006; Bartsch *et al.*, 2008). In China and Japan *Laminaria* cultivation from zoospores relies exclusively on naturally occurring sori restricted to a certain part of the year, so that hatchery facilities are not in use throughout the whole year. However, previous work in France on the free-living stage of the gametophytes (Pérez *et al.*, 1984, 1992) and recent developments in China in mass culture of filamentous gametophytes with establishment of vegetative gametophyte stocks in the range of kilo-grams of fresh weight has changed this concept. (*e.g.* Kaas, 2006;

Xu *et al.*, 2008; Zhang *et al.*, 2008). Using the so called free-living technique to keep male and female or mixed cultures free floating as stock cultures allows seeding of long-lines virtually any time of the year. Although successful growth at sea is still dependant on environmental conditions, such as light and temperature (Bartsch *et al.*, 2008).

Processing and Fermentation of Macroalgae Biomass

The water content of macroalgae is higher than for terrestrial biomass (80-85 %), making sea-weeds more suited for microbial conversion than for direct combustion or thermo-chemical conversion processes, which is an alternative for land-based biomass (Horn *et al.*, 2000a; Ross *et al.*, 2008). The carbohydrates from seaweeds may be used as substrates for microbial production of a wide range of fuels and chemicals. For a cost-efficient production the seaweed biorefinery concept should be applied (Fig. 1) to produce several products, *e.g.* by using different processes of conversion of the different substrates in the most efficient way, and by co-production of compounds with higher price than fuels. Target products, such as energy carriers like ethanol and butanol or other materials like itaconic acid are base material for the biofuel and plastics industry, representing important industrial products that can replace or substitute products that currently are produced from oil sources.

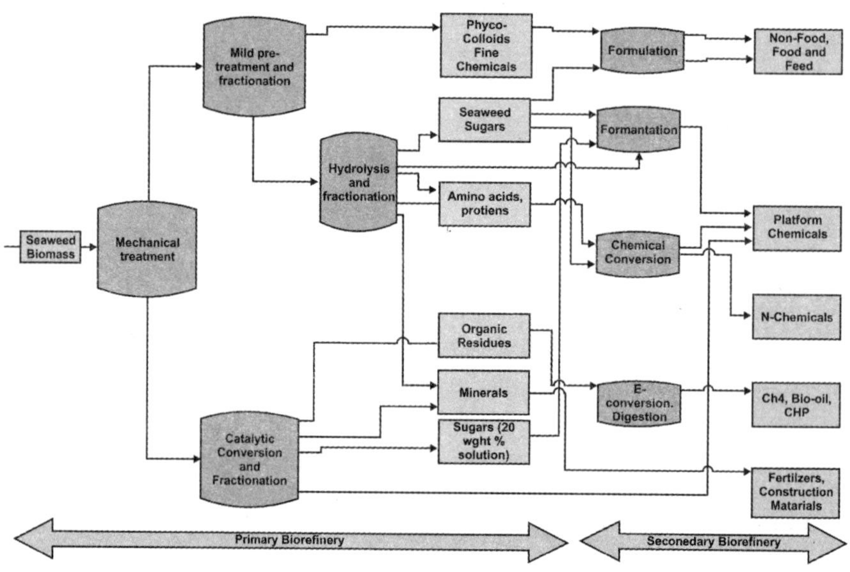

Fig. 1. A schematic representation of the seaweed biorefinery concept (Reith *et al.*, 2009 with permission)

Ethanol production from hexose sugars, derived from *e.g.* corn stover or sugar cane, is a well-known process. However, hexose-based poly-saccharides constitute only about 30-40% of the carbo-hydrates in kelp. The remaining fraction is composed of C-5 sugars that until now have not been applied as substrates for microbial production processes. However, recent breakthroughs have been made in C5 sugar fermentation technology allowing up to 90% of the available carbohydrates to be fermented (McCaughren, 2008, Trevors (Comm.).

Pre-Treatment of the Seaweed Biomass and Hydrolysis of the Polysaccharides

Fresh harvested seaweed contains about 15-20% carbohydrates of the total wet weight, which equals about 200 g carbohydrates per kg wet weight, an appropriate substrate con-centration for microbial conversion processes (Horn *et al.*, 2000a). The lack of lignin in seaweeds implies that the harsh pre-treatment applied for release of fermentable sugars from lignocellulosic biomass is not required. Laminaran and mannitol can easily be extracted by water (Horn *et al.*, 2000b). Alginates (consisting of polymer blocks of Uronic and guluronic acid) is present in the macroalgae biomass at 30-40% (Honya *et al.*, 1993; McHugh, 2003). Sodium alginate can be removed from the initial extraction solution by adding a calcium salt. This causes calcium alginate to form with a fibrous texture; it does not dissolve in water and can be separated from it. The separated calcium alginate is suspended in water and acid is added to convert it into alginic acid. This fibrous alginic acid is easily separated, placed in a planetary type mixer with alcohol, and sodium carbonate is gradually added to the paste until all the alginic acid is converted to sodium alginate. The paste of sodium alginate is sometimes extruded into pellets that are then dried and milled (McHugh, 2003). Due to the high viscosity, dilution with large amounts of water is required. The process is operating at alginate concen-trations in the order of ~2%. Such a dilution cannot be applied to processes aimed at use of alginate as fermentation substrate. Preferably, no water should be added, as it will increase the downstream processing costs (McHugh, 2003).

Bioethanol production from cellulosic materials can be achieved by running simultaneously enzymatic hydrolysis and fermentation. For the wet seaweed biomass it is not as easy due to the alginates which are harder to release from the biomass resulting that enzymatic degradation of un-treated biomass will be rate-limiting if combined with fermentation. Hydrolysis should therefore be a part of the biomass pre-treatment. This hydrolysis can be carried out mechanically through grinding and emulsifying equipment, chemically using acid or alkali, or enzymatic. Researchers at the Biochemistry Department of The National University of Ireland, Galway, in conjunction with AER Ltd in Ireland, have developed an enzyme cocktail with key enzymes isolated from a marine fungus which is able to degrade 90% of the poly-saccharides of *L. digitata*, leaving only some insoluble fibre and salts (M. Tuohy pers. Comm). Moreover, several alginate lyases have been isolated from marine bacteria that can degrade alginates (Doubet and Quatrano, 1982; Wang *et al.*, 2006). Moreover, the Norwegian institute SINTEF-MK has established large-scale processes for production of microbial alginate lyases (Aasen *et al.*, 1992; Dyrset *et al.*, 1994) although the technology has been available since the early 1980s (Doubet and Quatrano, 1982). Several methods for partial or complete degradation of alginate as an integrated part of the mechanical pre-treatment of the biomass are known. Chemical hydrolysis should follow existing technologies, *e.g.* by modification and adaptation of methods used for acid and alkali pre-treatment of wood biomass (*e.g.* Ballesteros *et al.*, 2002; Klinke *et al.*, 2001) and by combination of acid- or alkali with steam treatment. Other existing technologies make use of a brown-rot fungi and is based on the dihydroxybenzene driven Fenton reaction (Contreras *et al.,* 2007), which produce hydroxyl radicals and other oxidants that partially depolymerise carbohydrates and oxidize phenols and other fermentation inhibitors. Although the easiest and most environmental friendly way is through mechanical pre-treatment of algal biomass in combination with enzymatic hydrolysis (Doubet and Quatrano, 1982).

Ethanol and Butanol from Brown Seaweeds

Ethanol production from fermentation is the most obvious one as it has a direct application in the replacement of fossil fuels. However other products such as butanol and itaconic acid can be produced as well which can substitute and/or replace similar products produced from fossil resources. There are many microorganisms in the marine environment that can degrade and utilize algal carbohydrates as a carbon source for energy. Often these microorganisms are associated with the seaweeds being present on the blade surface or in tissue as many kelp species produce exo-polysaccharides as mucus layer or shed entire skin. This would imply that these organisms possess the necessary enzymes for cleavage of the algal polysaccharides. However, compounds such as ethanol and butanol are produced by anaerobic fermentation that require the presence of specific metabolic pathways generating these compounds as end products, *e.g.* yeast for ethanol production and *Clostridia* for butanol production. Limited information is available with respect to which degree these organisms are capable of converting the specific seaweed carbohydrates to the desired products (Horn *et al.*, 2000a, 2000b), although several break through have recently been made in respect of ethanol production (McCoughran, 2008)

Ethanol

The potential ethanol production from seaweeds can be calculated and is based on the following assumptions: A carbohydrate content of 60% of dry weight and a 90% conversion ratio to ethanol. Through fermentation one gram of sugar can yield 0.4 g ethanol. This will yield 0.22 kg or 0.27 l ethanol from 1 kg dry weight seaweed biomass, corresponding to approximately 0.05 l ethanol per kg wet weight.

Several companies and institutes can produce ethanol through carbohydrate fermentation including alginates using a marine fungus; however, no studies on production of ethanol from alginate or its monomers, mannuronic acid and gulu-ronic acid have been published as this information is often confidential and is not published in the public domain. In bacteria uronic acids can be meta-bolized to pyruvate and glyceraldehyde-3-P, which may then be fermented to ethanol by the glyco-lytic pathway (Van Maris *et al.*, 2006). In anaerobic fermentation processes, as ethanol and butanol production, oxygen is not available for removal of excess hydrogen generated. This implies that the conversion reaction from sub-strate to product must be red-ox balanced. Many bacteria have the enzyme "trans-hydrogenases", omitting this problem, but yeasts lack this enzyme. Ethanol-production from hexose sugars is redox balanced, while production from pentoses or mannitol generates excess hydrogen. For yeasts, the problem may be solved by a small, controlled supply of oxygen. However, oxygen leads to complete oxidation of the substrate to CO_2 and water, and reduced ethanol yields. Another strategy is introduction of transhydrogenase into strains that lack this, through genetic engineering (Fortman *et al.*, 2008; Lee *et al.*, 2008).

Butanol

Butanol could be another possible route as a viable biofuel and an alternative to ethanol, since it has a higher energy content (butanol 29.2 MJ/l, ethanol 19.6 MJ/l, compared to gasoline 32 MJ/l). Butanol can be used to supplement both gasoline and diesel fuels and can be supplied using existing infrastructures (Fortman *et al.*, 2008). Butanol is an important industrial chemical and is currently produced via petrochemical processes. In the last century butanol was produced through bacterial fermentation of starch rich compounds using *Clostridia* strains (Zverlov *et al.*, 2006). These bacteria can use hexose as well as pentose sugers.

Current Situation

Recently an exponential growth of research, technological development and demonstration initiatives has kick started the exploration of aquatic biomass as alternative source for the biofuel industry. Scientific research has been initiated on the production of various species and families of algae including micro-algae, macroalgae, plankton, and other aquatic biomass in fresh water, *e.g.*, duckweed. in order to transform them in biofuels using very different technology pathways ranging from the esterification of fats and oils (biodiesel) to Fischer-Troops (BTL), anaerobic digestion (biogas) or ethanol production. Industrial pilots are also being developed and built and some companies already claim to be able to provide large-scale production in the short term. Though, such initiatives are not coordinated neither at a European level, nor at a national level. The enthusiasm expressed by the scientific community, the industry and also by political representatives have created the conditions for further development but also the risk of confusion and overlapping R&D areas. Additionally, a disconnection between lab and field results was one of the most important issues underlined by U.S. Department of Science (DOE) scientists on their report in 1998, after a twenty year study on algae at U.S. National Renewable Energy Laboratory (Chynoweth, 2002; Chynoweth *et al.*, 1981, 1987).

The Seaweed Biorefinery Concept

The replacement of fossil-fuel based products and energy by bio-based products and bioenergy is an important route for the reduction of CO_2 emissions. The development of biorefinery concepts in order to enable the manufacturing of bio-based is an essential part for successful implementation. At the same time, the targets set for bio-based products and bioenergy increase the demand for biomass as feedstock.

Several initiatives in Europe aim to further develop the bio-refinery concepts for a still unexploited and large volume of biomass, seaweed from the Atlantic Area. Some currently running large scale projects are: BioMara (EU Programme for Cross Border Territorial Cooperation INTERREG IVA; Supergen II (EPSRC, UK: Potential of Marine Biomass for Energy, Fuels and Chemicals); Biorefinery and Biofuel Competency Centre, Enterprise Ireland; EOS LT Seaweed Bio-refinery project, SenterNovem. The Netherlands; and recent initiatives undertaken by SINTEF Norway to start large scale seaweed cultivation projects for biofuels.

Due to the different biochemical composition of seaweed compared with land plants, seaweed offers a vast potential of new chemical components. Bio-refinery of seaweed has a large potential both in quantity as in quality for the production of a wide variety of unique chemical components and bio-energy. Previous studies (Reith *et al.*, 2003) have shown that growing seaweed can be made economic viable if the production of bio-based products, biofuels and bio-energy in combined. Growing seaweeds with the current technologies available for energy purposes only is not attractive on the short term from an economic point of view. Certain products from the seaweed industry have long been used for the production of phycocolloids such as alginates, carrageenans or agars (McHugh, 2003). These polymers are either located in cell walls or within the cells serving as storage materials. Characteristic for seaweed is the abundance of sulphated polyssacharides in the cell walls. From the cell walls of green seaweeds the polysaccharide ulvan can be produced (Lahaye and Robic, 2007). This type of components exist of varying proportions of repeating sequences of rhamnose, glucuronic acid, iduronic acid, xylose and sulphate. Possible applications of these components (after modification) are as pharmaceuticals, gelling agents and other chemicals (Lahaye and Robic, 2007).

Ulvans (and its derivates) are hydrocolloids and components not found in terrestrial plants. Using seaweed as feedstock will add new classes of components for several traditional and novel applications.

Before a successful commercial implementation of a bio-refinery plant based on seaweed processing can be realized a number of questions and uncertainties need to be answered and clarified. These questions concerns the cultivation and harvesting aspects of seaweed in Europe and also the subsequent processing of the harvested seaweed into bio-based products and bio-energy. Another very important issue is Marine licenses to be obtained adequate ocean surface to be able to grow the required biomass. Currently obtaining an aquaculture license is a huge stumbling block *e.g.*, see Regulation (EC) No 889/2008 laying down detailed rules for the implementation of Council Regulation (EC) No 834/2007, as regards laying down detailed rules on organic aquaculture animal and seaweed production.

Other Environmental and Social Benefits

The primary production of kelp per unit area is amongst the highest known in aquatic Ecosystems (Birkett *et al.*, 1988). Kelp primary production results in the production of new biomass, detrital material shed from the blade tip, mucus and other dissolved inorganic material and spores (Kelly, 2005). The holdfast, stipe and fronds of kelp plants present available substratum for colonisation by marine flora and invertebrates. The holdfasts tend to host strongly associated communities of epiphytes and marine invertebrates. Kelp contributes directly and indirectly to the food resource of suspension and deposit feeding invertebrates that in turn serve as prey to more mobile invertebrates such as polychaetes, cnidaria and larger decapods (Steneck *et al.*, 2004). The rich fauna of mobile invertebrates in kelp beds makes this an important habitat in the diet of fish species.

Kelp forests provide a foraging habitat for birds due to the associated and diverse invertebrate and fish communities present. Some invertebrate and fish species exhibit egg attachment and nest-building, respectively, in kelp habitats while others such as juvenile gadoids and salmon utilise kelp habitats as important nursery and refuge grounds. A number of studies have indicated the importance of kelp forests as habitat for spawning and reproduction in fish species. The importance of kelp habitat as a nursery area for the development of juvenile fishes has been widely recognized (Carr, 1983; Shaffer, 2003; Lorentsen *et al.*, 2004). The effective extension of the substratum into the water column increases shelter, or refugia, available to fishes while also providing habitat for the prey species used as a forage base by reef fishes (Kelly, 2005).

Sea-weed farms function in a similar way and create high biodiversity areas as the kelp farmed will attract other animals such as fish, invertebrates and benthos. These animals on their turn attract birds and seals. These kelp farms also create good habitats for fish to spawn and shelter for juvenile fish and could act as nursery areas, and may thereby enhance locally the recruitment of fish stocks under pressure such as gadoid species. Other researchers have highlighted the importance of the habitat for particular species, *e.g.*, Sjøtun and Lorentsen (2003) recently demonstrated the importance of *Laminaria hyperborea* forests as important habitats for juvenile gadoids. Development of future large-scale ocean farming of seaweeds will contribute to the growth of maritime sectors and can be a viable alternative for fisherman as existing infrastructure can be applied. It further might enhance employment opportunities in local rural coastal areas in the form of seed hatcheries, seeding units and processing units and create employment opportunities that otherwise would not exist.

Eutrophication and Bioremediation

Inputs of biodegradable organic matter and inputs deriving from fertiliser run-off together with run off or dilution from finfish and shellfish rearing in near-shore waters and land based activities have many effects on the quality of coastal inshore waters and are a primary cause of eutrophication due to increased availability of nutrients (EPA, 2003). In estuaries and shallow coastal embayments, this can lead to the proliferation of vast green algal mats, known as 'green tides' (Fletcher, 1996). Aesthetically unpleasant, noxious-smelling agglomerations of algae can influence local fisheries and tourism industries and can persist for years and may pervasively and fundamentally alter ecosystems (Valiela *et al.*, 1997).

Kelp farms (inshore and nearshore) are able to act as bio-filters and are able to remove nitrates and phosphates from the surrounding eutrophic inshore waters. This allows for increased production of farmed seaweed as demonstrated by (Chopin *et al.*, 2001; Chopin, 2008). Eutrophe waters are high in ammonia and phosphorous which can be stripped from the water by seaweed at rates varying from 60% up to 90% of the nutrient input. Macroalgae are able to take up nitrogen from seawater with rates to allow for a biomass increase of ca. 10% day-1. It is well documented that in tank systems the green alga Ulva is able to remove 90% of the nitrogenous compounds and the red alga Gracilaria up to 95% of dissolved ammonium from fish effluent (Neori *et al.*, 2000). Moreover, primary production in costal waters is about 10 times higher than that of phytoplankton on an area basis. These data indicate that macroalgae have a greater potential to remove nutrients compared to phytoplankton in inshore waters. By sea cultivating and harvesting macroalgae as biofilters integrated with other shellfish or fish production systems, nutrient polution from these aquaculture systems could be alleviated through a process called IMTA (Integrated Multitrophic Aquaculture) while increasing production and carrying capacity (Chopin *et al.*, 2001; Chopin, 2008; Troell *et al.*, 2009). Production of macroalgae in near-shore sea cultivation can be harvested for the bioethanol market to produce a value added marketable product acting as both an economic incentive and environmental incentive.

References

Aasen, I.M., Folkvord, K. and Levine, D.W. (1992). Development of a process for large-scale chromatographic purification of an alginate lyase from *Klebsiella pneumonia*. *Applied Microbiology and Biotechnology*, **37**: 55-60.

Annon, (2006a). Biofuels in the European Union - A vision for 2030 and beyond, 2006.

Annon, (2006b). Commission of the European Communities, Communication from the Commission: An EU Strategy for Biofuels. Brussels 8.2.2006

Annon, (2008a). European Biofuels Technology Platform: Strategic Research Agenda & Strategy Deployment Document, January 2008 p. 42.

Annon, (2008b). http://www.euractiv.com/29/images/UEPA%20Balanced%20 Approach_tcm29 172883.pdf

Arzel, P. (1998). Les laminaires sur les côtes bretonnes. Evolution de l'exploitation et de la flottille de pêche, état actuel et perspectives. Editions IFREMER BP 70 – 29280 Plouzané –France. p. 139.

Ballesteros, I., Oliva, J.M., Negro, M.J., Manzanares, P. and Ballesteros, M. (2002). Enzymic hydrolysis of steam exploded herbaceous agricultural waste (*Brassica carinata*) at different particle sizes. *Process Biochem*. **38(2)**: 187-92.

Bartsch, I., Wiencke, C., Bischof, K., Buchholz, C.M., Buck, B.H., Eggert, A., Feuerpeeil, P., Hanelt, D., Jacobsen, S., Karez, R., Karsten. U., Molis, M., Roleda, M.Y., Schubert, H., Schumann, R., Valentin, K., Weinberger, F. and Wiese, J. (2008). The genus *Laminaria sensu lato*: Recent insights and developments. *Eur. J. Phycol.*, **43**: 1-86.

Birkett, D.A., Maggs, C.A., Dring, M.J., Boaden, P.J.S. and Seed, R. (1988). Infralittoral reef biotopes with kelp species (volume VII). An overview of dynamic and sensitivity characteristics for conservation management of marine SACs. Scottish Association of Marine Science (UK Marine SACs Project). p. 174.

Brinkhuis, B.H., Levine, H.G., Schlenk, G.G. and Tobin, S. (1987). *Laminaria* cultivation in the Far East and North America. *In*: Bird, K.T. and Benson, P.H. (*Eds*.), Seaweed cultivation for renewable resources. Elsevier, Amsterdam, pp. 107-146.

Buck, B.H. and Buchholz, C.M. (2004). The offshore-ring: A new system design for the open ocean aquaculture of macroalgae. *J. Appl. Phycol.* **16**: 355–368.

Buschmann, A.H., Correa, J.A., Westermeier, R., del Carmen Hernandez-Gonzalez, M. and Norambuena, R. (2001). Red algal farming in Chile: a review. *Aquaculture* **194**: 203-220.

Carmona, R., Kraemer, G.P. and Yarish, C. (2006). Exploring Northeast American and Asian species of *Porphyra* for use in an integrated finfish–algal aquaculture system. *Aquaculture* **252**: 54-65.

Carr, M. (1983). Spatial and temporal patterns of recruitment of young of the year rockfishes (genus *Sebastes*) into a central Californian kelp forest. MSc. Thesis: San Francisco State University, San Francisco, California. p. 104.

Chapman, A.R.O. (1987). The wild harvest and culture of *Laminaria longicruris* de la Pylaie in Eastern Canada. *In*: Doty, M.S., Caddy, T.F. and Santelices, B. (*Eds*.). Case studies of seven commercial seaweed resources. *FAO Fisheries Technical Paper*, **181**: 193-238.

Chynoweth, D.P. (2002). Review of biomethane from Marine Biomass. Review of history, results and conclusions of the "US Marine Biomass Energy Program" (1968-1990). p. 194.

Chynoweth, D.P., Ghosh, S. and Klass, D.L. (1981). Anaerobic digestion of kelp. *In*: Sofer, S.S. and Zaborsky, O.R. (*Eds.*), Biomass Conversion Processes for Energy and Fuels, Plenum Press, New York, pp. 315-338.

Chynoweth, D.P., Fannin, K.F. and Srivastava, V.J. (1987). Biological gasification of marine algae. *In*: Bird, K.T. and Benson, P.H. (*Eds.*), Seaweed Cultivation for Renewable Resources, Elsevier, Amsterdam, pp. 285-304.

Chiu, Y.-W., Walsh, B. and Suh, S. (2009). Water embodied in bioethanol in the United States. *Environ. Sci. Technol.*, **43**: 2688-2692.

Chopin, T., Buschmann, A.H., Halling, C., Troell, M., Kautsky, N., Neori, A., Kraemer, G.P., Zertuche-Gonzalez, J.A., Yarish, C. and Neefus, C. (2001). Integrating seaweeds into marine aquaculture systems: a key towards sustainability. *Journal of Phycology* **37**: 975-986.

Chopin, T., Robinson, S., Sawhney, M., Bastarache, S., Belyea, E., Shea, R., Armstrong, W., Stewart, I. and Fitzgerald, P. (2004). The AquaNet integrated multi-trophic aquaculture project: rationale of the project and development of kelp cultivation as the inorganic extractive component of the system. *Bull. Aquac. Assoc. Can.* **104(3)**: 11-18.

Chopin, T. (2008). Mussels and kelps help salmon farmers reduce pollution. SeaWeb Vol 1, No 2, 3 July Département d'Ecologie Côtière, B.P. 70, 29280 Plouzané http://www.ifremer.fr/envlit/documentation/documents.htm

Contreras, D., Rodrýguez, J., Freer, J., Schwederski, B. and Kaim, W. (2007). Enhanced hydroxyl radical production by dihydroxybenzene driven Fenton reactions: implications for wood biodegradation. *J. Biol. Inorg. Chem.* **12**: 1055-1061.

Copenhagen Climate Conference, (2009). http://www.copenhagenclimatecouncil.com/Critchley, A., Ohno, M. and Largo, D. (2006). The seaweed resources of the world., A CD-rom project *Eds.* Alan Critchley, Masao Ohno and Danilo Largo, ETI, Amsterdam The Netherlands.

Doubet, R.S. and Quatrano, R.S. (1982). Isolation of marine bacteria capable of producing specific lyases for alginate degradation. *Applied and Environmental Microbiology*, **44**: 3754-756.

Druehl, L.D., Baird, R., Lindwall, A., Lloyd, K.E. and Pakula, S. (1988). Longline cultivation of some Laminariaceae in British Columbia, Canada. *Aquacult. Fish. Manag.* **19**: 253-263.

Dyrset, N., Lystad, K.Q. and Levine, D.W. (1994). Development of a fermentation process for the production of an alginate G-lyase from *Klebsiella pneumoniae*. *Appl. Microbiol. Biotechnol.*, **41**: 523-530.

EPA, (2003). National Environmental Monitoring Programme for Transitional, Coastal and Marine Waters, Environmental Protection Agency.

Fletcher, R.L. (1996). The occurrence of "green tides" - a review. Marine benthic vegetation: recent changes and the effects of eutrophication. W.S.A.P.H. Nienhuis. Berlin, Springer. pp. 7-43.

Fortman, J.L., Chhabra, S., Mukhopadhyay, A., Chou, H., Lee, T.S., Steen, E. and Keasling, J.D. (2008). Biofuels alternatives to ethanol: pumping the microbial well. *Trends Biotechnol*, **26**: 375-381.

Gao, K. and McKinley, K.R. (1994). Use of macroalgae for marine biomass production and CO_2 remediation: a review. *J. Appl. Phycol.*, **6**: 45-60.

Global Warming, (2009). Http:\\www.globalwarming.org. Accessed 14 September, 2009.

Guiry, M.D. and Guiry, G.M. (2009). *AlgaeBase*. World-wide electronic publication, National University of Ireland, Galway. http://www.algaebase.org; searched on 20 December 2009.

Hernández-González, M.C., Buschmann, A.H., Cifuentes, M, Correa, J.A. and Westermeier, R. (2007). Vegetative propagation of the carrageenophytic red alga *Gigartina skottsbergii* Setchell et Gardner: Indoor and field experiments. *Aquaculture* **262**: 120-128.

Honya, M., Kinoshita, T., Ishikawa, M., Mori, H. and Nisizawa, K. (1993). Monthly determination of alginate M/G ratio, mannitol and minerals in cultivated *Laminaria japonica*. *Nippon Suisan Gakkaishi* **59(2)**: 295-299.

Horn, S.J., Aasen, I.M. and Østgaard, K. (2000a). Production of ethanol from mannitol by *Zymobacter palmae*. *J. Ind. Microbiol. Biotechnol.*, **24**: 51-57.

Horn, S.J., Aasen, I.M. and Østgaard, K. (2000b). Ethanol production from seaweed extract. *J. Ind. Microbiol. Biotechnol.,* **25**: 249-254.

ISC, (2009). Seasonal analysis in 10 commercial edible sea vegetables of the Irish west coast. Internal report. Irish Seaweed Centre, National University of Ireland, Galway.

Kaas, R. (2006). The seaweed resources of France. *In*: Critchly, A.T., Ohno, M. and Largo, D.B. The seaweed resources of the world. Interactive CD ROM, ETI. The Netherlands.

Kain, J.M. and Dawes, C.P. (1987). Useful European seaweeds: past hopes and present cultivation. *Hydrobiologia* **151/152**: 173-181.

Kelly, E. (*Ed.*) (2005). The role of kelp in the marine environment. Irish Wildlife Manuals, No. 17. National Parks and Wildlife Service, Department of Environment, Heritage and Local Government, Dublin, Ireland.

Klinke, H.B., Thomsen, A. and Ahring, B.K. (2001). Potential inhibitors from wet oxidation of wheat straw and their effect on growth and ethanol production by *Thermoanaerobacter mathranii*. *Appl Microbiol Biotechnol* **57**: 631-638.

Kraan, S. and Guiry, M.D. (2001). Strain selection in the edible brown seaweed *Alaria esculenta*: Genetic fingerprinting and hybridization studies under laboratory conditions, Marine Resource Series No. 18, Marine Institute, Dublin 2001.

Kraan, S. and Guiry, M.D. (2006). The seaweed resources of Ireland. *In*: Seaweed resources of the world, A CD-rom project *Eds*. Alan Critchley, Masao Ohno and Danilo Largo, ETI, Amsterdam, The Netherlands.

Lahaye, M. and Robic, A. (2007). Structure and functional properties of Ulvan, a polysaccharide from green seaweeds. *Biomacromolecules*, **8(6)**: 1765-1774.

Lee, S.K., Chou, H., Ham, T.S., Lee, T.S. and Keasling, J.D. (2008). Metabolic engineering of microorganisms for biofuels production: from bugs to synthetic biology to fuels. *Current Opinion in Biotechnology*, **19**: 556-563.

Licht, F.O. (2009). *World Ethanol and Biofuels Report*, **7(18)**: 365.

Lorentsen, S.-H., Gremillet, D. and Nymoen, G.H. (2004). Annual variation in diet of breeding Great Cormorants: Does it reflect varying recruitment of Gadoids? *Waterbirds*, **27**: 161-169.

Lüning, K. (1990). Seaweeds. Their environment, Biogeography and Ecophysiology. *Eds*. Yarish, C. and Kirkman, H., John Wiley & Sons, Inc., New York, p. 527.

Lûning, K. and Pang, S. (2003). Mass cultivation of seaweeds: current aspects and approaches. *J. Appl. Phycol.*, **15**: 115-119.

Mann, K.L. (1982). Ecology of coastal waters: a systems approach. University of California Press: Berkeley, CA.

McCaughren, S. (2008). Sunday Business Post. http://www.sbpost.ie/post/pages/p/story.aspx-qqqt=IRELAND-qqqm=news-qqqid=36121-qqqx=1.asp

McHugh, D.J. (2003). A guide to the seaweed industry. *FAO Fisheries Technical Paper* **441**: 105.

Mohammed, A.I. and Fredriksen, S. (2004). Production, respiration and exudation of dissolved organic matter by the kelp *Laminaria hyperborea* along the west coast of Norway. *J. Mar. Biol. Ass.* U.K., **84**: 887-894.

Neori, A., Shpigel, M. and Ben-Ezra, D. (2000). A sustainable integrated system for culture of fish, seaweed and abalone. *Aquaculture* **186**: 279-291.

Ohno, M. and Largo, D.B. (2006). The seaweed resources of Japan. *In*: Critchly, A.T., Ohno, M. and Largo, D.B. (Eds), The seaweed resources of the world. Interactive CD ROM, ETI, The Netherlands.

Peak, Oil (2009). Http:\\www.peakoil.com. Accessed 14 September, 2009.

Pérez, R., Kaas, R. and Barbaroux, O. (1984). Culture expérimentales de l'algue *Undaria pinnatifida* sur les côtes de France., Science et Pêche, **(343)**: 3-15.

Pérez, R., Kaas, R., Campello, F., Arbault, S. and Barbaroux, O. (1992). La culture des algues marines dans le monde. *In*: Service de la Documentation et des Publications (SDP). IFREMER, Plouzane.

Reddy, C.R.K., Jha, B., Fujita, Y. and Ohno, M. (2008). Seaweed micropropagation techniques and their potentials: an overview. *J Appl Phycol* **20**: 609-617.

Reith, J.H., Deurwaarder, E.P., Hemmes, K., Curvers, A.P.W.M., Kamermans, P., Brandenburg, W. and Zeeman, G. (2003) BIO-OFFSHORE, Grootschalige teelt van zeewieren in combinatie met offshore windparken in de Noordzee. ECN-C—05-008 ECN projectnummer 8.20300. p. 137.

Reith, H., Huijgen, W., van Hal, J. and Lenstra, J. (2009). Seaweed potential in the Netherlands. ECN Biomass, Coal and Environmental Research Kick-off Meeting EOS-LT project Seaweed Biorefinery 24 Sept. 2009, Amsterdam.

Ross, A., Jones, J.M., Kubacki, M.L. and Bridgeman, T.G. (2008) Classification of macroalgae as fuel and its thermochemical behaviour. *Bioresource Technology*, pp. 6494-6504.

Schuenhoff, A., Shpigel, M., Lupatschb, I., Ashkenazib, A., Msuyab, F.E. and Neori, A. (2003). A semi-recirculating, integrated system for the culture of fish and seaweed. *Aquaculture* **221**: 167-181.

Shaffer, S. (2003). Preferential use of nearshore kelp habitats by juvenile salmon and forage fish. Proceedings of the Georgia Basin/Puget Sound Research Conference. p. 11.

Sjøtun, K. and Lorentsen, S-H. (2003). Kelp forest (*Laminaria hyperborea*) as habitat for juvenile gadoids. Poster presented at the 3rd European Phycological Congress, Belfast, North-Ireland, 21-26 July, 2003.

Steneck, R., Vavrinec, J. and Leland, A. (2004). Accelerating trophic-level disfunction in kelp forest ecosystems of the western north Atlantic. *Ecosystems*, **7**: 323-332.

Titlyanov, E.A., Titlyanova, T.V., Kadel, P. and Lüning, K. (2006). New methods of obtaining plantlets and tetraspores from fragments and cell aggregates of meristematic and submeristematic tissue of the red alga *Palmaria palmata*. *Journal of Experimental Marine Biology and Ecology* **339**: 55-64.

Thomas, D.N. (2002). Seaweeds, Natural History Museum, London.

Troell, M., Joyce, A., Chopin, T., Neori, A., Buschmann, A.H. and Fang, J.-G. (2009). Ecological engineering in aquaculture — Potential for integrated multi-trophic aquaculture (IMTA) in marine offshore systems. *Aquaculture* **297**: 1-9.

Trevors, J. CEO of AER Ltd 162 Clontarf Road, Dublin 3, Ireland.

Tseng, C.K. (1987). Some remarks on the kelp cultivation industry of China. *In*: Bird, K.T. and Benson, P.H. 1987. (*Eds.*). Seaweed cultivation for renewable resources. Developments in aquaculture and fisheries science, 16. Elsevier, Amsterdam. pp. 147-156.

Tuohy, M. Department of Biochemistry, National University of Ireland, Galway.

Valiela, I., McClelland, J., Hauxwell, J., Behr, J., Hersh, D. and K., F. (1997). Macroalgal blooms in shallow estuaries: controls and ecophysiological and ecosystem consequences. *Limnology and Oceanography* **42**: 1105-1118.

Van Maris, A.J.A., Abbot, D.A., Bellisimi, E., van den Brink, J., Kuyper, M., Luttik, M.A.H., Wisselink, H.W., Scheffers, W.A., van Dijken, J.P. and Pronk, J.T. (2006). Alcoholic fermentation of carbon sources in biomass hydrolysates by *Saccharomyces cerevisiae*: current status. *Antonie van Leeuwenhoek*, **90**: 391-418.

Wang, Y., Han, F., Hu, B., Li, J. and Yu, W. (2006). *In vivo* prebiotic properties of alginate oligosaccharides prepared through enzymatic hydrolysis of alginate. *Nutrition Research* **26**: 597-603.

Werner, A. and Kraan, S. (2004). Review of the potential mechanisation of kelp harvesting in Ireland. Marine Environment & Health Series, No.17 pp. 1-52, Marine Institute, Galway.

Wu, C.Y. and Pang S.J. (2006). The seaweed resources of China. *In*: Critchly, A.T., Ohno, M. and Largo, D.B. The seaweed resources of the world. Interactive CD ROM, ETI, The Netherlands.

Xu, B., Zhang, Q.S., Qu, S.C., Cong, Y.Z. and Tang, X.X. (2008). Introduction of a seedling production method using vegetative gametophytes to the commercial farming of *Laminaria* in China. *J. Appl. Phycol.* **20**: 9.

Yarish, C. and Pereira, R. (2008). Mass production of marine macroalgae. *In* Sven Erik Jørgensen and Brian D. Fath (*Editor-in-Chief*), Ecological Engineering. Vol. [3] of Encyclopedia of Ecology, 5 vols. pp. [2236-2247] Oxford: Elsevier.

Zhang, Q.S., Qu, S.C., Cong, Y.Z., Luo, S.J. and Tang, X.-X. (2008). High throughput culture and gametogenesis induction of *Laminaria japonica* gametophyte clones. *J. Appl. Phycol*. 20, 205-211.

Zverlov, V.V., Berezina, O.V., Velikodvirskaya, G.A. and Schwarz, W.H. (2006). Bacterial acetone and butanol production by industrial fermentation in the Soviet Union: use of hydrolyzed agricultural waste for biorefinery. *Applied Microbiology and Biotechnology* **71**: 587-597.

10

OUR TRYST WITH MICROALGAE HAVING POTENTIAL FOR BIOPIGMENTS, BIOPOLYMER, BIOFUEL AND BIOREMEDIATION

Sandhya Mishra*

Abstract

Microalgae including Cyanobacteria are the largest primary producers in any aquatic ecosystem. Their metabolic functions have developed during the evolutionary process for adapting to diversified & extreme environments. With the development of marine biotechnology, microalgae from unique & extreme marine environments are attracting attention as a novel biological sources for superfine chemicals to be employed in varied applications of pharmaceuticals, neutraceuticals, cosmeceuticals, etc. where biopigments like Phycobiliproteins, Carotenoids; biopolymers like Polyhydroxyalkanoates (PHAs), Exo-polysaccaharide (EPS) have an indispensable role to play. Due to the present day health awareness amongst the human beings, the gradual increase in demand of the natural / herbal products have rightly evaluated the treasure of our planet Earth i.e. Microalgae. Since, a decade we have been exploring and exploiting the potential of Microalgae for C-PC, C-PE, R-PE, Beta-carotene, PHAs, EPS, Lipids, Biosequestration of metals and Bioremediation, though there are still many unexploited and unexplored potentials hidden within the vast reign of Microalgae. Why only few microalgal species made their way into successful Biotechnology? Only three main Genus Arthrospira (Spirulina), Dunaliella and Chlorella have been grown at large scale by several companies for neutraceuticals.

Marine Biotechnology and Ecology Discipline, Central Salt & Marine Chemicals Research Institute, CSIR G.B. Marg, Bhavnagar – 364 002, India.

* *Corresponding author*: E-mail: smishra@csmcri.org, smishracsmcri@rediffmail.com

Spirulina and Dunaliella thrive in extreme conditions of alkalinity and salinity respectively, while, Chlorella is endowed with a relatively high growth rate, which makes them obey the" Law of survival of the Fittest "avoiding the possible contaminants while growing in open pond culture. There are many challenges to be faced by the Microalgae to be a commercially promising candidate in future, amongst which 4 main points are: (a) proper strain selection and further genetic improvement, (b) rate of biomass production, (c) efficient harvesting and (d) prevention of the product from denaturation. As per our gradually growing needs and the concern for the better LIFE with SAFE ENVIRONMENT, the known potentials of microalgae should be scaled up and the latent potentials should be explored and exploited in a balanced and tactful manner.

Key words: Biopigment, Biopolymer, Biofuel, Bioremediation, *Arthrospira (Spirulina), Dunaliella, Chlorella*

Introduction

Some of the potentials of Microalgae which have been studied at CSMCRI, Bhavnagar for a decade are the attempts to enter the Plant Kingdom of prokaryotic and eukaryotic algal inhabitants from Cyanophyceae, Chlorophyceae and Rhodophyceae families and the glimpse is provided over here. Developments in the field of algal biotechnology have yielded a few excellent contributions dealing with different aspects of the technology and some related problems in the mass cultivation of Microalgae. The concept of biotechnology involved is the same as in conventional agriculture, *i.e.* the utilization of the photosynthetic machinery for the production of biomass to be used as a source of food, feed, chemicals and energy.

Biopigments

The Microalgae are unique in the Plant Kingdom having variation of the photosynthetic pigments in each phylum and in many of the classes, for the remainder of the kingdom are monotonously similar. This diversity of the light-harvesting pigments implies that the common ancestor was primitive and that no close affinity exists between blue-green (cyanobacteria), red, brown, golden-brown and green algae, to use their common names. This diversity of pigments has led to their recognition as taxonomic and phylogenetic markers. Measuring the relative and absolute concentration of light-harvesting pigments in particular, chlorophyll-a is widely used to study the density of marine algae in relation to depth and area. Remote sensing by satellite is now being employed for this purpose. All chlorophylls, phycobiliproteins, and carotenoids found in algae are designated as of photosynthetic pigments.

Cyanobacteria and microalgae were evaluated for the production of other biological compounds, including pigments such as phycoerythrins, which has potential as a fluorescent probe (Glazer, 1994). Another biliprotein, phycocyanin, is being extracted from *Spirulina platensis* for the health-food and cosmetic market in Japan, â-carotene, a vitamin A precursor which has uses as a food colorant, has attracted much attention. The commercial potential of *Dunaliella salina* for â-carotene and glycerol production has been well established (Ben-Amotz and Avron, 1973; Ben-Amotz *et al.*, 1982). At high salinities (250–320 g l^{-1}) *Dunaliella salina* contains high levels of â-carotene (approximately 10% of dry wt.). Furthermore, *Dunaliella salina* has the advantage of remaining the dominant alga during outdoor mass culture and does not suffer from predation. A range of other carotenoids such as astaxanthin is produced in significant quantities by certain species of *Euglena, Haematococcus* and *Chlorella* which are also of commercial interest (Chen and Chi, 1981; Borowitzka *et al.*, 1984). Among the different pigments which are characteristic of cyanobacteria, the carotenoids and the phycobiliproteins have high commercial value. The carotenoids are used as "natural" food colorings, as feed additives to enhance the color of the flesh of Salmanoid fish, to enhance the color of egg yolks and to improve the health and fertility of cattle. Among the various types of carotenoids, â-carotene is known to be the precursor of vitamin A and hence, is of immense nutritional value, *Spirulina* is reported to be rich in this pro-vitamin A. The first full-scale research and development attempt towards the mass production of green microalgae was started in Czechoslovakia around 1960, when the laboratory for Experimental Algology was founded at Trebon. Intensive outdoor production of algal biomass is limited by several major factors: minerals, CO_2, temperature, light and pH. Minerals and CO_2 can be supplied to the culture at a reasonable cost but temperature and light limitations confine algae-mass culture to the tropical countries while in temperate countries the expenditure with the glass house on sauna and temperature (30 ± 5°C) increases the cost of cultivation (outdoor).

In the Rift Lakes (East Africa) and in Lake Texaco, Mexico (Clement, 1967; Durand-Chastel, 1980) grew *Spirulina* in highly alkaline natural and artificial lakes respectively. Bicarbonate concentrations as high as 200 gm/lit has been reported by Zarrouk (1966). Others have been able to grow various *Spirulina* species in different growth media ranging from waste-water effluents (Kosaric, 1974) to cow manure (Oron *et al.*, 1984), but always with bicarbonate supplements, which could be eliminated by vigorous aeration. However, *Spirulina* has never been the dominant species in large-scale outdoor marine cultures, as it clearly cannot compete with weed species such as *P. tricornatium* (Goldmann and Ryther, 1976). Besides, the difficulties in mass culturing unique species such as *Spirulina* and *Dunaliella* under conditions suitable for yield optimization, there is virtually no information on how the chemical composition of the algae can be varied to enhance maximum production of certain organic compounds.

Simply, the conditions favoring lipid production are not compatible with those required for yield optimization. Similarly, the physiological conditions (*e.g.* temperature, light quality, N_2 and phosphorus, etc) for biomass productivity may be optimized but, which may not be the parallel optimized conditions for the yield of a desired organic product; Hence, growth kinetics with respect to the biomass as well as yield is necessary. According to E. Jel-OR, S. Boussiba and A.E. Richmond, Institute for Desert Research, Ben-Gurion University of the Negev, Sede Boger Campus, Israel different products and chemicals from *Spirulina platensis* can be obtained by a combined process for the extraction of several components in addition to bulk protein *e.g.* enzymes, photosynthetic pigments and fine chemicals.

Spirulina is rich in proteins, which amount to 60-70% of its dry weight (Belay *et al.*, 1993). The proteins having the highest economic potential are the biliproteins *i.e.* C-phycocyanin, allophycocyanin and phycoerythrin. The protein fraction may contain up to 20% of phycocyanin *i.e.* 10-14% of the dry wt. Phycocyanin is mainly used as a food pigment; however, small quantities are used as a biochemical tracer in immunoassays, microscopy and cytometry owing to the fluorescent properties of the pigment. It could be utilized as a natural pigment for drug, food and cosmetic industries to replace the currently used synthetic pigments which are carcinogenic. The phycobiliproteins were introduced as a novel class of fluorescent tags in 1982 which was followed by diagnostic clinical assays in histochemistry and on diverse research application (Glazer, 1994). Cyanobacteria are considered to be very efficient biological system for harvesting solar energy for the production of various organic compounds via. Photosynthesis where the accessory pigments (Phycobiliproteins ® C-PC, APC, PE). Such a docile organism if cultivated in a well balanced and planned way could yield really high valued commercial products. At a time and the same batch it yields different valued products. The product which is desired the most can be induced to yield high by either physicochemical means or through genetic manipulation.

At CSMCRI, Bhavnagar we developed an integrated process for the production of different grade of C-Phycocyanin from *Spirulina platensis* aligned with other byproducts.

Besides, a process for the highly pure C-Phycocyanin also developed (Patel *et al.*, 2005; Mishra, 2006) Simultaneously, Single crystals of C-PC developed from cyanobacteria of different habitat also compared through XRD (Satyanarayana *et al.*, 2005) Further, their antioxidant potential (*in vitro* and *in vivo*) also compared (Mishra *et al.*, 2005; Patel *et al.*, 2006; Paul *et al.*, 2006). In order to increase the stability of food grade C-Phycocyanin a process with edible preservatives was developed (Mishra *et al.*, 2008). Although it is impossible to predict the economic feasibility of the superfine product like phycocyanin but still it could definitely replace the synthetic pigments as they are carcinogenic and people have become more health conscious now, they would not risk their lives with the cheaper and synthetic products.

Apart from imparting blue color to the food and beverages it has several beneficial physiological effects *e.g.* antidiabetic, hypolipidemic effect, anti-inflammatory, anti-oxidant and anti-cancer. *In vitro* assay of *Spirulina* C-PC and *Porphyra* R-PE was also done with "HERBOPRINT" at IICT, Hyderabad (Mishra *et al.*, 2008).

Single product process is not economically feasible; hence an integrated process was developed for the production of highly pure C-PC aligned to various byproducts like chlorophyll-a, carotenoids, plant growth promoters and aquafeed (fish feed) from the Cyanobacteria (*S. platensis*) cultivated in different ways selected for different grades of C-PC with maximum utilization of the raw material (biomass) and minimum loss of the product with almost zero effluent discharge, through the indigeneously developed ultra-filtration unit and the membrane also.

Biopolymers

The potential for biopolymers have been explored from *Spirulina* spp. for both intracellular (PHA) and extracellular (EPS) biopolymers. The yield of PHA can further be increased by recombinant strains and metabolic/genetic engineering required to improve the expression level of PHB in cyanobacteria. Under extreme/stress conditions cyanobacteria are reported to produce several biopolymers like EPS, PHA etc., which can be produced extracellularly and intracellularly respectively. PHAs (Poly hydroxyl alkanoates) are of biological origin, they are also capable of being completely broken down to water and carbon dioxide by microorganisms found in a wide range of environments, such as soil, water, and sewage.

PHA is polyester like synthesized and stored intracellular by bacteria as energy and carbon storage materials having structural properties similar to polypropylene. PHAs have immense applications in packaging films, disposable items, bone replacements, blood vessel replacements and scaffold material in tissue, engineering of heart valves etc.

Cyanobacteria however, are indigenously the sole prokaryotes that accumulate PHB by oxygenic photosynthesis, and are of particular interest because of their minimal requirements for their growth and biomass production. Due to their minimal nutrient requirement and ability to grow even in wastewaters in presence of CO_2 and sunlight these photoautotrophs (cyanobacteria) could be explored as an alternative source for PHB production, as the biomass could be inexpensively converted into biodegradable plastics by solar energy which can aid in overall reduction of the production cost of biodegradable plastics, which is the limiting factor for the replacement of synthetic polymers by such biodegradable biopolymers (Shrivastav *et al.*, 2008).

Biofuel

Marine or freshwater micro-algae can be a potential source of future fuel and chemical feedstock for India, besides acting as a 'Thermostat'. Micro-algae have higher photosynthetic efficiency (higher biomass productivity) than higher plants. Unlike terrestrial plant biomass, the carbon in algal biomass is easily convertible and recoverable as oil or chemical feedstock such as methane since it is not lignified. About 25% of algae are extractable oil which can be converted to biodiesel. Other components can be converted to biogas through biomethanation. The mineralized residue after biomethanation can be recycled as nutrient source for algal growth. The intensive cultivation of micro-algae requires addition of CO_2 since natural rate of CO_2 mass transfer from atmosphere is insufficient to support the high growth rates. Flue gases from thermal power plants are an attractive source of carbon dioxide for micro-algae production. Tropical country like India, especially Gujarat state has natural advantage for algal cultivation because of the high insolation throughout the year, besides having large coastline with several million hectares of salt marshes and coastal backwaters which can be utilized for algal cultivation. No diversion of agricultural land or freshwater resources is required for algal cultivation. Algal cultivation will require only 3% of the land area required for *Jatropha* to produce equivalent quantity of biodiesel. Energy security and environment are key factors for the future growth of India's economy. Two important issues that need technological change are: (a) The development of biodiesel for transportation so as to reduce the dependence on imported hydrocarbon and to reduce fossil carbon emissions (b) The capture of CO_2 from point sources such as power plants in order to meet India's commitments towards global reduction of greenhouse gas emissions. Biodiesel from oil seeds for complete replacement of mineral oil is severely constrained by land and water resource availability and yield and cost issues. It is therefore, necessary to consider development of various parallel approaches to biofuels.

We don't require technological breakthroughs for the conversion of algal biomass to these fuels but pond productivity needs incremental improvement to achieve competitive energy production over alternate energy crops. Since algal biomass rather than defined algal species cultivation is all that is desired in the system, issues related to maintenance of strain purity in open pond cultures are irrelevant. Algal cultivation is carried out commercially in India for production of speciality food materials (neutraceuticals). This requires narrowly defined algal cultures and the value of products is high. On the other hand, energy production requires high biomass productivity and low cost of cultivation rather than purity of culture. Algal cultivation for food and neutraceuticals suffers from issues such as contamination, maintenance of algal strain purity faced with infections, wild growth etc. that to an extent can be overcome by using sufficiently large quantity starting (seed) cultures prepared under highly controlled conditions. Mixed biomass representing a naturally stable ecosystem for a given environmental condition in the culture pond, generally avoid such problems, and may be considered as an alternate strategy for energy production through open algal ponds.

The concept of mixed algal cultures for biomass production has not reached commercial production anywhere. Internationally algal biomass production as a fuel source is being researched in USA (*e.g.*, http://www.unh.edu/p2/biodiesel/pdf/algae_salton_sea.pdf) and in other countries uses the approach of mixed cultures (IEA Greenhouse Issues Newsletter Dec http://www .ieagreen.org.uk/newsletter/80Dec.pdf)

Algal biofuel concept has not been examined in India, although India is climatically well suited because of high insolation available throughout the year and because of long coastlines. India also has several areas such as tidal mudflats of Kutch in the west and lagoons and tidal deltas in the east that can be profitably used for such cultivation.

CSMCRI has done some exploratory studies with the existing cultures for Biofuel. We are at present involved in preliminary stages of selecting the microalgal strain and optimizing its growth conditions for high yield of good quality oil required for the biodiesel production in future. Simultaneously, chemical characterization of the oil extracted from the lab grown culture of the selected strain of microalgae is also done. Mixotrophic (heterotrophic and autotrophic) growth conditions is approached for the originally Autotrophic (photosynthetic) microalgal strains under lab. controlled conditions.

Biological management of solar salt production for which US patent has been granted no. 6, 812, 011, 2004. Developed the process for the removal of Ca^{++} impurity from the brine by marine cyanobacteria. The cyanobacterial strains exploited for the above purpose were isolated from the west coast of Arabian Sea and have shown good adaptability to brine compositions varying from sea brine to subsoil brine. They are *Euryhaline* spp. having salinity tolerance range from 3°Be to 29°Be. They do not require external energy or chemical for growth and activity and perform their task under ambient conditions. They can be reused also for the removal of calcium from the fresh brine. The 42 hrs young cultures take up Ca^{++} upto 90% within 48 hrs. The lab scale experiments have shown reproducibility in the field too.

Calcium impurity in sodium chloride is detrimental in certain industrial applications such as production of caustic soda and chlorine from brine through Nafion membrane cells. Thus, price of salt is strongly dictated by the calcium ion impurity level in salt. Normally, calcium is removed from brine through a process of fractional crystallization wherein the major fraction of the less soluble calcium carbonate and calcium sulphate salt is precipitated prior to crystallization of salt. However, some calcium remains in the brine at the time of crystallization of salt and this precipitates out along with salt since its solubility is also decreased abruptly with crystallization of salt. The process of fractional crystallization is strongly influenced by the composition of brine, namely its sulphate content. With a view to enhance production of salt, CSMCRI has been developing a process of salt manufacture from sub-soil brine which is 4-5 times more concentrated than sea brine.

Unfortunately, the calcium impurity in the salt produced from sub-soil brine is higher than from sea brine which makes it unacceptable for industrial applications. Solutions to these problems which were identified include: external addition of sodium sulphate to reduce the solubility of calcium sulphate. This is, however, not practical and also the results are less than satisfactory unless large quantities of sulphate are added. The latter makes the process economically unviable besides increasing magnesium sulphate concentration in salt which is more difficult to remove. The process is cost effective, eco-friendly, more efficient, requires no extra chemical, energy or labour, increases the rate of evaporation in the brine by forming bluish green carpet like, cyanobacterial mat leading to greater salt productivity, saves the salt from additional chemical impurities, reduces the seepage of the brine and also removes the undesirable bacteria *i.e.* acting as a, depurating system, the process is recyclable, the extra biomass if occurs and is not utilized in the process can be used as an aqua-feed or cattle-feed as a by-product.

The halophilic microalgae *Dunaliella salina* (pleomorphic) possesses capacity to accumulate high quantities of β-carotene (more than 10% of the algal dry weight) for which it is cultured massively for commercial purposes. The carotenes of *D. salina* are a mixture of β-carotene and its isomers, 9-cis and all-trans, while synthetic β-carotene is composed only of all-trans isomer. The high liposolubility of the 9-cis β-carotene permits its great absorption and storage in animal tissues showing its therapeutic effect through its antioxidant properties. The carotenoids from selected microalgal strain of *Dunaliella salina* and *Spirulina* spp. from the Salt farm of CSMCRI, Bhavnagar were investigated (Joshi *et al.*, 2008).

Bioremediation

The conventional methods used for the effluent treatment are precipitation as hydroxides or sulphides, oxidation or reduction & ion-exchange, though the biochemical methods involving the microbes is better alternate to conventional physicochemical ones. Biochemical methods involve number of advantages such as rapid kinetics of removal, capability of biomass to remove contaminants at lower concentration, the utilization of naturally abundant renewable biomaterials, possibility to treat the effluent containing multiple contaminants, eco-friendly & cost effective, reusability of biomass through immobilization or dead biomass usage, generates minimum secondary effluent, so, it is less polluting. Some cyanobacteria are extremophilic microbes surviving under any adverse conditions. Highly saline effluents containing complex organic & inorganic compounds having very high COD was treated with different biomass such as cyanobacteria, macro algae, fungi & drumstick. The cyanobacterial consortium in the system pretreated with EDTA showed maximum removal of phosphates *i.e.* 65.03%. The effects seen in the cases of macro algae *Ulva lectuca* and fungal biomass of *Aspergillus niger* was 55.90% and 53.99% respectively. The significant reduction in the COD was observed with cyanobacterial biomass (70-80%) as compared to macroalgal, fungal and plant biomass (Mishra, 2004; Patel *et al.*, 2001; Mishra *et al.*, 2001; Sonawane *et al.*, 2002). Besides, uranium sequestering potential of *Spirulina platensis* was studied in batch mode & its equilibrium was established in about 60 min having, maximum sorption of uranium at pH 4.0-5.0. The desorption results show that sodium citrate solution is very effective with 83% of uranium being recovered effectively. The mineral acids though effective in desorption cause destruction of the biomass. The uranium binding is also studied through fluorescence studies (Dheetcha and Mishra, 2008).

Conclusions

The first practical use of microalgae dates back 2000 years to the Chinese, who used *Nostoc* to survive famine. More than 10,000 species are believed to exist in nature, out of which few thousands are in culture collections, few hundreds are explored and just few are mass cultivated for industrial applications in tonnes (Spolaore *et al.*, 2006). The cultivation itself needs to be improved for being economically feasible and more competetive, through either genetic improvement of desired strains or metabolic engineering. India, being a tropical country has a boon of immense solar energy and vast coast line for the much better opportunity than western countries for outdoor mass culture of the extremophilic microalgae. If integrated process is developed for any desired microalgal product it would become commercially viable. Let us make Microalgae meet all the challenges and compete with synthetic products in order to preserve and treasure the life of future generation.

Acknowledgements

I am highly grateful to the project staff & students. Namely Dr. Yogi Badheka, Dr. Vimal Ramani, Mrs. Chanchal Raviya, Dr. Amit Dubey, Dr. Anamika Patel, Dr. Richa Pawar, Mrs. Purvi Shukla, Mr. Shailendra

Sonawane, Mr. Bhavin Raval, Mr. Sanjiv Mishra, Ms. Anupama Shrivastav, Ms. Deepti Jain, etc and others. Author is also thankful to the Department of Biotechnology, New Delhi, M/s. Strides Arco Lab, Bangalore and CSIR, India for financial support. I am highly thankful to Mr. M.R. Gandhi for his help in all the processes development. Thanks are also due to Analytical Division of CSMCRI for instrumental analysis (FTIR, NMR, GC, SEM, DSC, TGA, etc). Grateful thanks are due to Dr. P.K. Ghosh, Director, CSMCRI, Bhavnagar for conceptualizing various novel ideas and fruitful discussions while interpreting the results from Microalgal research and developing different processes. Last but not the least I would like to pay my gratitude to Mr. A. Tewari (Ex-Scientist'G', CSMCRI, Bhavnagar) for his moral support and constant encouragement during the initial stages of the work at CSMCRI, Bhavnagar.

References

Belay, A., Ota, Y., Miyakawa, K. and Shimamatsu, H. (1993). Current knowledge on potential health benefits of *Spirulina. Journal of Applied Phycology,* **5**: 235-241.

Ben-Amotz, A. and Avron, M. (1973). The role of glycerol in the osmotic regulation of the halophilic alga *Dunaliella parva. Plant Physiology,* **51(5)**: 875-878.

Ben-Amotz, A., Katz, A. and Avron, M. (1982). Accumulation of carotene in halotolerant algae: purification and characterization of â-carotene globules from *Dunaliella salina* (Chlorophyceae). *Journal of Phycology,* **18**: 529-537.

Borowitzka, L.J., Borowitzka, M.A. and Moulton, T.P. (1984). The mass culture of *Dunaliella salina* for fine chemicals: from laboratory to pilot plant, *Hydrobiologia,* **116(117)**: 115-121.

Chen, B.U. and Chi, C.H. (1981). Process development and evaluation for algal glycerol production. *Biotechnology & Bioengineering,* **23:** 1267-1287.

Clément, G., Giddey, C. and Menzi, R. (1967). Amino acid composition and nutritive valve of the alga *Spirulina maxima. Journal of the Science Food & Agriculture.,* **18(11)**: 497-501.

Dheetcha, A. and Mishra, S. (2008). Biosequestering potential of *Spirulina platensis* for Uranium. *Current Microbiology,* (In press).

Durand-Chastel, H. (1980). Production and use of *Spirulina* in Mexico. *In*: Shelef, G. and Soeder, C.J., *Editors, Algae Biomass,* Elsevier/North Holland Biomedical Press, Amsterdam, pp. 51-64.

Glazer, A.N. (1994). Phycobiliproteins — a family of valuable, widely used fluorophores. *Journal of Applied Phycology,* **6(2)**: 105-112.

Goldman, J.C. and Ryther, J.H. (1976). Waste reclamation in an integrated food chain system. *In*: *Biological Control of Water Pollutio,. Eds.* Toubier, J. and Pierson, R.W., Jr., University of Pennsylvania press, Philadelphia, PA. pp. 197-214.

Joshi, V., Desai, P. and Mishra, S. (2008). Preliminary studies on comparison between â-carotene of prokaryotic and eukaryotic algae: *Spirulina platensis*

SM-Pc1 & Dunaliella salina SM-Bc1. Journal of Cell and Tissue Research, **8(1)**: 1263-1270.

Kosaric, N. (1974). Growth of *Spirulina maxima* algae in effluents from secondary waste water treatment plants. *Biotechnology & Bioengineering,* **16**: 881.

Mishra, S. (2004). Microbial degradation of Phenol in sea water sediments of Arabian Sea. *Nature, Environment and Pollution Technology,* **3(4)**: 549-552.

Mishra, S. (2006). "Phycobiliproteins from Microalgae" *Recent advances on applied aspects of Indian marine algae with reference to Global scenario, CSMCRI* **2**: 271-284.

Mishra, S., Chundawat, R.S., Sood, P.P. and Mishra, S. (2005). Efficacy of C-phycocyanin against Cr (IV) toxicity induced lipid peroxidation. *Journal of Cell and Tissue Research,* **5(2)**: 13-514.

Mishra, S., Ray, P., Trivedi, G.S., Adhikary, S.K., Sravankumar, V.G., Tewari, A. and Ghosh, P.K. (2001). Application of electrodialysis and blue green algae for removal of inorganic salt and color from the fungicide manufacturing industrial waste. *Research Journal of the Chemistry & Environment,* **5(2)**: 17-20.

Mishra, S.K., Shrivastav, A. and Mishra, S. (2008). Effect of preservatives for food grade C-PC from *Spirulina platensis. Process Biochemistry,* **43**: 339-345.

Mishra, S.K., Shrivastav, A., Kumar, D.V. and Mishra, S. (2008). Evaluation of *Spirulina* & C-Phycocyanin as medicine by "Herboprint". *Journal of Ethnopharmacology* (under review).

Oron, G., Wildschut, L.R. and Porath, D. (1984). Waste water recycling by duckweed for protein production and effluent renovation. *Water Science and Technology,* **17**: 803-817.

Patel, A., Mishra, S. and Ghosh, P.K. (2006). Antioxidant potential of C-Phycocyanin isolated from three cyanobacterial species: *Lyngbya, Phormidium and Spirulina* spp., *Indian Journal of Biochemistry and Biophysics* **43**: 31-35.

Patel, A., Mishra, S., Pawar, R. and Ghosh, P.K. (2005). Purification and characterization of C-Phycocyanin from three cyanobacterial species of different habitat. *Protein Expression & Purification,* **40(2)**: 248-255.

Patel, A., Pawar, R., Mishra, S. and Tewari, A. (2001). Exploitation of marine cyanobacteria for removal of colour from distillery effluent. *Indian Journal of Environmental Protection,* **21(12)**: 1118-1121.

Paul, B.T., Selvam, G., Murugesan, R., Patel, A., Mishra, S. and Ghosh, P.K. (2006). Photodynamic action of C- phycocyanins obtained from marine and fresh water cyanobacterial cultures: A comparative study using EPR- spin trapping technique. *Free Radical Research,* **40(8)**: 821-825.

Satyanarayana, L., Suresh, C.G., Patel, A., Mishra, S. and Ghosh, P.K. (2005). X-ray crystallographic studies on C-phycocyanin from cyanobacteria of different habitats: marine & fresh water. *Acta Crystallographica Section, F* **61**: 844-847.

Shrivastav, A., Mishra, S.K. and Mishra, S. (2008). Polyhydroxyalkanoate (PHA) synthesis by *Spirulina subsalsa* from Gujarat coast of India. *Process Biochemistry* (under review).

Sonawane, S.K., Pawar, R., Patel, A. and Mishra, S. (2002). Effect of biophysico-chemical factors on biosequestering of chromium from the tannery effluent by marine cyanobacteria. *Research Journal of Chemistry and Environment,* **6**: 43-47.

Spolaore, P., Joannis-Cassan, C., Duran, E. and Isambert, A. (2006). Commercial applications of microalgae 101. *Journal of Bioscience and Bioengineering* **2**: 87-96.

Zarrouk, C. (1966). Contributión à l'étude d'une Cyanophycée. Influence de divers facteurs physiques et chimiques sur la croissance et la photosynthèse de *Spirulina maxima* (Setch et Garner) Geitler. Ph.D. Thesis, Univ. Paris, France.

11

IMPACT OF RAW MATERIAL COMPOSITION ON BIODIESEL: A CRITICAL STUDY

Sukumar Puhan[1*], A. Gopinath[2] and G. Nagarajan[2]

Abstract

Biodiesel is an oxygenate fuel made from vegetable oils and animal fats by conversion of triglycerides esters through transesterification. In the present work, the potential availability of non-edible oil in India, their chemical composition, physical and chemical properties, transesterification reaction and biodiesel properties comparison with ASTM standards were given. The scope of the present work also includes raw material selection criteria, impact of raw material on their physical and chemical properties, impact of raw material on transesterification process and biodiesel quality. The economic issues of biodiesel commercialization and employment potential in India, the biodiesel initiatives in India were also presented. In addition, thirteen biodiesel fuels including five blends were prepared and the properties were measured experimentally following the methods specified in the ASTM standards. Using these biodiesel fuels, experiments were conducted on a single cylinder, four stroke, air cooled, direct injection, 4.4 kW compression ignition engine at full load. The effects of fatty acid ester composition on biodiesel properties, the effect of biodiesel properties in relation with fatty acid ester composition on performance, and emission parameters were investigated. The biodiesel properties were significantly influenced by unsaturation percentage in biodiesel. Gradients are proposed between different parameters and percentage of unsaturation. From the present

1 Anna University, Chennai - 6000 25, Tamil Nadu, India
2 Internal Combustion Engineering Division, Department of Mechanical Engineering, Anna University, Chennai - 25, India
* *Corresponding author*: E-mail: sp_anna2006@yahoo.co.in.

investigation it is concluded that the raw material for biodiesel production has a substantial impact on transesterification process and biodiesel fuel properties in relation with biodiesel performance and emission in DI diesel engine.

Key words: Biodiesel, Emission, Fatty acid composition, Performance, Properties, Raw material, Unsaturation

Introduction

Biodiesel

The basic composition of any vegetable oil is triglyceride, which is of three fatty acids and one glycerol molecule. Biodiesel is a renewable diesel fuel that can be obtained when a vegetable oil is chemically reacted with an alcohol to produce mono alkyl esters (biodiesel) [1, 2]. In the process, glycerol is obtained as co-product. Methanol has been the most commonly used alcohol in the commercial production of biodiesel. Biodiesel is also made through simple processes such as dilution, micro emulsification, and pyrolysis [3, 4, 5, 6]. In Europe, biodiesel is widely available in both in neat form (100% biodiesel, also known as B100) and in blends with petroleum diesel. Most European biodiesel is made from rapeseed oil (a cousin of canola oil). In the United States, initial interest in producing and using biodiesel has focused on the use of soybean oil as the primary feedstock, mainly because it is the world's largest producer of soybean oil. In India the strong hold of biodiesel is yet to take place, but a lot of research and projects works are going on to bring the Indian status of biodiesel on par with other parts of established nations.

For Biodiesel, Its Benefits and Drawbacks

Need for biodiesel

- It can reduce independence on foreign petroleum.
- It can leverage limited supplies of fossil fuels.
- It can help reduce greenhouse gas emissions.
- It can help reduce air pollution and related public health risks.
- It can benefit our domestic economy.

Benefits of biodiesel

- Biodiesel runs in any conventional, unmodified diesel engine.
- Biodiesel can be used alone or mixed in any amount with petroleum diesel fuel.

- Biodiesel is biodegradable and non-toxic.
- The higher cetane number of biodiesel compared to petro-diesel indicates potential for higher engine performances
- The superior lubricating properties of biodiesel increases functional engine efficiency.
- Biodiesel has a high flash point of about 149°C compared to petroleum diesel fuel, which has a flash point of 52°C. The higher flash point makes them safer to store and to transport.
- They contain higher amount of oxygen (up to 10%) that ensures more complete combustion of hydrocarbons.
- Reduced exhaust emissions such as sulfur dioxide, hydrocarbons, particulates, carbon monoxide and carbon dioxide by 100, 56, 55, 43, and 78 per cent respectively as compared to conventional diesel fuel.
- The production of biodiesel has a net positive energy balance (as much as 2.5:1) due to high-energy value of ester based food stocks, the low energy requirements of the conversion process.

Drawbacks

- Slight increase in NO_X emissions compared to diesel fuels (up to 10 %)
- Slight decrease in fuel economy on energy basis (about 10% for pure biodiesel)
- Thickness more than diesel fuel in cold weather, may need to use blends in sub freezing conditions.
- More expensive

Raw Material for Biodiesel Production

Selection criteria for raw materials

The following characteristics must be focused to a greater extent when selecting vegetable oil seeds for biodiesel production.

- Non-edible and availability
- High yield per hectare and low gestation period
- Competitive cost
- Safety in transportation
- Better engine performance
- Conformity to the environmental standards
- Higher commercial benefits

- Production of the fuel must be energy positive – that is, the production process should not consume more energy than that produced at the wheels, other than to utilize an existing supply surplus.

Tree borne oil seeds as a source of biodiesel in India

Biodiesel can be produced from a large variety of vegetable oils. India has about 86 different oil tree species. However the review of available

Table 1. Potential tree borne oil seeds in India and their characteristics

S. No.	*Oil seed*	*Characteristics*
1	Jatropha	• Seed yield per shrub low· • Low storing period for seeds· • Potential to grow as weed in uncultivated land· • Difficulty in small scale chemical extraction of oil· • Toxic seeds and cake
2	Jojoba	• Maintenance of male and female tree ratio· • Self polymerize in the presence of sunlight· • Production of seed varies greatly from plant to plant in a stand and from year to year for a particular plant.
3	Karanja	• Long maturing time for pods (nearly 10 months)· • Varying yield (2–90 kg seed per tree)· • Longer time for germination· • Lower oil content (25 percent)
4	Kusum	• Fruit harvest coincides with monsoon· • Hard oil· • Lye content causing problem in effluent treatment
5	Mahua	• Sharp increase FFA (free fatty acids) in the oil of poorly stored kernels· • Utility as virgin fat in soap manufacturing limited· • Presence of toxic and bitter saponins
6	Neem	• Poor yield (< 60 kg/tree/year)· • Stunted growth· • Rapid deterioration of seeds and contamination with aflatoxin· • Harvesting during wet season
7	Simarouba	• Low percentage of high yielder in total population (5–10 %)· • Wide variation in yield· • Limited genetic base
8	Sal	• Very long gestation period· • Kernels extremely susceptible to attack by weevils. • Low oil content (< 15 %)

literature suggests that the commercial and academic interest veers around a limited indigenous and exotic species those exhibit some promising potential. The potential tree borne oil seeds in India and their characteristics are given in Table 1.

Production of Vegetable of Oils in India

Depending upon climate and soil conditions, different nations are looking into different vegetable oils for diesel fuels. For example, soybean oil in the United States, rapeseed and sunflower oils in Europe, palm oil in Southeast Asia (mainly Malaysia and Indonesia), and coconut oil in the Philippines are being considered as substitutes for diesel fuels. India has rich and abundant forest resources with a wide range of plants and oilseeds. The production of these oilseeds can be stepped up many folds if the government takes the decision to sue them for producing diesel fuels. India has rich and abundant forest resources with a wide range of plants and oilseeds. The cost of biodiesel can be reduced if non-edible oils are considered instead of edible oils. Non-edible oils such as neem, mahua, karnaja, sal, jatropha, etc are easily available in many parts of the world including India, and are very cheap compared to edible oils. Moreover the demand for edible oils in a highly populated country like India is high. The potential availability of some non-edible oils in India is given in Table 2.

Table 2. Non-edible oil resources of India

Oil	*Potential (million tonnes)*	*Utilised (million tonnes)*	*% Utilization*
Mahua	400,000	24,500	6
Sal	720,000	23,000	3
Neem	400,000	20,000	6
Karanja	135,000	8,000	6

Composition and Properties of Vegetable Oils

Vegetable oils comprise 90 to 98% tri-glycerides and small amounts of mono- and di-glycerides. Table 3 summarizes the fatty acid composition of some vegetable oils.

The fatty acids, which are commonly found in vegetable oils, are stearic, palmitic, oleic, linoleic and linolenic. The properties of different vegetable oils are listed in Table 4.

Table 3. Composition of vegetable oils (Hideki *et al.*, 2001; Mandpe *et al.*, 2005; Agrawal, 2003)

Vegetable oil	***Fatty acid composition, wt.%***									
	14:0	16:0	18:0	18:1	18:2	18:3	20:0	22:0	22:1	24:0
Corn	0	12	2	25	6	Tr	Tr	0	0	0
Cottonseed	0	28	1	13	58	0	0	0	0	0
Crambe	0	2	1	19	9	7	2	1	59	1
Jatropha	–	14.2	6.9	43.1	34.4	Tr	–	–	–	–
Karanja	–	9.8	6.2	72.2	11.8	–	–	–	–	–
Linseed	0	5	2	20	18	55	0	0	0	0
Palm		40.3	3.1	43.4	13.2	–	–	–	–	–
Peanut	0	11	2	48	32	1	1	2	0	1
Rapeseed	0	3	1	64	22	8	0	0	0	0
Safflower	0	9	2	12	78	0	0	0	0	0
H.O. Safflower	Tr	5	2	79	13	0	Tr	0	0	0
Sesame	0	13	4	53	30	0	0	0	0	0
Soya bean	0	12	3	23	55	6	0	0	0	0
Sunflower	0	6	3	17	74	0	0	0	0	0

Where, Tr = Traces

Performance of Vegetable Oils as Diesel Fuels

It has been found that the use of vegetable oils as diesel fuels in conventional diesel engines leads to a number of problems that are related to the type and grade of oil and local climatic conditions. The injection, atomization and combustion characteristics of vegetable oils in diesel engines are significantly different from those of hydrocarbon-based diesel fuels. The high viscosity of vegetable oils interfere with the injection process and leads to poor fuel atomization. The inefficient mixing of oil with air contributes to incomplete combustion. The high ash point attributes to its lower volatility characteristics. This leads to more deposit formation, carbonization of injector tips, ring sticking and lubricating oil dilution and degradation. Table 5 summarizes the known problems, probable cause and potential solutions for using straight vegetable oils in diesel engines.

The combination of high viscosity and low volatility of vegetable oils causes poor cold engine start up, misfire and ignition delay. Oxidative and thermal polymerization of vegetable oils causes a deposition on the injectors forming a film that will continue to trap fuel and interfere with combustion. In the long-term operation, vegetable oils normally develop gumming, injector coking and ring sticking. Thus two strategies Mandpe *et al.* (2005) to overcome the

Table 4. Properties of vegetable oils (Babu & Devaradjane, 2003; Mandpe *et al.*, 2005; Raheman & Phadatare, 2004; Ghadge *et al.*, 2005)

Vegetable	*Kinematic viscosity (mm^2/s)*	*CN*	*HV (MJ/Kg)*	*CP (°C)*	*PP (°C)*	*FP (°C)*	*Density (kg/l)*	*Carbon residue (wt %)*	*Ash (wt %)*	*Sulfur (wt %)*	*IV (g Iodine/ 100g oil)*
Corn	34.9[a]	37.6	39.5	–1.1	–40.0	277	0.9095	0.24	0.010	0.01	103 – 140
Cottonseed	33.5[a]	41.8	39.5	1.7	–15.0	234	0.9148	0.24	0.010	0.01	90 – 140
Cramble	53.6[a]	44.6	40.5	10.0	–12.2	274	0.9044	0.23	0.050	0.01	93
Linseed	22.2[a]	34.6	39.3	1.7	–15.0	241	0.9236	0.22	< 0.01	0.01	168 – 204
Peanut	39.6[a]	41.8	49.8	12.8	–6.7	271	0.9026	0.24	0.005	0.01	80 – 106
Rapeseed	37.0[a]	37.6	39.7	–3.9	–31.7	246	0.9115	0.30	0.054	0.01	94 – 120
Safflower	31.3[a]	41.3	39.5	18.3	–6.7	260	0.9144	0.25	0.006	0.01	126 – 152
H.O. Safflower	41.2[a]	49.1	39.5	–12.2	–20.6	293	0.9021	0.24	< 0.001	0.02	–
Sesame	35.5[a]	40.2	39.3	–3.9	–9.4	260	0.9133	0.25	< 0.01	0.01	104 – 120
Soybean	32.6[a]	37.9	39.6	–3.9	–12.2	254	0.9138	0.27	< 0.01	0.01	117 – 143
Sunflower	33.9[a]	37.1	39.6	7.2	–15.0	274	0.9161	0.23	< 0.01	0.01	110 – 143
Palm	39.6[a]	42.0	–	31.0	–	267	0.9180	–	–	–	35 – 61
Babassu	30.3[a]	38.0	–	20.0	–	150	0.9460	–	–	–	–
Tallow	–	–	40.0	–	–	201	–	6.21	–	–	–
Jatropha	49.9[b]	45	39.774	–	–	–	–	–	–	–	93
Mahua	24.6[a]	–	36.0	–	15.0	232	0.960	3.70	0.90	–	53
Karanja	27.8[a]	–	34.0	–	–	205	0.912	–	–	–	89.1
Neem	30.0[b]	47	39.399	–	–	180	–	–	–	–	69.2

Where, CN = Cetane Number, HV = Heating Value, CP = Cloud Point, PP = Pour Point, FP = Flash Point and, IV = Iodine Value. Superscript "a, b" indicates viscosity at 40°C and 30°C respectively.

Table 5. Known problems, probable cause and potential solutions for using straight vegetable oil as engine fuels (Harwood, 1984; Ma & Hanna, 1999)

Problem	*Probable cause*	*Potential solution*
Short-term		
1. Cold weather starting	High viscosity, low cetane, and low flash point of vegetable oils	Preheat fuel prior to injection. Chemically alter fuel to an ester
2. Plugging and gumming of filters, lines and injectors	Natural gums in vegetable oil.	Partially refine the oil to remove gums. Filter to 4-microns.
3. Engine knocking	Very low cetane of some oils. Improper injection timing.	Adjust injection timing. Use higher compression engines. Preheat fuel prior to injection. Chemically alter fuel to an ester
Long-term		
4. Coking of injectors on piston and head of engine	High viscosity of vegetable oil, incomplete combustion of fuel. Poor combustion at part load with vegetable oils	Heat fuel prior to injection. Switch engine to diesel fuel when operation at part loads. Chemically alter the vegetable oil to an ester
5. Carbon deposits on piston and head of engine	High viscosity of vegetable oil, incomplete combustion of fuel. Poor combustion at part load with vegetable oils	Heat fuel prior to injection. Switchengine to diesel fuel when operation at part loads. Chemically alter the vegetable oil to an ester
6. Excessive engine wear	High viscosity of vegetable oil, incomplete combustion of fuel. Poor combustion at part loads with vegetable oils. Possibly free fatty acids in enginevegetable oil. Dilution of lubricating oil due to blow-by of vegetable oil.	Heat fuel prior to injection. Switch engine to diesel fuel when operation at part loads. Chemically alter the vegetable oil to an ester. Increase motor oil changes. Motor oil additives to inhibit oxidation.
7. Failure of engine lubricating oil due to polymerization	Collection of polyunsaturated vegetable oil blow-by in crankcase to the point where polymerization occurs	Heat fuel prior to injection. Switch engine to diesel fuel when operation at part loads. Chemically alter the vegetable oil to an ester.

problem of using direct vegetable oil in diesel engines that can be inferred from the table above are that:

- Modification of the engine to suit the fuel
- Modification of the fuel to suit the existing engine

The first strategy only appears attractive when using neat vegetable oil in stationary engines, as the price of the engine is too costly due to low production numbers. Therefore, even in developing countries, the second strategy seems to be more practicable. Considerable efforts have been made to develop vegetable oil derivatives that approximate the properties and performance of the hydrocarbon-based diesel fuels. The problems with substituting triglycerides for diesel fuels are mostly associated with their high viscosity, low volatility and polyunsaturated character. The second strategy involving the viscosity and other problems can be approached by the following routes:

- Micro emulsification with short chain alcohols (*e.g.* ethanol or methanol)
- Pyrolysis/Cracking
- Transesterification

Methods of Producing Bio-Diesel

Micro-emulsions

Micro-emulsions with solvents such as methanol and ethanol can also be used to resolve the problem of the high viscosity of vgetable oil. A micro-emulsion is defined as a collodial equlibrium dispersion of optically isotropic fluid microstructures with dimension generally in the 1-150 nm range, formed spontaneously from two normally immiscible liquids (Agarwal, 2003).

Pyrolysis (thermal cracking)

Pyrolysis is the conversion of one substance into another by means of heat or by heat in presence of catalyst. The paralzed material can be vegetable oils, animal fats, natural fatty acids, or methyl fatty esters fatty acids (Agarwal, 2003). Thermal decomposition of triglycerides produces alkanes, alkenes, alkadines, aromatics and carboxylic acids (Pramanik, 2003).

Transesterification process

The main objective of the transesterification process is to reduce the viscosity of vegetable oil. Transesterification is a process of producing a reaction in triglyceride and alcohol in the presence of a catalyst to produce alkyl ester and glycerol. Transesterification process involves the transformation of the large, branched, triglycerides molecules of vegetable oils into smaller, straight chain molecules, similar in size to components of diesel fuel. Fatty esters form vegetable oils and animal fats have shown promise as alternative diesel fuels as a result of improved viscosity and volatility relative to the triglycerides. A simple representation of transesterification process is given in Fig. 1.

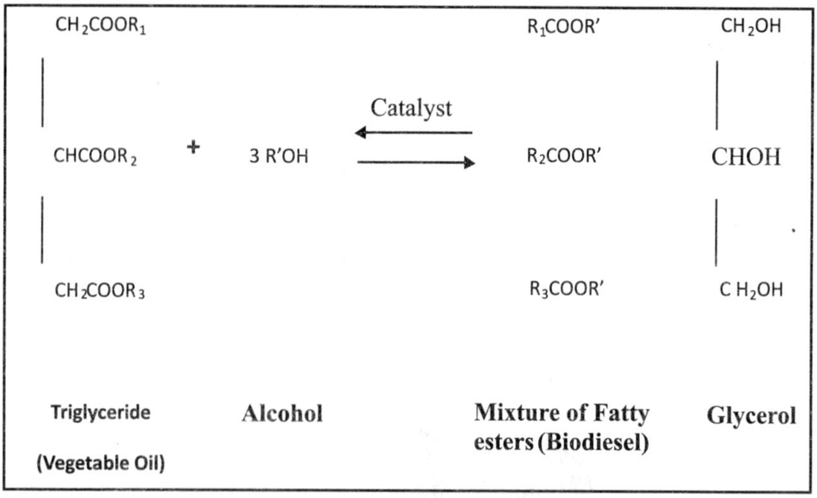

Fig. 1. Simple representation of transesterification process

A catalyst is usually used to improve the reaction rate and yield. Because the reaction is reversible, excess alcohol is used to shift the equilibrium to the products side. Among the alcohols that can be used in the transesterification process are methanol, ethanol, propanol, and butanol. Methanol and ethanol are used more frequently, especially methanol because of its low cost and its physical and chemical advantages (polar and shortest chain alcohol). It can easily react with triglycerides and sodium hydroxide is easily dissolved in it. Alkali (Sodium hydroxide, pottasium hydroxide), acids (Sulphuric acid and Hydrochloric acid) catalyze reaction (Kumar *et al.*, 2003; Hass, 2005). Alkali catalyzed transesterification is faster than acid catalyzed transesterification and is most used commercially (Puhan *et al.*, 2005). The reaction to produce biodiesel consumes the following reactants with approximate amounts shown:

- Triglyceride (vegetable oils) (*e.g.* 100 kg of vegetable oil)
- Primary alcohol (*e.g.* 10 kg methanol)
- Catalyst (*e.g.* 0.3 kg sodium hydroxide)
- Neutralizer (*e.g.* 0.25 kg sulfuric acid)

After transesterification of triglycerides, the products are a mixture of esters, glycerol, alcohol, catalyst and tri-, di-, and monoglycerides. Glycerol is separated by gravitational settling or centrifuging. Transesterification consists of a number of consecutive, reversible reactions. The triglycerides are converted stepwise to diglyceride, monoglycerides and finally glycerol. A mole of ester is liberated at each step. The reactions are reversible, although the equilibrium lies towards the production of fatty acid esters and glycerol.

Biodiesel Initiatives in India

India has great potential for production of bio-fuels like bio-ethanol and biodiesel from non-edible oil seeds. Wild crops cultivated in the wasteland also form a source of biodiesel production in India and according to the Economic Survey of Government of India, out of the cultivated land area; about 175 million hectares are classified as waste and degraded land. Thus, given a demand-based market, India can easily tap its potential and produce biodiesel in al large scale. The government has set up a group (August 26, 2003) to monitor the programme to promote biodiesel. Meanwhile Indian Oil Corporation (IOC) has already identified 200 acres of land for the plantation of jatropha plant. In IOC's pilot project, experimentation on both edible and non-edible plants is under progress According to a study conducted by government and non-government agencies, the cost of one liter of biodiesel has been calculated between Rupees 17 and Rupees 21 at a plant with a capacity of 1 tonne per day.

India will introduce biodiesel, a mix of gas oil and 20 percent plant oil, by 2012 and India has already launched a programme to blend ethanol with petrol to reduce hydrocarbon consumption in the energy-hungry country. Indian Railways are experimenting with the new eco-friendly biodiesel fuel to run passenger trains. The first successful trial run of a passenger train was conducted on December 31, 2002 when the Delhi-Amritsar Shatabdi Express used five percent of biodiesel as fuel. In 2007, The Energy and Resources Institute (TERI), an Indian research group, launched a 10-year, $9.4 million project to research issues involved in taking jatropha from seed to filling station. One challenge is growing the plant in poor soil.

Reliance Industries Ltd. to enter Bio-Fuel: Reliance Industries Ltd (RIL) is planning to enter the bio-fuel segment in a big way. Godrej Agrovet plans Indian Rupees 5 billion for bio-fuel projects: Godrej Agrovet Ltd is planning to invest over Indian Rupees 5 billion, for jatropha and palm oil cultivation in the states of Gujarat and Mizoram. Emami group is planning to set up a bio-diesel plant at Haldia, at cost of Rs 1.5 billion. Several smaller Indian companies are already working towards developing bio-diesel. Companies like Nandan Bioagro and Labland Biotech have tied up with British oil company D1 Oils to produce jatropha and trade in it. The company will encourage hundreds of farmers to cultivate the crop under an arrangement with the company.

Economic Issues of Biodiesel Commercialization and Employment Potential

The major hurdle for commercial Bio-diesel production as of now is its higher cost of production compared to mineral based petroleum fuels. The actual cost of Bio-diesel production depends on the cost of procurement of seeds, scale of manufacturing, government policies related to taxation, utilization of by-products, utilization of seed cake etc. However, while considering the cost factor, it is essential to look at the rural employment generation, energy security, carbon trading issues, and savings of foreign exchange. The estimated cost of biodiesel per kg is shown in Table 6.

Table 6. Estimated cost of biodiesel

	Rate (Indian rupees/kg)	*Quantity (kg)*	*Cost (Indian rupees)*
Seed (after collection)	8.00	3.28	26.24
Cost of decortications and oil extraction	3.00	1.05 kg oil from 3.28 kg seed	3.15
Less cake produced	3.00	2.23 kg cake from 3.28 kg seed	–6.69
Transesterification cost	8.00	1.00	8.00
Cost of glycerol produced	35.00	0.08	–2.80
Cost of Biodiesel			27.90

The glycerin and seed cake will be additional valuable by-products, which will cross-subsidize Bio-diesel production. Seed cake can be used as manure, as feedstock for biogas production, animal feed etc., whereas glycerin can be utilized in pharmaceutical as well as cosmetic industries. This can also be used in defense, soap manufacture, and in the chemical industry. However, government will have to provide tax benefits as well as subsidies, if Bio-diesel has to be a commercial success in the short run in this country.

Likely demand of petro diesel by 2006-07 will be 52 MMT and by 2011-12 it will increase to 67 MMT. 5% blend of Bio-diesel with petro diesel will require 2.6 MMT of Bio-diesel in 2006-07. By 2011-12, for 20% blend with Petro-diesel, the likely demand will be 13.4 MMT. To meet the requirement of 2.6 MMT of bio-diesel, plantation of Jatropha should be done on 2.2-2.6 million ha area. 11.2-13.4 million ha of land should be covered by 2011-12 for 20% bio-diesel blending. The employment potential in different areas is listed in Table 7.

Table 7. Employment potential (as per Planning Commission report on bio-fuels, 2003)

Year	*No. of jobs in plantation*	*In maintenance*	*Operation of biodiesel units*
2006-07	2.5 million	0.75 million	0.10 million
2011-12	13.0 million	3.9 million	0.30 million

Biodiesel Chemistry

All vegetable oils consist primarily of triglyceride molecules such as three fatty acids and one glycerol molecule. Biodiesel can be obtained from the process called transesterification in which a vegetable oil is chemically reacted with an alcohol to produce mono alkyl esters. In the process, glycerol is obtained as a co-product. A simple molecular representation of a triglyceride is shown in Fig. 2.

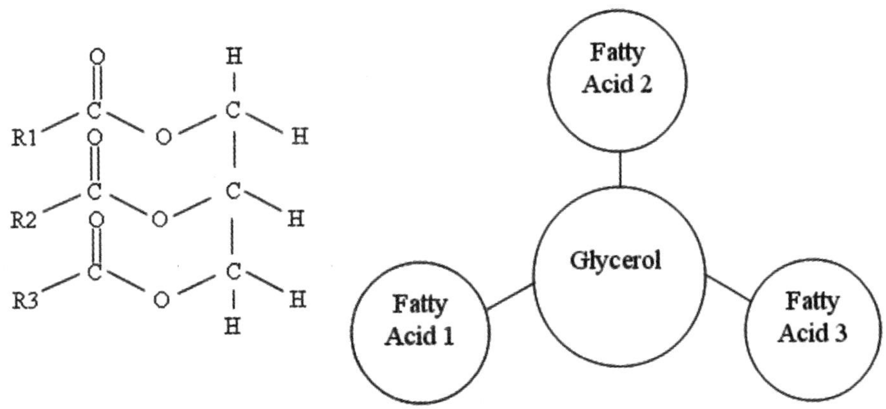

Fig. 2. Structure of a triglyceride

R1, R2, and R3 are the long hydrocarbon chains, some times called fatty acid chains. The fatty acids vary in their carbon chain length and in number of double bond of carbons (Babu and Devaradjane, 2003). Fatty acids may be saturated or unsaturated. A saturated fat is one that cannot chemically add hydrogen and contains only single bonds. An unsaturated fat can be hydrogenated; for each degree of unsaturation, that is each double bond, one hydrogen molecule can be added per fatty acid Michael and Robert (1998). The chemical composition of fatty acids is dependent upon the length and degree of unsaturation of the fatty acid alkyl chains. The structural formula for fatty acids is listed in Table 8.

Table 8. Structural formula for fatty acids

Acid Chain	*C : N*	*Type*	*Structure*
Caprylic	C8 : 0	S	$CH_3\ (CH_2)_6\ COOH$
Capric	C10 : 0	S	$CH_3\ (CH_2)_8\ COOH$
Lauric	C12 : 0	S	$CH_3(CH_2)_{10}\ COOH$
Myristic	C14 : 0	S	$CH_3(CH_2)_{12}\ COOH$
Palmitic	C16 : 0	S	$CH_3(CH_2)_{14}\ COOH$
Palmitoleic	C16 : 1	US	$CH_3(CH_2)_5\ CH{=}CH(CH_2)_7\ COOH$
Stearic	C18 : 0	S	$CH_3(CH_2)_{16}\ COOH$
Oleic	C18 : 1	US	$CH_3(CH_2)_7\ CH{=}CH(CH_2)_7COOH$
Linoleic	C18 : 2	US	$CH_3(CH_2)_4CH{=}CHCH_2CH{=}$ $CH(CH_2)_7COOH$
Linolenic	C18 : 3	US	$CH_3CH_2CH{=}CHCH_2CH{=}CHCH_2CH{=}$ $CH\ (CH_2)_7COOH$
Arachidic	C20 : 0	S	$CH_3(CH_2)_{18}COOH$
Eicosenoic	C20 : 1	US	$CH_3(CH_2)_7CH{=}CH(CH_2)_9COOH$
Behenic	C22 : 0	S	$CH_3(CH_2)_{20}COOH$
Erucic	C22 : 1	US	$CH_3(CH_2)_7CH{=}CH(CH_2)_{11}COOH$

Where, C:N, indicates C the number of carbons and N the number of double bond of carbons S = Saturated fatty acids, US = Unsaturated fatty acids

Biodiesel Quality and ASTM Specification

The parameters those are selected and established to define the quality of biodiesel can be divided into two groups. One group contains general parameters which are also used for mineral-oil-based fuels, and the other group especially describes the chemical composition and purity of fatty acid alkyl esters.

Some of the general parameters, like density, cetane number sulfur content mainly depend on the choice of vegetable oil and cannot be influenced by different production methods or purification steps. The flash point strictly corresponds to methanol content and the viscosity correlates with content of unreacted triglycerides.

The Conradson carbon residue is one of the most important indicators of biodiesel quality. It can be shown in different investigations that this parameter is an excellent indicator for the quality of biodiesel. The Conradson carbon residue corresponds strictly to the content of glycerides as well as free fatty acids, soaps, remaining catalyst and other impurities.

The content of inorganic matte, like impurities of remaining catalyst, is limited by the values of the ash content. A higher content of phosphorous in the fuel can lead to higher particulate emissions. The phosphorous content depends mainly on the grade of refined oil. Fully refined oils have phosphorus content only a several ppm phosphorus, whereas unrefined oil can have phosphorus content over 100 ppm.

Higher water content in biodiesel might reduce the storage ability or lead to the separation of water when blending with mineral diesel. The iodine value, which describes the content of unsaturated fatty acids, is dependent only on the origin of the vegetable oils. A limitation of unsaturated fatty acids is necessary, due to the fact that heating higher unsaturated fatty acids results in polymerization of glycerides. This can lead to formation deposits or to deterioration of the lubricating oil. This effect increases with the number of double bonds (unsaturation) in the fatty acid chain.

The ASTM system is the basis for defining product specifications and measurement methods for most segments of the fuels and industrial products market in the U.S. ASTM D 6751 – 02 sets forth the specifications that must be met for a fatty acid ester product to carry the designation "biodiesel fuel" or "B100" (Table 9).

Table 9. ASTM D 6751 – 02 Requirements

Property	***Method***	***Limits***	***Units***
Acid number	D 664	0.80 max	mg KOH/g
Carbon residue	D 4530	0.050 max	wt. %
Cetane number	D 613	47 min	–
Copper strip corrosion	D 130	No. 3 max	–
Flash point, closed cup	D 93	130 min	°C
Free glycerin	D 6584	0.02	wt. %
Kinematic viscosity @ 40 °C	D 445	1.9–6.0	mm^2/s
Phosphorus	D 4951	0.001	wt. %
Sulfated ash	D 874	0.020 max	wt. %
Total glycerin	D 6584	0.24	wt. %
Total sulfur	D 5453	0.05 max	wt. %
Vacuum distillation end point	D 1160	360°C max, at T-90	% distilled
Water and sediment	D 2709	0.050 max	% volume

Effects of Structure of Fatty Acids on Biodiesel Properties and its Emission Parameters: An Overview

The densities of pure FAMEs at 80°C are reported in (Ejim *et al.*, 2005) From the reported values, it was observed that the density of FAMEs generally increases with the degree of unsaturation. In other words, the density of FAMEs increases with increase in number double of bonds. The data points and the trend curve of density of methyl esters are shown in Fig. 3. The data points were taken from (Ejim *et al.*, 2007).

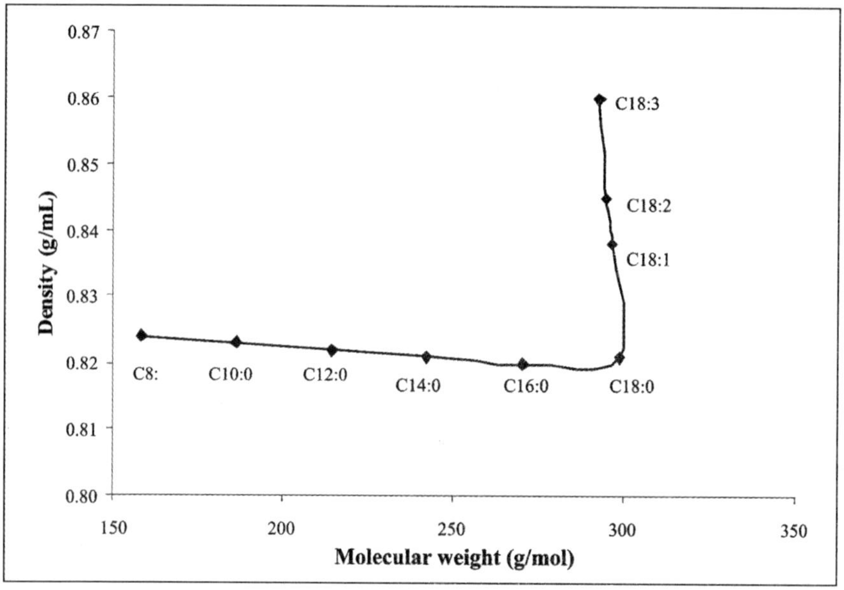

Fig. 3. Density trend lines for methyl esters at 80°C

The viscosity of a transesterified vegetable oil (biodiesel), is about an order of magnitude lower than that of the parent oil. Viscosity of fatty ester mixtures/biodiesel increases with chain length (number of carbon atoms) and with increasing degree of saturation. The configuration of double bond influence viscosity (cis double bond configuration offers lower viscosity than trans) while double bond position affects viscosity less Gerhad (2005). Branching in the ester moiety, however, has little or no influence on viscosity Gerhad (2005).

Dynamic viscosities of pure FAMEs are reported in Allen *et al.* (1999). It was stated in the literature that the dynamic viscosity at 40°C increased with carbon number in a curvelinear (second order polynomial) trend for saturated esters. It was also stated that a complete different curve was observed as the degree of unsaturation progressed from 18:1 to 18:3. As the number of double bonds increased, a non-linear (second order polynomial) decrease in viscosity (at 40°C) was observed Allen *et al.* (1999). The data points and trend curve of viscosity of methyl esters are shown in Fig. 4.

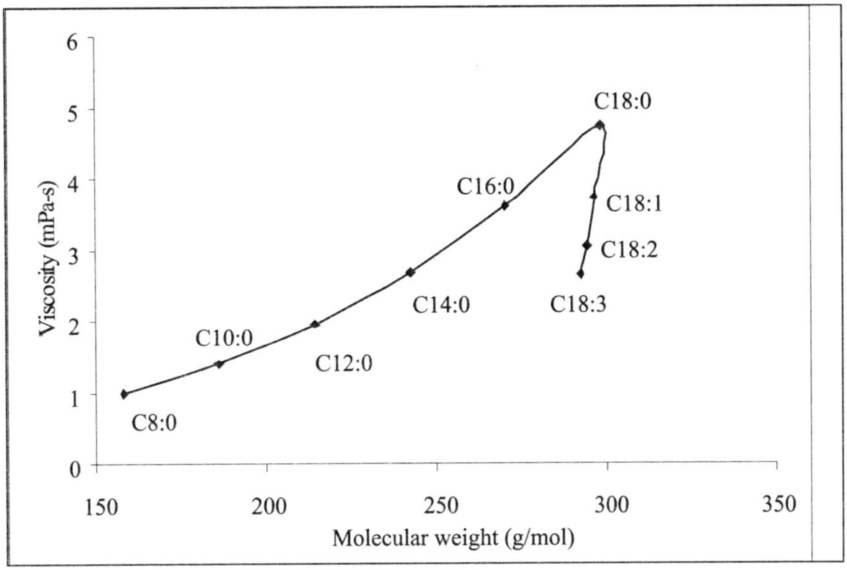

Fig. 4. Viscosity trend lines for methyl esters at 40°C (Allen *et al.*, 1999)

The cetane number of the fuel, specified by ASTM D-613, is a measure of ignition quality. Higher cetane number indicates shorter ignition delay. Cetane number is the percentage of cetane (hexadecane with cetane number of 100) which must be mixed with heptamethylnonane (cetane number = 15) to give the same ignition performance under standard conditions Heywood (1998). The cetane number of biodiesel depends on the parent oil source. However, both moieties, the fatty acid chain and alcohol functionality, contribute to the ignition quality of biodiesel fuels Gerhad (2005). Cetane number increases with chain length, decreases with number of double bonds and decreases as double bonds and carbonyl group move toward the centre of the chain [21]. Cetane numbers of fatty esters generally increases with [26]:

i) the number of methylene groups (CH_2) in the chain of fatty compound
ii) number of CH_2 groups in the ester moiety
iii) with increasing saturation of the fatty compound.

In addition it was quoted that the cetane number also depends on the position of the double bond in monounsaturated fatty compounds and increases if the double bond move towards one end of the molecule. For methyl esters, the cetane numbers were found to increase in a nonlinear relationship with the chain length (molecular weight) of the fatty acid Klopfenstein (1985). It was also cited that an increase in the molecular weight of the fatty acid portion of the ester produce greater increases in cetane number than the same change in molecular weight in the alcohol portion of the ester. The data points and trend curve of cetane number of methyl esters are shown in Fig. 5. The data points were taken from Klopfenstein (1985) and [28].

Heating value of a fuel is the magnitude of the heat of reaction at constant pressure or at constant volume at a standard temperature (usually at 25°C) for the complete combustion of unit mass of fuel Heywood (1998). Heating value of vegetable oils increases with increasing chain length and decreases with increase in number of double bonds. The increase in heat content may be believed to be due to the increase in number of carbon and hydrogen atoms, as well as increase in the ratio of these elements relative to oxygen Ayhan (1998). The possible cause for the decrease in heat content may be the deficiency of hydrogen atoms (greater unsaturation).

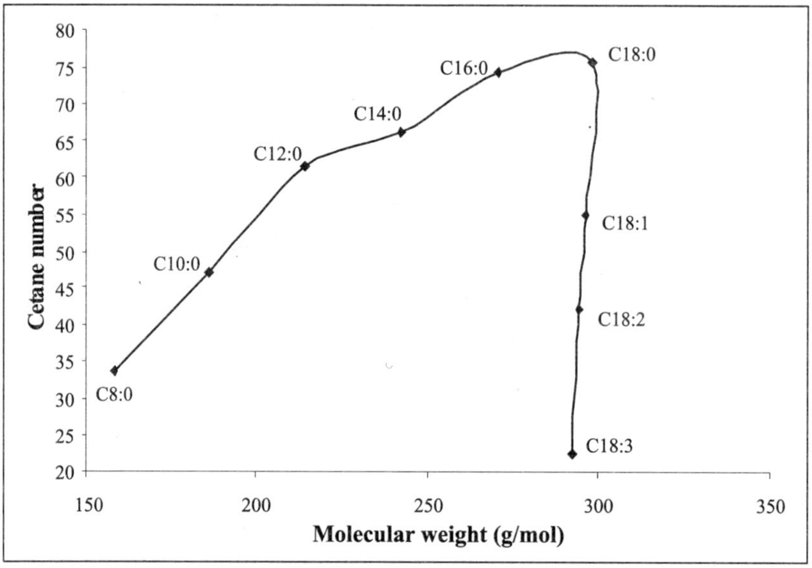

Fig. 5. Cetane number trend lines for methyl esters

Iodine value of an oil can be considered as a direct measure of unssaturation and may be defined as the amount of iodine (in grams) necessary to saturate hundred grams of oil sample. Iodine value of vegetable oil increases with increase in unsaturation (number of carbon double bonds) Ayhan (1998). While saponification value of oils can be defined as milligrams of pottosium hydroxide (KOH) necessary to saponify one gram of oil sample. The saponification value of an oil decreases with increase of its molecular weight Ayhan (1998). In other words, higher saponification value could be expected for lower chain fatty acids (lower molecular weight).

The increase in unsaturation (and hence increase in iodine number and density) of biodiesel fuels can be proposed as factors for increase in NO_X emissions. The cetane number decreases with increase in iodine number, thus, excessive ignition delay and poor combustion performance may also be proposed as a cause of higher NO_X. The differences in the speed of sound and isentropic bulk modulus of biodiesel relative to petroleum diesel can advance the effective injection timing and thereby cause NO_X to increase Robert *et al.* (2001). The fuel injection timing is mainly influenced by the bulk modulus and viscosity of fuel (the higher the bulk modulus and viscosity is, the faster the injection timing is). The bulk modulus of saturated fatty acid methyl esters is lower than that of unsaturated fatty acid methyl esters. PME, and RME reported as 90%, 50%, and 6% respectively. The saturated fatty acid methyl esters have lower NO_X emissions compared with the unsaturatedfatty acid methyl esters. The shorter ignition delay may have an effect on the reduction of HC emissions Eiji *et al.* (2006). The overall biodiesel properties are strongly influenced by the properties of individual fatty acid esters in biodiesel. It therefore looks sensible to supplement particular fatty ester(s) with desirable properties in the fuel to upgrade the properties of the whole fuel. It may be attainable in the future to improve the properties of biodiesel by means of genetic engineering of the parent vegetable oils, which could ultimately lead to a fuel enriched with certain fatty acids, those exhibit a combination of improved fuel properties Gerhad (2005).

The Present Work

In the present work, thirteen biodiesel fuels including five blends were prepared and the properties were measured experimentally following the methods specified in the ASTM standards. Using these biodiesel fuels, experiments were conducted on a single cylinder, four stroke, air cooled, direct injection, 4.4 kW compression ignition engine at full load. The effects of fatty acid ester composition on biodiesel properties, the effect of biodiesel properties in relation with fatty acid ester composition on performance, and emission parameters were investigated. Gradients are proposed between different parameters and percentage of unsaturation.

Experiments

Biodiesel preparation

The biodiesel fuels were prepared in the authors' laboratory. According to the quality of the parent oil, the selection of transesterification process was made as shown in Fig. 6. If the free fatty acid (FFA) content and moisture content are lesser than 0.5%, good quality of biodiesel can be produced. The objective of the transesterification process is to reduce the viscosity of vegetable oil. In the present work, biodiesel fuels were produced by single stage transesterification process. The step-by-step procedures to produce biodiesel fuels are mentioned below.

1. The parent vegetable oil was poured in a three necked flask.
2. The oil was heated using heating mantle.
3. In Parallel, methoxide (methanol + sodium hydroxide (NaOH) or methanol + potassium hydroxide (KOH), in the present work, NaOH used as catalyst) was prepared. The mixing ratio was taken as 1 kg oil: 200 g of methanol: 5 g of NaOH (generally the alcohol and the catalyst quantity are 20% and 0.3 to 0.5% of oil respectively). The mixture of methanol and NaOH stirred manually till all the NaOH diffused in the methanol.
4. The prepared methoxide was then poured in the three necked flask that contains the heated oil when it reached to 50°C. The whole mixure in the flask was continuosly stirred by a magnetic stirrer. The speed of the stirred was maintained at about 250 m/s (speed can be varied according to the mixture quantity).
5. One neck of the flask closed using thermowell and a thermometer inserted along with the thermowell. The second neck closed with a stopper to avoid the contact between system and atmosphere. The last neck closed with a condenser. The cooling water temperature maintained at dew point temperature of alcohol (about 20°C).
6. Every 10 min, the temperature of the reactant was measured with the help of thermometer. The regulator in the heating mantle

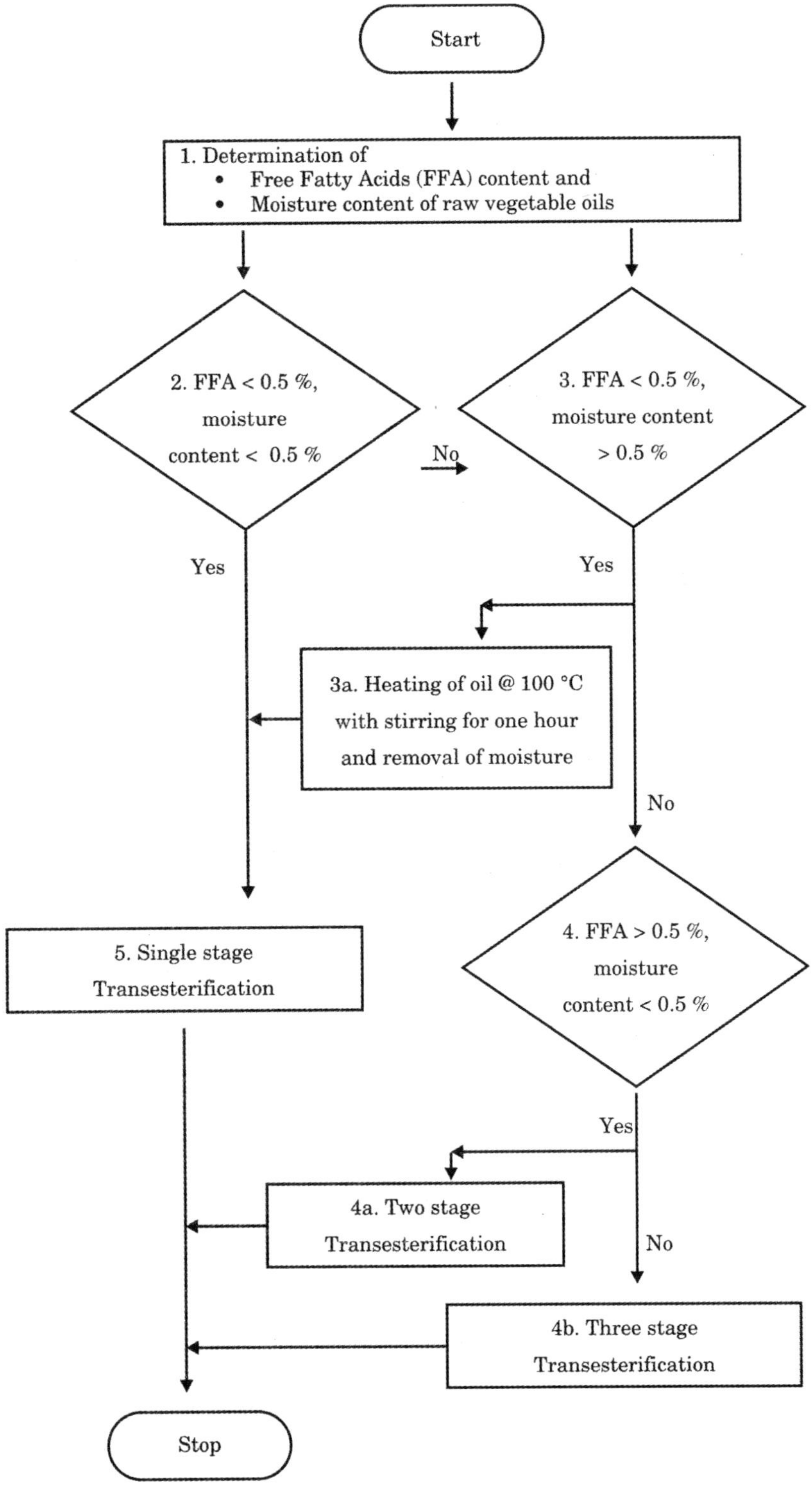
Start
1. Determination of
• Free Fatty Acids (FFA) content and
• Moisture content of raw vegetable oils
2. FFA < 0.5 %, moisture content < 0.5 %
No
3. FFA < 0.5 %, moisture content > 0.5 %
Yes
Yes
3a. Heating of oil @ 100 °C with stirring for one hour and removal of moisture
No
4. FFA > 0.5 %, moisture content < 0.5 %
5. Single stage Transesterification
Yes
4a. Two stage Transesterification
No
4b. Three stage Transesterification
Stop

was adjusted to maintain the reactant temperature at 60°C (the boiling point of methanol is 60°C).

7. The reaction process was allowed for about two hrs. After two hrs, the heating mantle was switched off to stop the reaction process.
8. All the three necks of the flask opened and the mixture was transferred from the flask to a two liter separating funnel.
9. The mixture was allowed in the funnel for about eight hrs. The mixture was separated into two layers as mono alkyl ester (biodiesel) and glycerol.
10. Due to the higher density of glycerol, it was settled at the bottom.
11. The glycerol alone was removed out through the opening of the funnel.
12. To remove the alcohol and catalyst present in the ester, the ester was washed with distilled water.
13. 120 g (generally 10% of ester by mass) of distilled warm water at 40°C (generally at 35–40°C) poured in the ester.
14. The ester and the water were allowed in the funnel to eight hours. The water (milky white colour) was settled at the bottom due to higher density.
15. The water was removed out through the opening of the funnel.
16. The ester (biodiesel) was heated again and maintained at 100°C about an hour to remove the moisture content. And thus the required biodiesel was made ready.

Following the procedures mentioned above, all the other biodiesel fuels were produced. By mixing 50% by volume of two different biodiesel fuels, the relevant blends were prepared.

Biodiesel properties

The fuel properties were determined following the methods specified in ASTM standards as given in Table 10 for different biodiesel fuels. The fatty acid composition, free fatty acid content and moisture content were measured in Tamil Nadu Oil and Seeds Association, Chennai. The density, kinematic viscosity, cetane number, heating value, iodine value and saponification value were measured in ITALAB Private Limited, Industrial Testing and Analytical Laboratories, Chennai.

Performance and pollutants emission parameters

A single cylinder air-cooled direct injection CI engine developing a power output of 4.4 kW was run at the rated speed of 1500 rpm at its full load.

Table 10. ASTM methods for determination of fuel properties

Property	*Unit*	*ASTM standard*
Fatty acid composition	wt %	D 6584
Density @15°C	g/cc	D 1298
Kinematic viscosity @ 40°C	mm^2 /s	D 445
Cetane number	–	D 613
Heating value	MJ/kg	D 240
Iodine value	g Iodine/100 g oil	D 1510
Saponification value	mg KOH/g oil	D 5558

Table 11. Test engine specifications

Parameters	*Specification*
Make	Kirloskar
Model	TAF – 1
No. of Cylinder	Single Cylinder
Stroke	4 Stroke
Type of Cooling	Air Cooled
Ignition	Compression Ignition
Fueling	High Speed Diesel Oil
Bore	87.5 mm
Stroke	110 mm
Compression Ratio	17.5:1
Injection Pressure	200 bar
Injection Timing	23 deg. bTDC
Speed	1500 rpm
Brake Power	4.4 kW

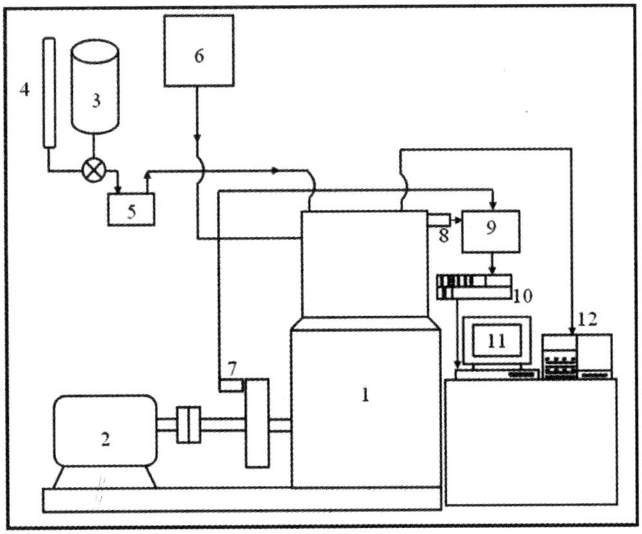

Fig. 7. Experimental setup

Engine details are given in Table 11. The experimental setup is depicted in Fig. 7.

1. Test engine
2. Dynamometer
3. Biodiesel tank
4. Burette
5. Fuel filter
6. Air surge tank
7. TDC pickup
8. Pressure transducer
9. Charge amplifier
10. Analog to digital converter card
11. Computer
12. Exhaust gas analyser

The engine was coupled to an electrical dynamometer and started with neat diesel fuel and warmed up. The engine was allowed to run for 10 min with biodiesel to attain steady condition at its rated speed of 1500 rpm. The engine was gradually loaded to full-load by switching on the load mains. The different biodiesel fuels and their blends were tested in a random order. The speed of the engine was maintained at 1500 rpm and the time taken for 10 cc of fuel consumption was measured using a digital stop watch. Chromel alumel thermocouple in conjunction with a digital temperature indicator was used for the exhaust gas temperature measurement. Exhaust gases (NO_X, CO, HC) from the engine were measured with the help of a QROTECH, QEO-402 gas analyzer. Smoke intensity was measured with the help of a Bosch smoke meter. The tests were repeated for five times and the average value of five readings was taken to eliminate the uncertainty.

The maximum uncertainty limits associated with different parameters are listed in Table 12.

Table 12. Maximum uncertainty limits associated with different parameters

S. No.	*Parameter*	*Unit*	*uncertainty*
1	Brake Power (kW)	kW	± 4.5 %
2	Fuel Volume Flow Rate	s	± 1 %
3	Total Fuel Consumption	kg/h	± 1 %
4	Brake Thermal Efficiency	%	± 1.4 %
5	Exhaust Gas temperature)	°C	± 2
6	Oxides of Nitrogen (ppm)	ppm	± 4 ppm
7	Hydrocarbons (ppm)	ppm	± 1 ppm
8	Smoke (BSU)	BSU	± 4 %

Results and Discussion

The results of fuel tests on different biodiesel fuels are summarized in Table 13 and Table 14.

A correlation analysis was made to find out the degree of linear association between different biodiesel properties and percentage of

Table 13. Properties of different biodiesel fuels.

Biodiesel	*Density (kg/m^3) @ 40°C*	*Kinematic viscosity (mm^2/s) @ 40°C*	*Cetane number*	*Heating value (MJ/kg)*	*Iodine value (g Iodine/ 100 g oil)*	*Saponification value (mg KOH / g oil)*	*Moisture content (wt %)*	*FFA content (wt %)*
Sunflower	841	4.87	47.8	39.5	136.0	193.0	0.03	0.01
Rubber seed	839	4.88	51.0	39.3	120.0	196.0	0.03	0.11
Jatropha	836	4.91	54.0	39.7	105.0	198.8	0.02	0.03
Cottonseed	837	4.95	52.1	39.4	113.2	202.7	0.02	0.02
Karanjia	837	5.00	52.0	39.5	92.0	198.0	0.02	1.09
JT 80:20	834	5.04	59.2	39.6	88.0	197.5	0.02	1.02
JP 50:50	834	5.10	59.0	39.6	84.0	198.0	0.03	0.44
Neem	832	5.16	58.7	39.8	83.2	201.0	0.02	0.03
JT 50:50	835	5.21	62.2	39.9	83.0	214.0	0.03	1.14
SFCt 50:50	834	5.27	54.6	39.9	81.5	210.0	0.03	1.04
Mahua	830	5.33	61.4	40.5	80.0	187.0	0.03	0.02
Palm	830	5.39	64.0	41.0	59.0	205.0	0.02	0.46
JCt 50:50	829	5.45	58.0	40.2	69.2	215.0	0.02	1.06

Where,
JCt 50:50 = Blend of 50 % *Jatropha* and 50 % Coconut biodiesel (by volume)
JP 50:50 = Blend of 50 % *Jatropha* and 50 % Palm biodiesels (by volume)
JT 50:50 = Blend of 50 % *Jatropha* and 50 % Tallow biodiesels (by volume)
JT 80:20 = Blend of 80 % *Jatropha* and 20 % Tallow biodiesels (by volume)
SFCt 50:50 = Blend of 50 % Sunflower and 50 % Coconut biodiesels (by volume)
FFA = Free Fatty Acid

Table 14. Composition of different biodiesel fuels

Biodiesel	*Lauric C12:0*	*Myristic C14:0*	*Palmitic C16:0*	*Stearic C18:0*	*Oleic C18:1*	*Linoleic C18:2*	*Linolenic C18:3*	*Others*	*% of US*
Sunflower	0.00	0.10	6.00	5.90	16.00	71.40	0.60	0.00	88.00
Rubber seed	0.00	0.24	12.46	8.32	27.78	37.65	13.44	0.11	78.87
Jatropha	0.00	0.10	14.90	9.50	40.50	34.70	0.30	0.00	75.50
Cottonseed	0.00	0.80	22.90	3.10	18.50	54.20	0.50	0.00	73.20
Karanjia	0.01	0.05	9.94	7.83	53.19	19.09	0.04	9.85	72.32
JT 80:20	0.12	0.64	16.58	11.48	40.88	28.34	0.42	1.54	68.18
JP 50:50	0.13	0.48	28.06	7.70	42.66	20.06	0.17	0.74	62.89
Neem	0.83	0.47	18.20	20.10	43.70	16.40	0.30	0.00	60.40
JT 50:50	1.22	1.72	19.94	15.44	41.64	15.62	0.66	3.76	57.92
SFCt 50:50	20.28	10.47	9.34	4.31	19.38	32.63	0.04	3.55	52.05
Mahua	0.00	0.00	24.20	25.80	37.20	12.80	0.00	0.00	50.00
Palm	0.21	1.30	43.90	4.90	39.00	9.50	0.30	0.89	48.80
JCt 50:50	20.88	10.40	13.71	7.16	26.11	18.21	0.12	3.41	44.44

Where, % of US = % of unsaturated fatty acids

unsaturation. The Pearson product moment correlation coefficient between different properties and percentage of unsaturation is shown in Table 15. The formula used to find out the Pearson correlation coefficient (r) is shown in equation (1).

$$r \quad \frac{(X \quad \overline{X})(Y \quad \overline{Y})}{\sqrt{X \quad \overline{X}^{2} \; Y \quad \overline{Y}^{2}}} \qquad (1)$$

Where, r = Pearson correlation coefficient X = % of unsaturation and Y = relevant properties.

Effects of fatty acid esters profile on biodiesel properties

The influences of fatty acid esters profile on biodiesel properties were investigated. From Table 13 and Table 14, the following points are observed.

1. Biodiesel of sunflower oil has higher density and is more unsaturated than the other biodiesel fuels. Also, the kinematic viscosity of sunflower oil methyl ester is lower than that of other biodiesel fuels.
2. JCt 50:50 blend has a lower density, a higher kinematic viscosity and is more saturated than the other biodiesel fuels.

The relationship between fatty acid methyl ester composition and the density of biodiesel fuels was investigated. The investigation reveals that the higher density of sunflower oil methyl ester may be due to that the contribution of unsaturated fatty acids in sunflower oil methyl ester is higher than that of other biodiesels. Similarly, the lower density of JCt 50:50 is believed due to the lesser contribution of unsaturated fatty acids in JCt 50:50. Table 15 provides a clear picture that a highly positive correlation exists between density and percentage of unsaturation. Therefore, it may be stated that the density of biodiesel fuels increases with increase in percentage of unsaturation or number of double bond of carbon.

Table 15. Pearson correlation coefficient (r) between biodiesel properties and percentage of unsaturation

	Density @ 40°C	*Kinematic viscosity @ 40°C*	*Cetane number*	*Heating Value*	*Iodine Value*	*Saponification value*
% of US	0.933	– 0.979	– 0.796	– 0.796	0.927	– 0.496

The scatter plot between density and percentage of unsaturation with fitted trend line equation is shown in Fig. 8. By differentiating the fitted line equation "y = 0.2535x + 818.23" with respect to x, *i.e.* % of unsaturation, the gradient between density and percentage of unsaturation was found as 0.2535. This means for every one per cent increase in unsaturation, it may result in 0.2535 units (kg/m^3) increase in density.

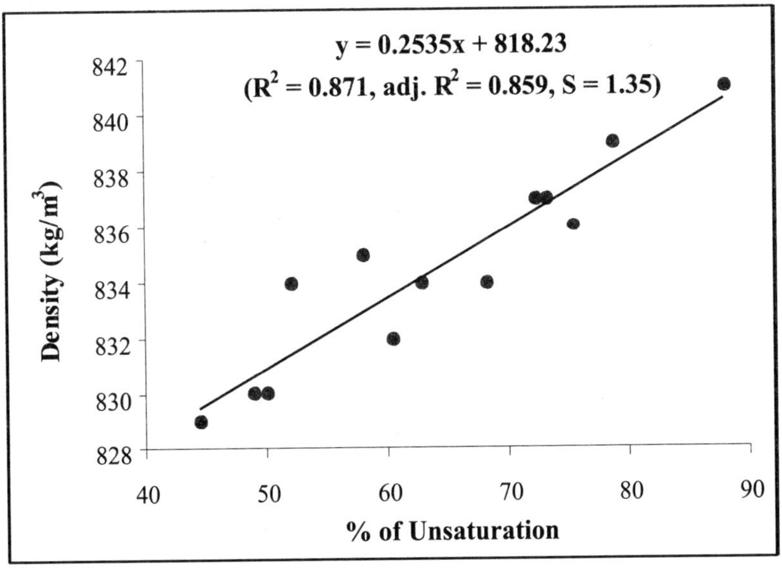

Fig. 8. Variation of density with percentage of unsaturation

The investigation on the relationship between kinematic viscosities and fatty acid composition exhibits an opposite trend as that of density. The viscosity decreases with increase in percentage of unsaturation in biodiesel fuels. This is due to the reason that the viscosity increases with increase in molecular weight and molecular weight decreases with increase in unsaturation. Table 15 shows that kinematic viscosity is highly negatively correlated with percentage of unsaturation.

Fig. 9 depicts the variation of kinematic viscosity with percentage of unsaturation. The fitted line possesses a good R^2 value. From the fitted line equation "y = – 0.0146x + 6.0536", the gradient between viscosity and percentage of unsaturation was found as – 0.0146. A decrease of 0.0146 units (mm^2/s) in viscosity may be expected for every one per cent increase in unsaturation.

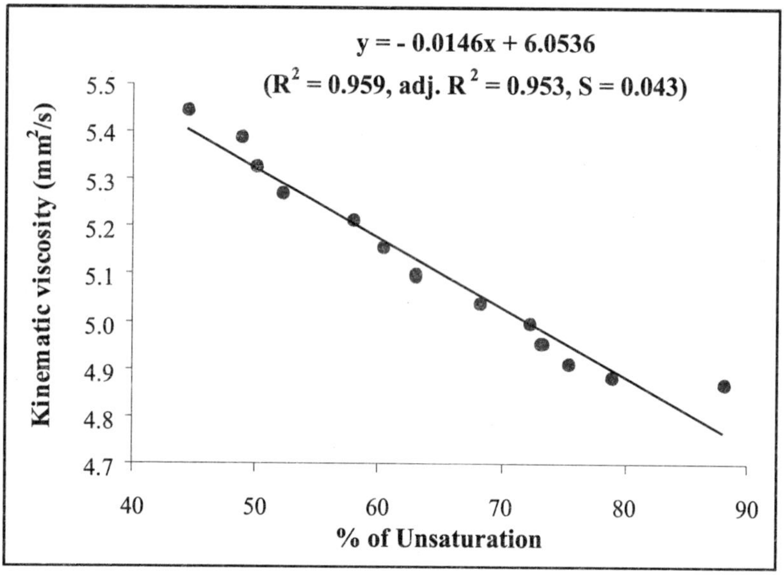

Fig. 9. Variation of kinematic viscosity with percentage of unsaturation

The relationship between fatty acid methyl ester composition and the cetane number of biodiesel fuels was investigated. Before getting into investigation findings, the following points can be recalled:

1. Cetane number increases with chain length, decreases with number of double bonds or unsaturation.
2. Cetane numbers of fatty esters generally increases with
   - the number of methylene groups (CH_2) in the chain of fatty compound

- number of CH_2 groups in the ester moiety
- increasing saturation of the fatty compound.

3. For the methyl esters, the cetane numbers were found to increase in a nonlinear relationship with the molecular weight (chain length).

It can be witnessed from Table 13 that the biodiesel of sunflower oil has a lower cetane number than that of other biodiesel fuels. This is due to the reason that the cetane number decreases with increase in number of double bonds (unsaturation). Although JCt 50:50 is more saturated than other biodiesel fuels, but has lower cetane number than palm, mahua and JT 50:50 blend. It may be believed due to the contribution of palmitic and stearic acids in palm, mahua and JT 50:50 are higher than that of JCt 50:50. From the above discussions it may be concluded that the overall percentage of unsaturation or saturation may not be the sole authority to decide the cetane number of a biodiesel. The contribution of long straight chain fatty acids may influence the cetane number of biodiesel to a large extent.

From Table 15, the correlation coefficient between cetane number and percentage of unsaturation can be observed as (–) 0.796. The correlation analysis reveals that the percentage of unsaturation exhibits a strong negative correlation with cetane number. Fig. 10 exemplifies the influence of unsaturation on cetane number.

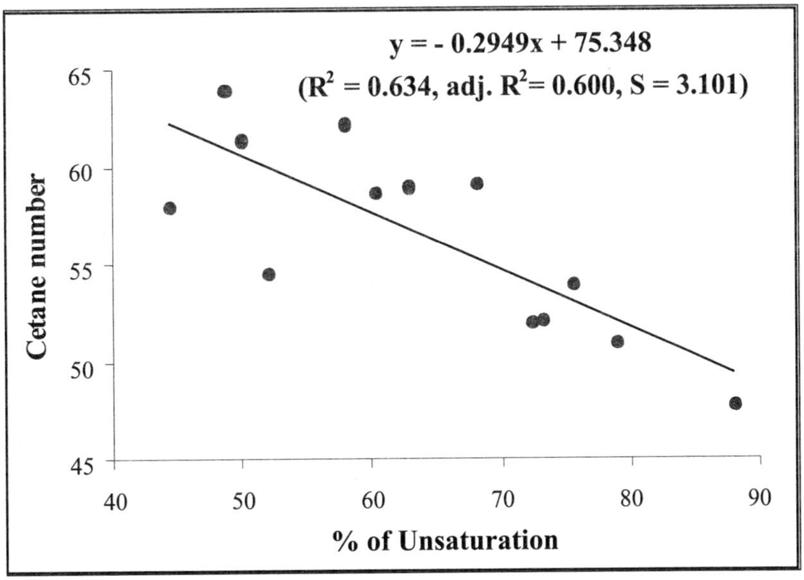

Fig. 10. Variation of cetane number with percentage of unsaturation

From the fitted line equation "y = – 0.2949x + 75.348", it can be noticed that a reduction of 0.2949 units in cetane number could be predictable for each percentage increase in unsaturation.

A more similar examination on the relationship between fatty acid profile and heating values of biodiesel was carried out. A strong negative correlation between heating value and percentage of unsaturation can be observed from Table 15. A recollection of following points will provide a more precise understanding on investigation findings:

- The molecular weight of fatty acids increases with increase in saturation.
- The heating value increases with increasing chain length (molecular weight) and decreases with increase in number of double bonds .
- The increase in heat content may be believed due to the increase in number of carbon and hydrogen. The possible cause for decrease in heat content may be the deficiency of hydrogen atoms (greater unsaturation).

From the investigation, it was found that the JCt 50:50 blend has a lower heating value than that of palm and mahua biodiesels although it has a higher amount of saturation than the other two. Yet again, it may be due to the contribution of palmitic acid and stearic acid in palm and mahua biodiesels. Hence a declaration can be made that the heating value increases with increasing chain length and decreases with increase in number of double bonds. The increase in heat content may be believed to be due to the increase in number of carbon and hydrogen. The possible cause for decrease in heat content may be the deficiency of hydrogen atoms (greater unsaturation). Eventually, it may be concluded that the overall percentage of unsaturation or saturation may not control solely the heating value of a biodiesel. The contribution of long straight chain fatty acids may influence the heating value of biodiesel to a greater extent. Fig. 11 illustrates the dependence of heating value on percentage of unsaturation.

From the fitted line equation "y = – 0.029x + 41.692", the projected decrease in heating value for every one per cent increase in unsaturation is 0.029 units (MJ/kg).

Iodine value which can be considered as a direct measure of unsaturation may be defined as the amount of iodine (in grams) necessary to saturate 100 g of oil sample. The influence of fatty acid ester composition on iodine value for biodiesel samples was scrutinized.

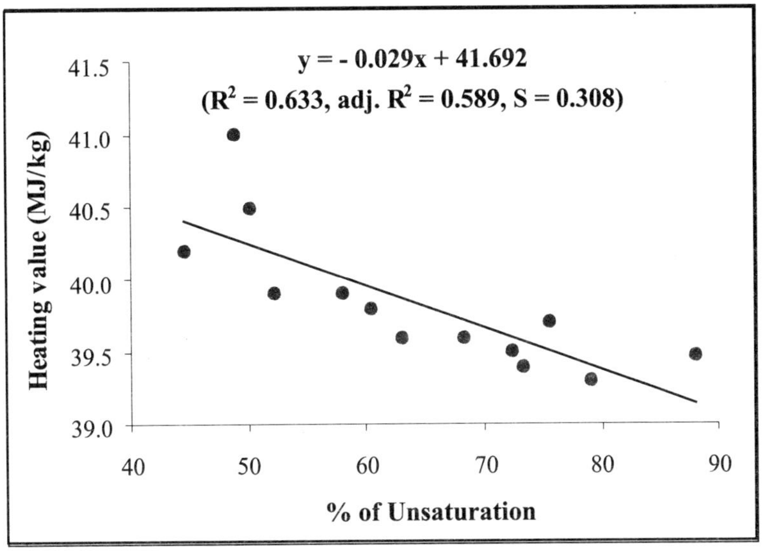

Fig. 11. Variation of heating value with percentage of unsaturation

The correlation analysis exhibits a strong positive correlation between iodine value and percentage of unsaturation. This was expected since from the definition, iodine value could be considered as a direct measure of unsaturation. Fig. 12 depicts the degree of variation in iodine value with unsaturation.

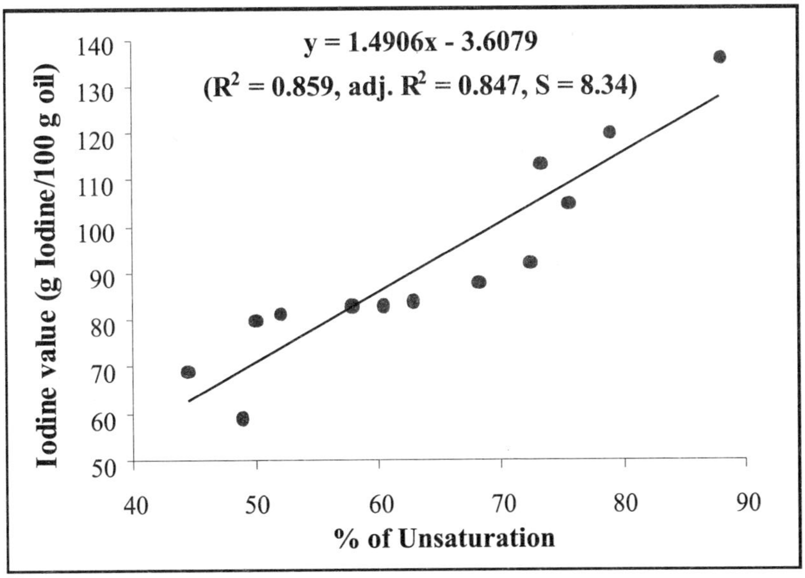

Fig. 12. Variation of iodine value with percentage of unsaturation

From the fitted line equation "y = 1.4906x – 3.6079", every one per cent increase in unsaturation may result in an increase of 1.4906 units (g Iodine/100 g oil) in iodine value.

However the proposed gradient between different properties and percentage of unsaturation was based only upon slope of the fitted lines. It may not be true for all cases, since the influence of each fatty acid ester on properties was not taken in to account. Also, if a large sample size would be taken for investigation, the gradient of aforesaid parameters may vary.

It may fairly be reasonable if the effect of lower fatty acids on saponification value would primarily be investigated. Because the saponification value of an oil decreases with increase of its molecular weight and hence the contribution of lower chain fatty acids may influence the saponification value to a larger extent. From Table 13, it can be seen that JCt 50:50 blend has a higher saponification value than that of other biodiesel fuels. This may be due to the contribution of lower chain fatty acids (especially, myristic and palmitic acids).

However the effect of unsaturation on saponification value can also be taken into account. Because biodiesel of palm oil possesses a higher amount palmitic acid (43.9%) than that of JCt 50:50 (13.71%), but it has a lower saponification value than that of JCt 50:50. From Table 15, it can also be observed that the saponification value is moderately negatively correlated with percentage of unsaturation.

It may be concluded that saponification value of biodiesel fuels could strongly be influenced by the lower chain fatty acids. In addition, saponification value of biodiesel fuels increases with increase in percentage of lower chain fatty acids and decreases with increase in unsaturation. It may further be proposed that the decrease in saponification value due to increase in percentage of unsaturation may not be significant as compared to the increase in saponification value due to increase in percentage of lower fatty acids. A pictorial representation of variation of saponification value with percentage of unsaturation is presented in Fig. 13.

The gradient between the saponification value and the percentage of unsaturation was not be proposed due to the lower R^2 value of the fitted line ($R^2 = 0.2462$).

Analysis of performance and pollutants emission parameters

In this section, the influence of biodiesel properties on the test engine's performance and emissions are discussed in detail. The following parameters were considered for the investigation.

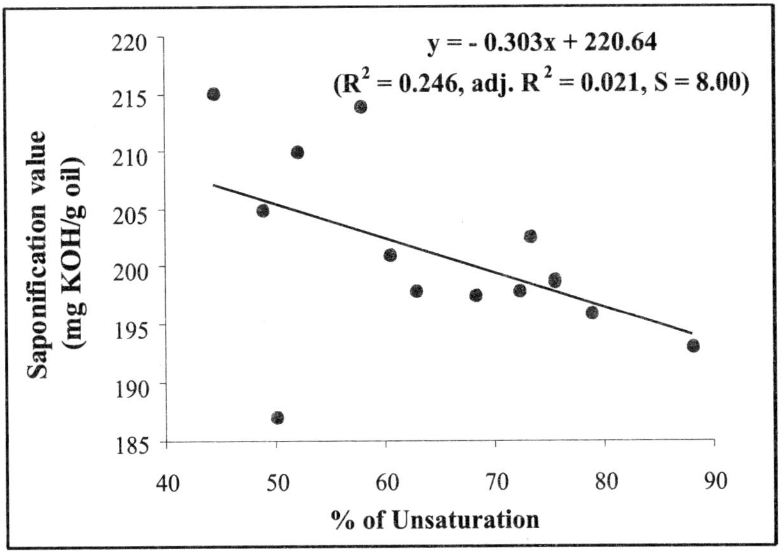

Fig. 13. Variation of saponification value with percentage of unsaturation

1. Brake Specific Fuel Consumption
2. Brake Specific Energy Consumption
3. Brake Thermal Efficiency
4. Exhaust Gas Temperature
5. Oxides of Nitrogen
6. Carbon Monoxide
7. Hydrocarbons
8. Smoke

The experimental values of all the above parameters at 100% load for various biodiesel fuels are listed in Table 16. The Pearson correlation coefficient between performance parameters and percentage of unsaturation is listed in Table 17.

Brake specific fuel consumption (BSFC)

The engine performance parameters could highly be influenced by the fuel properties such as mass based heating values and density. From Table 16, it can be observed that for a given operating conditions the engine consumes higher quantity of rubber seed oil methyl ester as compared to other biodiesel fuels. That is BSFC for rubber seed oil methyl ester is more than that of other biodiesel fuels. The investigation on the influence of fuel properties on engine's performance reveals that the increase in BSFC may be believed to be due to the increase in density. Increasing density may increase BSFC because the fuel injector injects a constant volume, but larger mass, of the more dense fuels. When

Table 16. Experimental values of performance and emission parameters of various biodiesel fuels

Biodiesel	*BSFC (kg/ kWh)*	*BSEC (MJ/ kWh)*	*Brake Thermal Efficiency (%)*	*Exhaust Gas Temperature (°C)*	*Brake specific NO_X (g/kWh)*	*Brake specific CO (g/kWh)*	*Brake specific HC (g/kWh)*	*Smoke (BSU)*
Sunflower	0.3342	13.20	27.27	365	14.38	3.62	0.59	0.8
Rubber seed	0.3345	13.15	27.38	360	14.14	2.58	0.56	0.9
Jatropha	0.3305	13.12	27.44	358	13.64	3.10	0.51	1.2
Cottonseed	0.3293	12.98	27.74	355	13.32	2.58	0.52	0.9
Karanjia	0.3276	12.94	27.82	350	13.40	2.58	0.52	1.2
JT 80:20	0.3259	12.91	27.89	348	12.94	2.58	0.48	1.3
JP 50:50	0.3254	12.89	27.93	343	12.68	2.07	0.43	1.3
Neem	0.3235	12.88	27.96	342	12.59	2.06	0.52	1.4
JT 50:50	0.3236	12.91	27.88	339	12.52	2.06	0.47	1.4
SFCt 50:50	0.3225	12.87	27.98	338	12.34	2.58	0.43	1.5
Mahua	0.3162	12.81	28.11	335	11.93	1.55	0.52	1.3
Palm	0.3110	12.75	28.23	334	11.70	1.06	0.43	1.4
JCt 50:50	0.3150	12.66	28.43	330	11.50	2.06	0.43	1.6

Table 17. Pearson correlation coefficient (r) between performance parameters and percentage of unsaturation

	BSFC	*BSEC*	*Brake Thermal Efficiency*	*Exhaust Gas Temperature*
% of US	0.938	0.939	– 0.940	0.989

compared to sunflower oil methyl ester, rubber seed oil methyl ester has a lower density but results in a higher BSFC. The possible reason for this increase may be due to the lower mass based heating value of rubber seed oil methyl ester. Because, as compared to sunflower biodiesel, the lower mass based heating value of rubber seed biodiesel requires larger mass of fuel to maintain constant energy input to the engine.

It can be stated that BSFC increases with increase in percentage of unsaturation, since an increase in percentage of unsaturation will result in an increase in density and in a decrease in heating value. From Table 17, a strong positive correlation can be observed between BSFC and percentage of unsaturation. Fig. 14 depicts the variation of BSFC with percentage of unsaturation. From the fitted line equation (0.0005x + 0.2921), every one per cent increase in unsaturation may result in an increase of 0.0005 units (g/kWh) in BSFC.

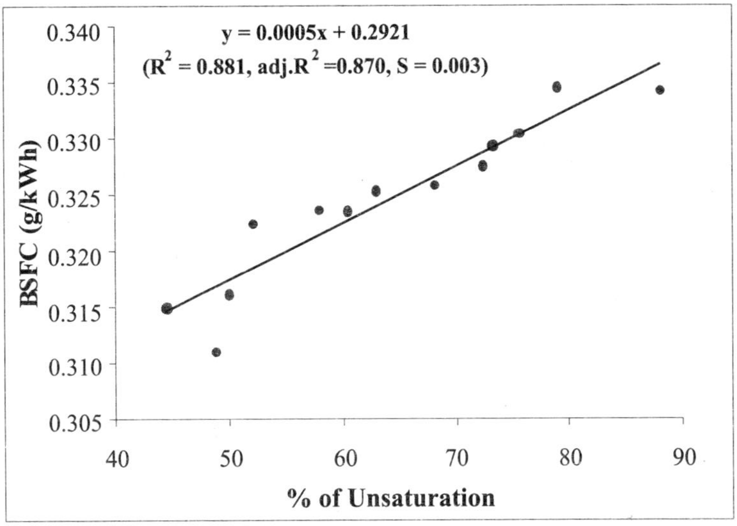

Fig. 14. Variation of BSFC with percentage of unsaturation

Brake specific energy consumption (BSEC)

In order to compare the performance of a given engine that is operated with different fuels, the term BSEC could provide clearer picture than the term BSFC. The BSEC can be obtained by multiplying the heating value with BSFC. From Table 16, it can be observed that the BSEC is higher for sunflower biodiesel and lower for JCt 50:50 biodiesel as compared to other biodiesel fuels.

It is obvious that the trend of BSEC should follow the same as that of BSFC. It can also be noted that the BSFC for sunflower biodiesel is lower than that for rubber seed biodiesel, but still the BSEC for sunflower biodiesel is higher than that for rubber seed biodiesel. Similarly the engine has a higher BSFC and a lower BSEC when it is operated with JCt 50:50 biodiesel as compared to palm biodiesel. The possible cause for the above findings may be due to the lower heating value of rubber seed biodiesel as compared to sunflower biodiesel and lower heating value of JCt 50:50 biodiesel as compared to palm biodiesel. From the correlation analysis, it was found that the BSEC was highly positively correlated with density and percentage of unsaturation. It is due to the fact the BSFC increases with increase in percentage of unsaturation and with density. It can also be noted that the BSFC increases with decrease in heating value. Fig. 15 depicts the variation of BSEC with percentage of unsaturation. From the fitted line equation $y = 0.011x + 12.224$, the gradient between BSEC and percentage of unsaturation can be proposed as 0.01. Every one per cent increase in unsaturation may result in an increase of 0.011 units (MJ/kWh) in BSEC.

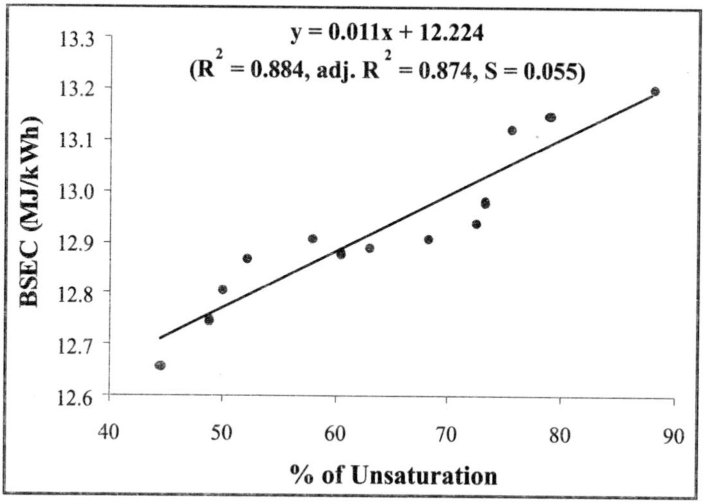

Fig. 15. Variation of BSEC with percentage of unsaturation

Brake thermal efficiency

Brake thermal efficiency shown in Table 16 is higher for JCt 50:50 blend and is lower for sunflower biodiesel. This shows that the order of magnitude of brake thermal efficiency for the biodiesel fuels are matched exactly with the reverse order of BSEC. From the correlation analysis, it was found that the brake thermal efficiency decreases with increase in percentage of unsaturation and density. This is due to the fact that the BSEC decreases with increase in percentage of unsaturation and density. The variation of brake thermal efficiency with percentage of unsaturation is illustrated in Fig. 16.

From the fitted line equation y = – 0.0236x + 29.366, a decrease of 0.0236 units (%) could be predictable for every one per cent increase in unsaturation.

Exhaust gas temperature

The exhaust gas temperature shown in Table 16 is higher for sunflower biodiesel and lower for JCt 50:50 blend. The order of magnitude of exhaust gas temperature matched with order of percentage of unsaturation. That is a more unsaturated biodiesel can produce higher value of exhaust gas temperature as compared to a less unsaturated

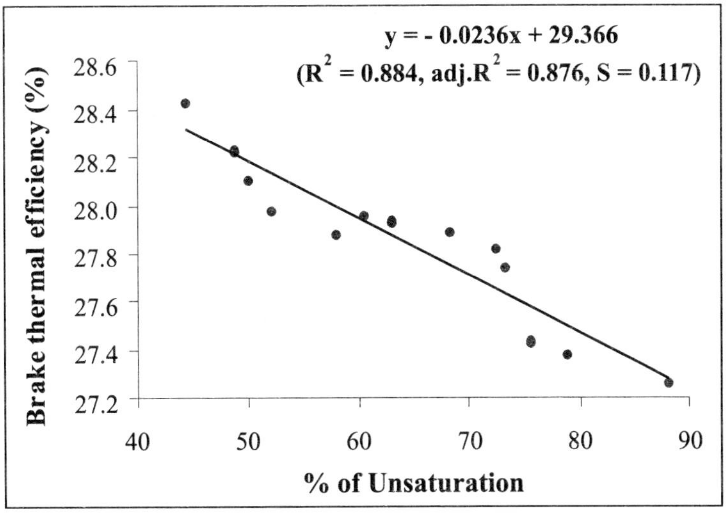

Fig. 16. Variation of brake thermal efficiency with percentage of unsaturation

one. This is believed to be due to the more afterburning stage for higher unsaturated biodiesel fuels which may increase the exhaust gas temperature. A high degree of positive correlation was found to exist between exhaust gas temperature and percentage of unsaturation from the correlation analysis. The variation of exhaust gas temperature with percentage of unsaturation is shown in Fig. 17.

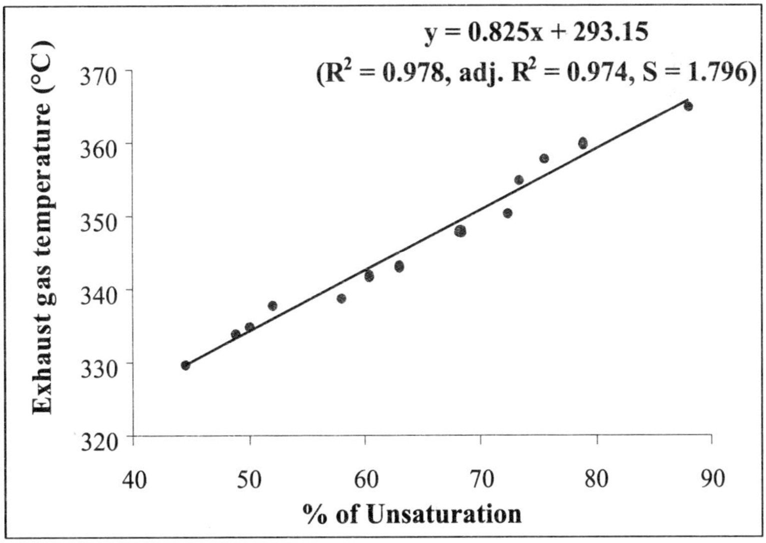

Fig. 17. Variation of exhaust gas temperature with percentage of unsaturation

From the fitted line equation y = 0.825x + 293.15, the gradient between exhaust gas temperature and percentage of unsaturation could be predicted as 0.825. In other words, every one per cent increase in unsaturation may result in an increase of 0.825 units (°C) in exhaust gas temperature.

Oxides of Nitrogen

Once again a correlation was performed (using equation 1) to allow a determination as to whether the measured emission values are correlated with biodiesel properties and unsaturation. The correlation coefficients are shown in Table 18.

Table 18. Pearson correlation coefficient (r) between emission parameters and biodiesel properties

"X" variable	*"Y" variable*	*Correlation coefficient*
% of US	brake specific NO_X	0.986
	brake specific CO	0.817
	brake specific HC	0.784
	Smoke	– 0.887

Understanding fuel spray regions is so important to understand the emissions formation. Fig. 18 shows the fuel spray regions.

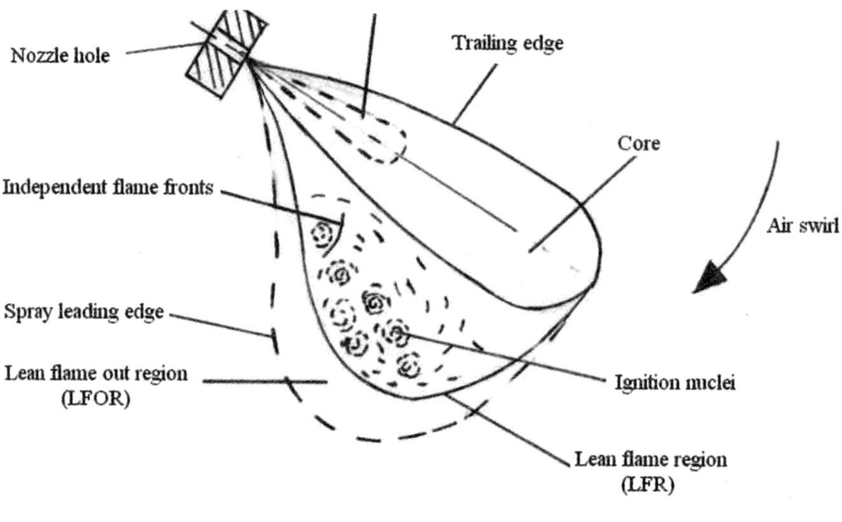

Fig. 18. Schematic diagram of a fuel spray injected in swirling air

The concentration of the vapour between the core and the leading edge of the spray is not homogeneous, and the local fuel-air ratio may vary from zero to infinity. Ignition nuclei would be formed at several locations where the mixture is most suitable for auto ignition. Once the ignition starts, small independent non-luminous flame fronts propagate from the ignition nuclei and ignite the combustible mixture around them. This mixture, on a mass average basis is lean. The region in which these independent flames propagate is referred to as the "lean flame region" (LFR). In this region, combustion is complete and nitric oxide is formed at high local concentrations. Near the far leading edge of the spray (downwind), the mixture is too lean to ignite or support combustion. This region is referred to as the "lean flame out region" (LFOR). Within this region, some fuel decomposition and partial oxidation products could be expected. It is believed that this region is one of the main contributors to the unburned hydrocarbons concentration in the exhaust. The effect of fatty acid ester composition on emissions is implicit in density, cetane number or iodine value. NO_X was found to be highly positively correlated with density, iodine value and percentage of unsaturation. Table 13 shows that the NO_X for sunflower biodiesel is higher and for JCt 50:50 are lower than the other biodiesel fuels. From the investigation, it was found that the NO_X concentration in the exhaust emissions increases with increase in biodiesel density, iodine value and percentage of unsaturation. This can be explained as follows.

- Increasing density may increase NO_X because the fuel injector injects a constant volume, but larger mass, of the more dense fuels. Since a larger mass of the fuel is burned more NO_X is produced.
- Another possibility is that the higher density which results in higher bulk modulus can advance the effective injection timing and thereby cause NO_X to increase. The injection advance results in longer ignition delay since the fuel is injected in air at lower temperatures and pressure.
- It could also be noted that the iodine value which is a direct measure of unsaturation is highly inversely correlated with cetane number. Thus, excessive ignition delay and poor combustion performance may also be proposed as a cause of the higher NO_X.

It may be expected that the lean flame region (LFR) is one of the major contributing regions to nitric oxide (NO) formation since it is the first part of the spray to burn and has the longer post-flame residence time. With lower cetane number fuel, the ignition delay is longer and more fuel is present in the LFR when combustion starts. The larger ignition delay produces a higher gas temperature upon combustion early in the cycle, and more nitric oxide (NO) could be formed in the LFR. It can be recalled that the cetane number decreases with increase in unsaturation. Hence, it may be concluded that, NO_X concentration in the exhaust emissions increases with increase in biodiesel density, iodine value, and percentage of unsaturation. Fig. 19 depicts the variation of NO_X with percentage of unsaturation.

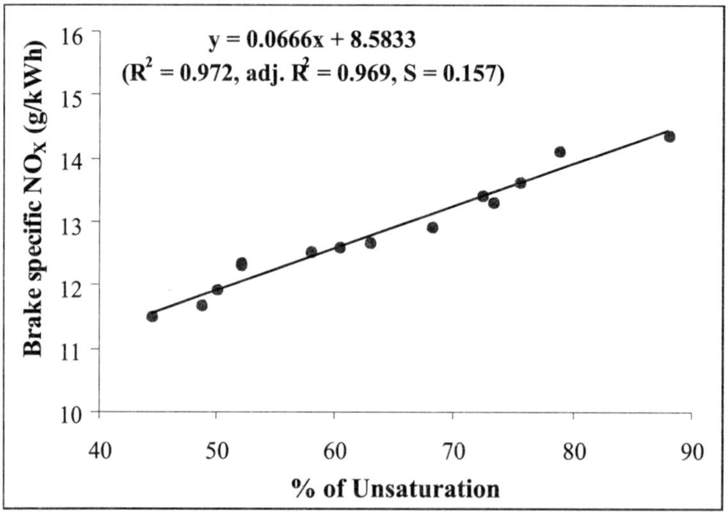

Fig. 19. Variation of NO_X with percentage of unsaturation

From the proposed fitted line equation y = 0.0666x + 8.5833, an increase of 0.0666 units (g/kWh) in brake specific NO_X might be expected for every one percent increase in unsaturation.

Carbon monoxide (CO) and Hydrocarbon (HC) Emissions

The carbon monoxide (CO) and hydrocarbon (HC) emission levels in diesel engines are small in absolute terms, so that they are of no real concern. But still it is fairly reasonable to discuss the effect of biodiesel properties and fatty acid composition on CO and HC emissions. The correlation analysis reveals that the CO and HC emissions are positively correlated with percentage of unsaturation. That is CO and HC emissions increase with increase in unsaturation. This may be believed to be due to the lower oxygen concentration in higher unsaturated biodiesel fuels. The oxygen content decreases with increase in unsaturation.

Carbon monoxide is believed to be formed at the borders between the lean flame out region (LFOR) and LFR during the early stages of spray combustion. At this stage, primary reaction can take place and the initial hydrocarbons may reduce to CO, H_2, and H_2O. As the local temperature is not enough at this stage, very little oxidation reactions take place. Increase in unsaturation may tend to decrease the gas temperature. The reduced gas temperature may not provide a positive situation for complete oxidation reaction thereby increase the CO concentration. The variation of CO with percentage of unsaturation is illustrated in Fig. 20.

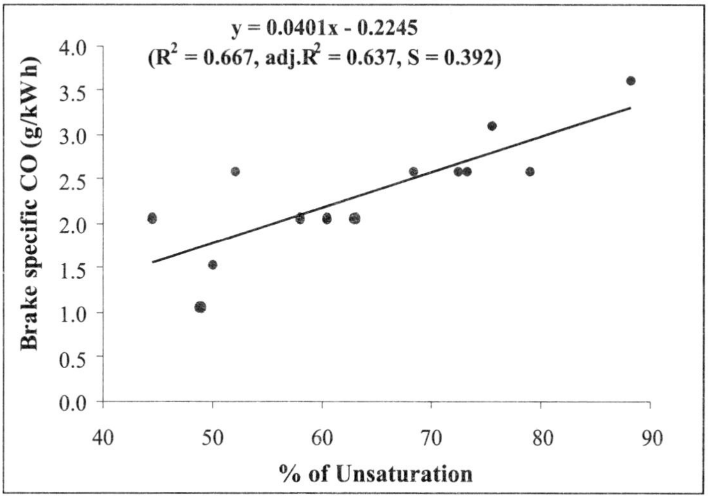

Fig. 20. Variation of CO with percentage of unsaturation

On the other hand, it may be believed that the LFOR is one of the main contributors to the HC concentration in the exhaust. LFOR is the region that is nearer to the leading edge of the spray (downwind). In this region the mixture is too lean to ignite or support the combustion. If the ignition delay is larger, the droplets and vapour will be carried farther away from the centre line of the spray in the downward direction, resulting in a wider LFOR. If unsaturation increases, the ignition delay increases which result in an increased width of LFOR. Increased width of LFOR eventually increases the HC concentration in exhaust. The variation of HC with percentage of unsaturation is shown in Fig. 20 and 21.

From the equation $y = 0.041x - 0.2245$, the probable increase in brake specific CO could be proposed as 0.0401 units (g/kWh) for every one per cent increase in unsaturation. The equation $y = 0.0031x + 0.2938$ provides the gradient between HC emission and percentage of

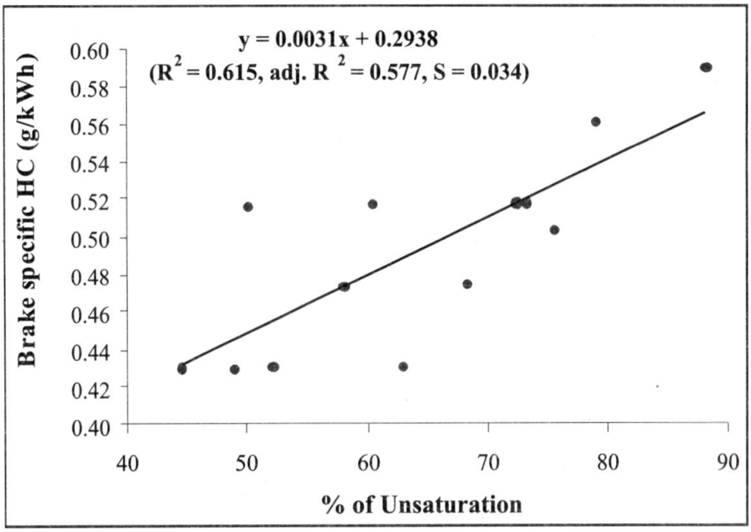

Fig. 21. Variation of HC with percentage of unsaturation

unsaturation as 0.0031. An increase of 0.0031 units (g/kWh) in brake specific HC could be expected for every one per cent increase in unsaturation.

Smoke

Smoke is emitted as a product of the incomplete combustion process, particularly at maximum loads. Smoke particles are formed from the fuel deposited on walls or in the spray core, especially under elevated loads. It was found from the investigation that the smoke intensity for biodiesel fuels decreases with increase in percentage of unsaturation. It is necessary to discuss the mechanism of smoke formation to justify the investigation findings. The smoke formation mechanism can be explained as follows.

- The evaporation, ignition and combustion of a fuel droplet begin at its outermost layer.
- As the outer layer of the fuel droplet burns, heat is produced and transferred to the inner layers. Therefore, the availability of oxygen for reaction is lesser at the interior layers.
- Hydrogen molecules have greater affinity towards oxygen compared to carbon molecules. As such hydrogen molecules combine with oxygen molecules more quickly leaving little or no oxygen for carbon molecules to react.

- The carbon molecules, therefore fail to burn within the limited time that is available for combustion. These unburned carbon molecules appear as smoke in the exhaust gases.

It is known that the unsaturation is nothing but the deficiency of hydrogen atoms. Greater unsaturation represents the greater deficit of hydrogen atoms. Therefore, for a given supply of air (and oxygen) to the engine, if unsaturation is more, carbon molecules could find more oxygen molecules to react with due to the less number of hydrogen molecules. Therefore, the degree of oxidation of carbon molecules may be higher in more unsaturated biodiesel fuels than that of biodiesel fuels with less unsaturation. Hence, it can be stated that the smoke intensity decreases with increase in unsaturation. Fig. 22 illustrates the variation of smoke with percentage of unsaturation.

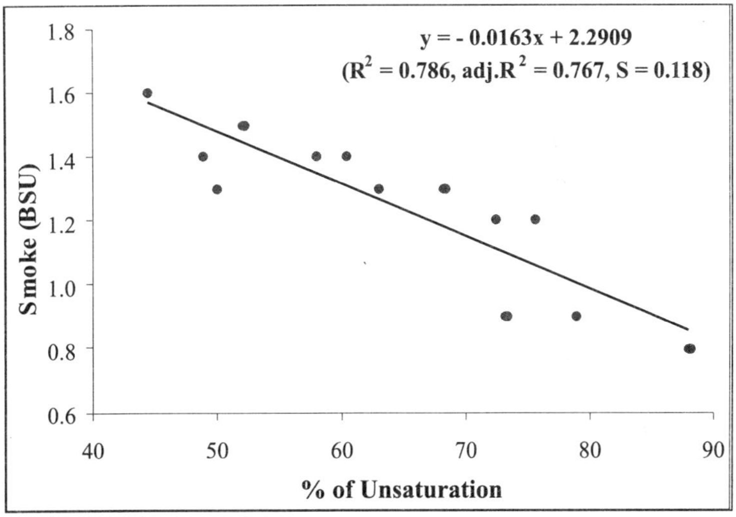

Fig. 22. Variation of smoke with percentage of unsaturation

The gradient between smoke and percentage of unsaturation can be found as – 0.0163 (from the equation y = – 0.0163 + 2.2909). It can be proposed that one per cent increase in unsaturation may result in a decrease of 0.0163 units (Bosch Smoke Units (BSU)) in smoke.

Conclusions

Thirteen biodiesel fuels including five blends were prepared and the properties were measured experimentally. Using these biodiesel fuels, experiments were conducted on a single cylinder, four stroke, air cooled, direct injection, 4.4 kW compression ignition engine at full load. The effects of fatty acid ester composition on biodiesel properties in relation with fatty acid ester composition on performance, and pollutants emission parameters were investigated. Gradients are proposed between different parameters and percentage of unsaturation. From the present work, it was found that the percentage of unsaturation in biodiesel fuels can have significant influence on fuel properties and therefore on performance and pollutants emission characteristics. Hence, all the properties and parameters were related to percentage of unsaturation. From the present work the following conclusions are arrived:

- The density of biodiesel fuels increases with increase in percentage of unsaturation. An increase of 0.2535 kg/m^3 in density could be expected for every one percentage increase in unsaturation.
- Kinematic viscosity of biodiesel fuels decreases with increase in percentage of unsaturation. One percentage increase in unsaturation may result in a decrease of 0.0146 mm^2/s units in kinematic viscosity.
- Cetane number of biodiesel fuels decreases with increase in unsaturation percentage and each percentage increase in unsaturation could reduce 0.2949 units in cetane number.
- Increase in percentage of unsaturation reduces the heating value of biodiesel fuels, and 0.029 MJ/kg reduction in heating value may occur for every one percentage increase in unsaturation.
- Iodine value which is a direct measure of unsaturation increases with increase in percentage of unsaturation. One percentage increase in unsaturation may increase 1.4906 g iodine/100 g oil in iodine value.
- Saponification value decreases with increase in unsaturation and increases with increase in percentage of lower chain fatty acid esters.
- Increase in unsaturation percentage result in an increase in BSFC and BSEC. Every one percentage increase in unsaturation may result in an increase of 0.0005 kg/kWh and 0.011 MJ/kWh in BSFC and BSEC respectively.
- Brake thermal efficiency decreases with increase in percentage of unsaturation. Every one percentage increase in unsaturation may result in an increase of 0.0236% in brake thermal efficiency.
- Exhaust gas temperature increases with increase in unsaturation percentage. An increase of 0.825°C may be predicted by increasing every one percentage of unsaturation.

- Oxides of nitrogen increases with increase in percentage of unsaturation, density and iodine value. An increase of 0.0666 g/kWh in brake specific NO_X can be expected for each percentage increase in percentage on unsaturation.
- Carbon monoxide emissions increased with increase in percentage of unsaturation. Every one percentage increase in unsaturation may cause an increase of 0.0401 g/kWh in brake specific CO.
- Hydrocarbon emissions increased with increase in percentage of unsaturation. Every one percentage increase in unsaturation may cause an increase of 0.0031 g/kWh in brake specific HC.
- Smoke decreases with increase in unsaturation percentage. A decrease of 0.0163 BSU in smoke may be expected for every one percentage increase in unsaturation.

The biodiesel fatty acid ester composition which is largely dependent on the raw material source (parent vegetable oil) can have considerable influence on biodiesel fuel properties with relation to performance and emission parameters. From the present study, it can be observed that almost all the properties/parameters are adversely affected by unsaturated fatty acids. But still unsaturated fatty acids should not completely be eliminated. Because some of the previous researches reveal that the unsaturated fatty acids can improve the cold flow properties of biodiesel. It therefore appears reasonable to enrich (a) certain fatty acid ester(s) with desirable properties in the fuel in order to improve the properties of the whole fuel. It may possible in the future to improve the properties of biodiesel by means of genetic engineering of the parent oils, which could eventually lead to a fuel enriched with (a) certain fatty acid (s) that exhibits a combination of improved fuel properties.

References

Agarwal Avinash Kumar (2003). Biofuels (alcohols and biodiesel) applications as fuels for internal combustion engines. *Progress in Energy and Combustion Science* **33**: 233-271.

Agarwal Deepak, Shailendra Sinha and Avinash Kumar Agarwal (2006). Experimental investigation of control of NO_X emissions in biodiesel fueled compression engine, *Renewable Energy* **31**: 2356-2369.

Allen, C.A.W., Watts, K.C., Ackman, R.G. and Pegg, M.J. (1999). Predicting the viscosity of biodiesel fuels from their fatty acid composition. *Fuel* **78**: 1319-1326.

Ayhan, Demirbas (1998). Fuel properties and calculation of higher heating values of vegetable oils, *Fuel* **77(9/10)**: 1117-1120.

Babu A.K. and Devaradjane, (2003). Vegetable oils and their derivatives as fuels for CI engines: An overview, Society of Automotive Engineers -01-0767.

Eiji Kinoshita, Thet Myo and Kazunori Hamasaki (2006). Diesel combustion characteristics of coconut oil and palm oil biodiesels, Society of Automotive Engineers 2006, Technical paper series, 2006 – 01 – 3251, Power Train and Fluid Systems Conference and Exhibition, Toronto, Canada.

Ejim, C.E., Fleck, B.A. and Amirfazli, A. (2007). Analytical study for atomization of biodiesels and their blends in a typical injector: Surface tension and viscosity effects. *Fuel* **86**: 1534-1544.

Fangrui, Ma, Davis Clements, L. and Milford, A. Hanna (1999). The effect of mixing on transesterification of beef tallow. *Bioresource Technology* **69**: 289-293.

Foidl, N, Foidl, G., Sanchez, M, Mittelbach, M. and Hackel, S. (1996). *Jatropha curcas* L. as a source for the production of biofuel in Nicaragua. *Bioresource Technology* **58**: 77-82.

Gerhad Knothe, (1997). Effect of structure of fatty compounds on their cetane numbers. The liquid biofuels news letter – 9, Information from the liquid biofuels activity. ISSN 1025-8698: p. 5.

Gerhad, Knothe (2005). Dependence of bio-diesel fuel properties on the structure of fatty acid alkyl esters. *Fuel Processing Technology* **86**: 1059-1070.

Ghadge Shashikant Vilas and Hifjur Raheman (2005). Biodiesel production from mahua (*Madhuca indica*) oil having high free fatty acids. *Biomass and Bioenergy* **28**: 601-605.

Harwood, H.J. (1984). Oleochemicals as a fuel: Mechanical and economic feasibility. *Journal of American Oil Chemists' Society* **61**: 315-324.

Hass, M.J. (2005). Improving the economics of biodiesel production through the use of low value lipids as feed stocks: Vegetable oil soap stock. *Fuel Processing Technology* **86**: 1087-1096.

Heywood, J.B. (1998). Internal combustion engine fundamentals, New York: McGraw Hill.

Hideki Fukuda, Akihiko Kondo and Hideo Noda (2001). Biodiesel fuel production by transesterification of oils. *Journal of Bioscience and Bioengineering* **92(5)**: 405-416.

Hideki Fukuda, Akihiko Kondo and Hideo Noda (2001). Review: Biodiesel fuel production by transesterification of oils. *Journal of Bioscience and Bioengineering*, **92(5)**: 405-416.

Klopfenstein, W.E. (1985). Effect of molecular weight of fatty acid esters on cetane numbers as diesel fuels. *Journal of American Oil Chemists' Society* **62(6)**.

Kumar Senthil, M., Ramesh, A. and Nagalingam, B. (2003). An experimental comparison of methods to use methonal and jatropha oil in a compression ignition engine. *Biomass and Bioenergy* **25**: 309-318.

Ma, F. and Hanna, M.A. (1999). Biodiesel production: a review. *Bioresource Technology* **70**: 1-15.

Mandpe Sanjeev, Suhas Kadlaskar, Winfried Degen and Stefan Keppeler (2005). On Road Testing of Advanced Common Rail Diesel Vehicles with Biodiesel from the Jatropha Curcas plant. *Society of Automotive Engineers*, 2005-26-356.

Marvin O. Bagby, Vegetable oils for diesel fuel: Opportunities for development, *American Society of Agricultural Engineers*, pp. 87-1588.

Marvin, O. Bagby, Bernard Freedman and Arthur, W. Schwab, Seed oils for diesel fuels: Sources and properties. *American Society of Agricultural Engineers*, pp. 87-1583.

Michael S. Graboski and Robert L. McCormick (1998). Combustion of fat and vegetable oil derived fuels in diesel engines. *Progress in Energy and Combustion Science* **24**: 125-164.

Pramanik. K., (2003). Properties and use of *Jatropha curcas* oil and diesel fuel blends in compression ignition engine. *Renewable Energy*, **28**: 239-248.

Puhan Sukumar and Nagarajan, G. (2008). Effect of biodiesel unsaturated fatty acid on combustion characteristics of a DI compression ignition engine. *Society of Automotive Engineers*, -28-41, Proceedings of 5th International Mobility Conference on Emerging Automotive Technologies Global and Indian Perspective (SIIMC 2008), 9th to 11th January 2008.

Raheman, H. and Phadatare, A.G. (2004). Diesel engine emissions and performance from blends of karanja methyl ester and diesel. *Biomass and Bioenergy*, **27**: 393-397.

Ramadhas, A.S., Jayaraj, S., Muraleedharan, C. and Padmakumari, K. (2006). Artificial neural networks used for the prediction of the cetane number of biodiesel fuel. *Renewable Energy* **31**: 2524-2533.

Robert L. McCormick, Michael S. Graboski, Teresa L. Alleman, Andrew M. Herring and Shaine Tyson K. (2001). Impact of biodiesel source material and chemical structure on emissions of criteria pollutants from a heavy–duty engine. *Environmental Science and Technology* **35(9)**: 1742-1747.

Schwab, A.W., Bagby, M.O. and Freedman, B. (1987). Preparation and properties of diesel fuels from vegetable oils, Fuel p. 66.

Sukumar Puhan, Vedaraman, N., Ramabrahamam, B.V. and Nagarajan, G. (2005). Mahua (*Madhuca indica*) seed oil: A source of renewable energy in India. *Journal of Scientific and Industrial Research* **64**: 890-896.

12

ALGAL TURF SCRUBBERS: CLEANING WATER WHILE CAPTURING SOLAR ENERGY

Walter Adey and Jeffrey Bannon[1*]

Abstract

Present paper deals with the algal turfs and Algal Turf Scrubbers (ATS) which clean water and remove Carbon dioxide more efficiently from water and the atmosphere and can be configured to remove CO_2 *from power plant stack gases. Algal turfs are communities of organisms dominated by aggregations of unicellular to branched filamentous algae and cyanobacteria (blue-green algae). Algal Turf Scrubber is applied biomimicry, an ecologically engineered, algal turf dominated mini-ecosystem. The applications of ATS in monitoring various environmental parameters have been discussed.*

Key words: Algal turf scrubbers, Solar energy, Cyanobacteria, Red algae, Environmental use, ATS engineering

Algal Turfs and Algal Turf Scrubbers (ATS)

Algal Turfs are biodiverse communities of unicellular to filamentous algae of all major algal phyla. Algal Turf Scrubbers (ATS) are bioengineered ecosystems dominated by algal turfs. They clean water to very high quality, and remove CO_2 from the atmosphere by capturing solar energy at rates 10 times that of agriculture and 50 times that of forestry.

ATS was invented at the Smithsonian Institution, by scientist, Walter Adey in the 1980s as a tool for controlling water quality in highly diverse model ecosystems. The technology received extensive R&D for aquacultural, municipal, and industrial water cleaning by Dr. Adey,

[1] Department of Botany, Smithsonian Institute, USA

* *Corresponding author*: Email : jbenemann@aol.com, adey.walter@nmnh.si.edu

using venture capital, through the 1990s. Later, HydroMentia, Inc., of Ocala, Florida, engineered ATS to landscape scale of 20-50 Mgpd (it is important to note that this is a modular system, capable of expanding to any size.) A 2005 independent study of ATS, by the South Florida Water Management District and the IFAS Institute of the University of Florida, certified ATS as 5-100 times more cost efficient at removing nutrients from Everglades canal waters than the next competitor, the STA, a managed marsh system. ATS and STA were the final contestants in a 15-year study of nine technologies, and ATS was the only technology that created a useable byproduct.

Water Quality and Energy Independence

ATS systems are capable of removing N and P from surface waters in the mid latitude US at \$0.60/kg and \$10.60/kg respectively, and independently producing an energy product consisting of biodiesel and Butanol. Since they are controlled ecosystems, using local algae, ATS does not suffer the major disadvantages of agricultural crops, which for maximum efficiency require fertilizers, herbicides and pesticides. ATS removes CO_2 from water and the atmosphere, and can be configured to remove CO_2 from power plant stack gases. As a normal part of operations, ATS removes heavy metals, break down toxic hydrocarbons, and oxygenates treated waters.

The Competition

Many technologies are under development for solving the U.S.'s energy dependency. Some, like artificial photosynthesis require major technological breakthroughs; others, like agricultural crops, *i.e.*, bioenergy, require intensive application. Switchgrass, a new hope for solving these problems, requires much more R&D and great land area. There are also microalgal techniques that require genetic engineering of single species of special characteristics. All of these techniques have environmental consequences that are potentially very significant. All are monocultural and agro-industrial by their nature, unlike ATS, which uses a biodiverse ecosystem of local organisms.

Challenges

Population growth, modernization, globalization, and the subsequent burden and stress they continue to place on the planet have created immense socioeconomic, health, and environmental challenges for society. Unfortunately, solutions receive different priorities depending on the scientist, NGO, or governmental agency in charge; efforts toward one solution are often piecemeal and only mildly effective, and unfortunately, often exacerbate other problems. As an example, current energy solutions are multifold, with solutions ranging from a combination of nuclear, coal, natural gas, oil, and recently biofuels. All however present numerous problems: nuclear has the problem of safety and security concerns; oil and natural gas have issues of over demand, supply, accessibility, and pollutant release; and coal, although readily and cheaply abundant, places an immense burden on the environment due to mining effects, and the release of carbon dioxide and other harmful pollutants. Assuming that sustainable solutions are even approached, all of those on the table today involve methodologies that will have serious impacts on the atmosphere, water quality, and biodiversity.

Biofuels are strong contenders to overcome some of the problems of fossil fuels; yet approaches that use corn, soy, require excessive land and rely heavily on fertilization. This would place a tremendous burden on food-cropland, streams, rivers, and coastal waters, and would lead to eutrophication, organic chemical toxicity and a decline in biodiversity. Switchgrass still requires decades of R&D. The problems are as numerous as the solutions. The key is to find solutions that have multiple benefits: they must be economically competitive and biological in nature, but ultimately lead to environmental restoration.

The Solution

What is an algal turf?

Algal turfs are communities of organisms dominated by aggregations of unicellular to branched filamentous algae and cyanobacteria (blue-green algae). They are attached to hard substrates, rock, wood or plant stems. Most major groups of algae provide species that occur in turfs: green algae in fresh waters, brown and red algae in marine waters and diatoms and blue greens universally. In addition, they support numerous epiphytes and enmeshed algal unicells, and like grasslands and lawns, are often maintained by fish and invertebrate grazers in the face of competition by larger algae. It is likely that there are tens of thousands to hundreds of thousands of algal species that can form algal turfs. While algae provide algal turf's primary structure, by definition, these are biodiverse communities containing numerous protozoans and small invertebrates. They are sometimes referred to by other names *e.g.*: periphyton, aufwuchs, algal mats, and biofilms.

Why algal turfs are so productive

Given strong current, surge or wave action, sunlight and regular grazing, algal turfs can be highly efficient at capturing solar energy. Most individual cells of algal turfs are photosynthetic; however, the extraordinary level of efficiency of algal turfs is also partly the result of high levels of mixing; flowing water, forced against cells by surge, greatly increases chemical exchange. In addition, photosynthesis in most higher plant and planktonic algal cells is biochemically inhibited in full sunlight, especially at high temperatures. Rarely are water temperatures as high in aquatic situations as in terrestrial environments, and of course, the typical problem of terrestrial plants: water loss, stomate closure and CO_2 cut off does not occur. Most important, however, because of the back and forth swashing of filaments in wave surge, individual cells receive flashing light and no cells are fully shaded by others. This allows a very high level of light capture, and typically, as measured by oxygen release, there is no inhibition even in full tropical summer sun at mid day. Equally important, a very high proportion of light energy captured is transferred to chemical storage as added biomass.

For many of the same reasons, algal turfs are only very weakly inhibited by low nutrient levels. Individual cells are able to uptake carbon, nitrogen and phosphorus at fractions of ppb levels. Since the water film adjacent to each cell cannot be exhausted of nutrients in a surge and flow environment, relatively high levels of productivity occur even at very low nutrient concentrations.

For decades, the dream of microalgae, planktonic algae grown in large vats or even ponds, for food, biochemicals and more recently bioenergy, has lived on. Because it is difficult to efficiently achieve high productivity levels and to harvest planktonic algae, intensive searches for unique species, and more recently to alter species by biotechnology, have been undertaken to solve these problems. However, at the end of a major bioenergy project at the U.S. Department of Energy (DOE) in the 1990's, it was concluded that their greatest difficulties lay in trying to maintain a monoculture - algal weeds are everywhere. As biodiverse ecosystems, algal turfs are not subject to these serious problems.

Algal Turf Scrubbers

An Algal Turf Scrubber is applied biomimicry—an ecologically engineered, algal turf dominated, mini-ecosystem. An individual Algal Turf Scrubber is a sloping plastic membrane surface, square meters to hectares in dimension, and has a rough surface or is overlain by a mesh screen of several millimeters in pore size. A thin layer of water (1-2 cm thick) is pulsed across the screen. Algal spores from the waters being treated settle quickly on the screens and grow rapidly, taking up nutrients and carbon dioxide as they grow.

Depending upon season and temperature, the resulting algal turf must be harvested every 5-15 days to maintain high levels of productivity. Given this regime, Algal Turf Scrubbers are considerably more productive than terrestrial crops. Traditional crop production during the summer in the United States is 2-4 grams dry weight/m^2/ day. Switchgrass, given intense fertilization, can be higher than corn and soy, but year-round production in the central U.S. is only 3-6g/m^2/ day. Unlike algal turfs, switchgrass provides tough stalks that require specialized, and, as yet unproven at industrial scale, enzyme breakdown. Algal turfs have less recalcitrant fibers and no lignin, and are more easily refined than even corn and soy—at the same levels of light, they can photosynthesize and produce new tissue at roughly 10 times that of typical crops (and 50 times that of forests.) Moreover, in central to southern areas of the country, Algal Turf Scrubbers can be operated year round. A breakdown of algal productivity can be found in the table below.

Algal turf scrubbers do not require fertilization; they operate at high production levels on waste nutrients. Even if nutrient levels on the major river systems on which operation is planned were to be reduced to pre-Columbian levels, production would remain moderately high because of the great efficiency of nutrient capture. Fish are the primary grazers of algal turfs, but they cannot function in the thin water layer of the scrubber. There are micrograzers of algal turfs such as chironomid larvae, but they are naturally kept in check by frequent harvest, and the dense bird populations that flock to large scrubbers. Diseases are not destructive because of the high diversity of algal turfs; individual species come and go to little effect.

Mature Technology

The above information is supported by dozens of experimental studies and hundreds of years of operational time (*i.e.*, multiple ATS over 25 years of operation). Moreover, ATS systems have received over 15 years of research and development testing followed by a ten-year history of pilot plant construction and operation for water quality control. This not only includes ATS systems for water treatment, but also for aquaculture. Two full-scale plants of tens of millions of gallons per day were operated for finfish production in Florida and Texas (one for over ten years) with great technical success. Finally in 2005, the University of Florida provided proof of concept with respect to the economic viability of ATS. This study reported on a one hectare ATS pilot system, normalized to a 56-acre, 50-year operation, and was funded and monitored by the South Florida Water Management District and the IFAS Institute of the University of Florida.

The Competition

Terrestrial Plants: Corn and soy are currently cultivated to produce gasohol, biodiesel, and methane. To achieve even moderate productivity, they require fertilization and often treatment with herbicides and pesticides. Diseases can become serious hazards in these extensive monocultures. Even with best management practices, nutrients and toxic compounds will find their way into already highly overburdened lakes, rivers and coastal waters. Most of the production of these plants in temperate zones occurs in summer, with die off or aestivation in winter.

Microalgae: Most of the algal production work for the last half-century has been carried out on plankton (*i.e.*, in large vats or ponds). These "microalgae" systems have a disadvantage in that the cells are free in the water column. The cells move with the water and are difficult to "mix"; being unattached, they carry their local water film with them and that film acts as a barrier to metabolite exchange; shading and light inhibition are difficult to avoid. Moreover, harvest is difficult, in the past requiring expensive centrifugation. There are newer methods of precipitation for microalgae, but they require chemical treatment and a sludge that must be managed. These are batch processes requiring continual management.

For the last decade, considerable effort has been expended to improve microalgal production by searching for and culturing specialized species and by genetically altering species to provide greater solar capture or more desirable storage products. However, these types of approaches have two major drawbacks: the basic problems of invasive species and monoculture.

In terms of biomimicry, ATS systems are like membranes at the cellular level. Efficiency is likely to be greater, given the same solar input and driving energy, on an ATS, than in a pond or vat. Also, ATS systems are communities and ecosystems; they employ local pre-existing species; since the biodiversity is high, diseases are not a factor in long-term production. None of the primary challenges of modern agriculture: herbicides, pesticides, fertilizers, soil depletion, drought and diseases are significant factors with ATS.

It is our belief that not only are ATS systems economically more competitive than microalgal pond systems, they are easier to maintain, require less energy and land for production, and produce more useable biomass than competitors. More importantly, the Algal Turf Scrubber is based on a biodiverse ecosystem, not a monoculture, and therefore has the advantage of being "local", stable, and sustainable.

Environmental Benefits of Algal Turf Scrubbers

ATS systems, through the process of cleaning water, produce multiple environmental and social benefits. The ultimate result is clean, healthy water, as well as restored waterways. This is accomplished in several ways:

Environmental Benefits

As in most photosynthesis, algal turfs abundantly release oxygen; in ATS systems, oxygen is dissolved into the overflowing water. It is not unusual for water flowing off an ATS plant in the afternoon to be highly supersaturated. In waters needing extensive bioremediation, such as the Chesapeake Bay and the Gulf of Mexico, large ATS plants would have a significant and immediate impact on downstream dead zones.

Nitrogen and Phosphorous

ATS systems are well known for their abilities to "scrub" nitrogen and phosphorus. Their lack of sensitivity to nutrient concentration, until extremely low levels are achieved, provides the ability to accomplish high water purity. Equally important, when high biomass production is desired, it is unlikely that fertilization will ever be required, except for the most specialized of uses.

Carbon Dioxide

In some fresh waters, mostly those of moderate quality and higher pH, readily available CO_2 in the water can be exhausted during the afternoon, as pH in water flowing down a large ATS is driven to over 10, affecting productivity. A simple solution to this problem is to link CO_2 in power plant stack gases to large-scale ATS systems. While this can be done with today's technology, an insertion process under development would greatly improve transfer efficiency, and would increase the CO_2 scrubbing aspects of ATS from simple atmospheric carbon removal to pre-atmospheric, industrial removal. Theoretically, ATS systems, at very large scale, are capable of not only blocking power plant release of CO_2 to the atmosphere but of significant draw down of existing atmospheric CO_2.

Heavy Metals

It has been known for a half century that algal cell walls adsorb heavy metals, and it is a characteristic of ATS phytoremediation, that heavy metals are removed from treated waters and sequestered into the algal biomass. This is an added value of ATS water cleaning, especially in waters with an industrial or urban component.

Toxic Organics

Combined with solar ultraviolet, ATS systems with high oxygen supersaturation break down entrained hydrocarbons. There is an extensive general research literature on this process, and a single ATS research study in the late 1990's demonstrated that when combined with artificial ultraviolet, ATS systems have considerable capability of breaking down a variety of chlorinated hydrocarbons. Additional research is necessary in this area, as a single study, even with a considerable laboratory support, cannot provide the basis for systems engineering; however, it is likely that if ATS is carried out at very large scale to produce bioenergy, it will significantly reduce toxic hydrocarbons in aquatic environments and ultimately the ocean.

Byproduct Potential of Algal Turf Scrubbers

In addition to the environmental benefits of ATS systems, there is a large potential for valuable byproduct use and reuse to create new economic opportunities. The key to realizing the full potential of Algal Turf Scrubbers lies in approaching solutions from a whole-systems perspective. Using ATS interconnected to other existing technologies, such as fermentation systems or anaerobic digesters, the potential for byproduct reuse increases dramatically. Not only do you get clean water and nutrient removal, you have the potential to sequester industrial CO_2, develop new sources of renewable energy, and create new sources of fertilization and animal feed. This is accomplished by re-imagining how value is defined and produced. Trapped within the confines of one technology or solution, it is naturally impossible to extend past the core competency of any technology to create more abundant value. However, like nature, when connected to other technologies, emergent qualities in the system produce a cascade effect. One waste or byproduct becomes the input for another byproduct. In this way, we can wring every amount of value out of the system, while simultaneously eliminating waste and improving the environment. It is a win-win-win situation.

Biodiesel and Energy Production

Bioenergy, as biodiesel, gasohol and methane from corn, soy and waste biomass, has become an important component of U.S. energy use. However, bioenergy remains at only a few percent of total U.S. requirements, and the most optimistic estimates would put it in the maximum range of 15-20% of imported oil (see e.g., Lovins *et al.*, 2004. Winning the Oil Endgame). The primary difficulty with agricultural bioenergy is that at current corn, soy, switchgrass levels of productivity, to match imported oil, all the agricultural land in the U.S. would be required. That could perhaps be cut in half with extremely intensive culture, but the fertilizers and pesticides required to do that would provide an ecological disaster. At an MIT seminar in 2005 on the U.S. energy crisis and its relationship to global warming, the participants could ascribe only a minor role to bioenergy and suggested nuclear as the only, but admittedly difficult option. On the other hand, some major organizations, such as NRDC, support a bioenergy option.

The U.S. Department of Energy (DOE) funded a number of projects to use microalgae for oil production in the 1990's. Being planktonic, systems of moderate productivity and low economic efficiency, they were not competitive at then existing prices, but they did provide the basis for understanding the conversion of algal biomass to biofuels – mostly either biodiesel or methane. An analysis of the algal product from the S-154 test facility (www.hydromentia.com) provided 55% volatile solids of dry weight, which is more favorable for direct energy conversion than the algae of the DOE study. The likely range based on IFAS economics is $0.85-$1.00/gal FOB at an ATS-plant. At roughly $40/bbl, this is highly competitive with fossil fuels today. When coupled with potential nitrogen and phosphorous credits, the economic potential becomes compelling, with costs below $0.40-$0.50 cents a gallon.

This is an area where economic research is required – namely what is the cheapest way to convert ATS algal turf to a useable energy product? Microalgae researchers usually settle on methane, though many are searching for species high in oil, or genetically engineered species, to achieve greater productivity. ATS scrubbing of moderate nutrient waters, tend to produce a biomass that is rich in diatoms (which carry their storage products as oil), and diatoms were probably the source of most petroleum. This is likely to tip the scale to a biodiesel product, though several other options are also viable.

ATS System Engineering

During the period from 1980-1991, ATS analysis consisted primarily of basic laboratory research on bench scale to greenhouse-scale systems treating waters ranging from highly oligotrophic model ecosystems to highly loaded aquaculture, chicken manure and raw sewage. Some of this work continued through the 1990's, adding toxic chlorinated hydrocarbon breakdown and extending to treating mesocosm systems of several million gallons. This research was mostly carried out by Walter Adey's Ecological Systems Technology, Inc. and the S.I.'s Marine Systems Laboratory and appeared in many publications (see Bibliography).

Beginning in the early 1990's, ATS analysis expanded to small, pilot field systems, ranging from a Florida Everglades agriculture water treatment system, to secondary and tertiary treatment of sewage at high and low ammonium levels, to finfish aquaculture ranging up to 500,000 gpd. This began the engineering phase as it related to design for maximum efficiency. Most of this work was carried out by Aquatic BioEnhancement Systems, Inc.

In the late 1990's and through to the present day, formal engineering to maximize efficiency of the ATS process was undertaken by the engineering staff of HydroMentia, Inc. HydroMentia has designed, constructed and operated several full-scale systems ranging up to 40Mgpd for aquaculture, agricultural and municipal non-point source treatment. A number of lake and stream treatment systems are currently under construction. Based on an engineering report for one of those systems, written for the South Florida Water Management District (SFWMD), an economic analysis was prepared by the IFAS Institute of the University of Florida (http://edis.ifas.ufl.edu/FE576). Note that ATS are sometimes called Managed Aquatic Plant Systems (MAPS).

References

Algal Turf Scrubber. (1982). United States Patent No. 4,333,263. Assigned to the Smithsonian Institution.

Cherfas, J. (1988). The Ocean in a Box. New Scientist **1602**: 46-50.

Coral Reef (2000). Ecosystems and human health: Biodiversity counts! *Ecosystem Health* **6**: 227-236.

Coral Reefs (1998). Algal structured and mediated ecosystems in shallow, turbulent alkaline seas. *Review Article J. Phycology* **34**: 393-406.

Coral Reefs (1998). Conservation by valuation and the utilization of pharmaceutical potential. *In*: Coral Reefs: Challenges and Opportunities for Sustainable Management. 5th Annual World Bank Conference on Environmentally and Socially Sustainable Development, (1997). Washington D.C. pp. 72-75.

Genetic Resources. **109**: 33-42. *In*: Sheppard C. (*Ed.*), Seas at the Millenium. Elsevier.

Hackney, J. and Sze, P. (1988) Photorespiration and productivity rates of a coral reef algal turf assemblage. *Mar. Biol.* **98**: 483-492.

Lessons learned from the construction and operation of coral reef microcosms and mesocosms (2006). (*In*: Precht, B.) Restoration of coral reef ecosystems. Book for CRC Press (In Press).

Marine Microcosms (1987). *In*: Jordan, W., Aber, J. and Gilpin, M. (*Eds*). Restoration ecology: a synthetic approach to ecological research. Cambridge University Press. p. 342.

Miller, J. (1980). Reef Alive. *Science News* **118**: 250-251

Miller, J. (1985). Diet for a blue planet. *Science News* **127**: 220-221.

NMNH and Related Publications Relating to Algal Turfs, ATS and the Use of ATS in Ecosystem Modeling (Adey publications, except as noted).

Sears, J.J. and Brawley, S. (1982). Smithsoniella gen. nov. A possible evolutionary link between the multicellular and siphonous habits in the Ulvophyceae, Chloroplyta. *Am. J. Bot.* **69**: 1450-1461.

The microcosm: a new tool for reef research (1983). Coral Reefs **1**: 193-201.

Walter Adey and Jeffrey Bannon (with Small A. and Spoon D.) (1998) Are current estimates of coral reef biodiversity too low? The view through the window of a microcosm. *Atoll Research Bulletin* **458**: 1-20.

Walter Adey and Jeffrey Bannon (with A. Small) (2001) Reef corals, zooxanthellae and free-living algae: a microcosm that demonstrates synergy between calcification and primary production. *Ecological Eng.* **16**: 443-457.

Walter Adey and Jeffrey Bannon (with Small A., Spoon D. and McConnaughey T.) (2000). Coral Reefs: Endangered, Biodiverse.

Walter Adey and Jeffrey Bannon (with Brawley S.) (1981). Micrograzers May Affect Macroalgal Density. *Nature* **292**: 177.

Walter Adey and Jeffrey Bannon (with Brawley S.) (1981). The effect of micrograzers on algal community structure in a coral reef microcosm. *Marine Biology* **61**: 167.

Walter Adey and Jeffrey Bannon (with Brawley S.) (1982). *Coralliophila abbreviata*: A Significant Corallivore! *Bull. Mar. Sci.* **32**: 595.

Walter Adey and Jeffrey Bannon (with Luckett C. and Jensen K.) (1993). Phosphorus removal from natural waters using controlled algal production. *Restoration Ecology* **1**: 1-11. (1995) Controlled Ecologies. Entry in *Encyclopedia of the Environmental Sciences*. p. 6. 5 figs. Academic Press.

Walter Adey and Jeffrey Bannon (with Luckett C. and Smith M.) (1996). Purification of industrially-contaminated ground waters using controlled ecosystems. *Ecological Engineering* **7**: 191-212.

Walter Adey and Jeffrey Bannon (with C. Luckett, Morrissey J. and Spoon Don). (1996). Coral reef mesocosms and microcosms - Successes, problems and the future of laboratory models. *Ecological Engineering* **6(1-3)**: 57-72.

Walter Adey and Jeffrey Bannon (with Rogers C. and Steneck R.) (1979). The South St. Croix Reef: A Study of Reef Metabolism as related to Environmental Factors and an Assessment of Environmental Management. Report to the Virgin Islands Government. Charlotte Amalie. p. 95.

Walter Adey and Jeffrey Bannon (with Griffith P., Cubit J. and Norris J.) (1987). Computer automated flow respirometry: metabolism measurements on a coral reef flat and in a microcosm. *Limnology and Oceanography*, **32**: 442-451.

Walter Adey and Jeffrey Bannon (with Hackney J. and Carpenter R.) (1989). Characteristic adaptations to grazing among algal turfs on a Caribbean Coral Reef. *Phycologia* **23**: 224-250.

Walter Adey and Jeffrey Bannon (with Hackney J.M.) (1989). The composition and production of tropical marine algal turfs in laboratory and fields experiments. p. 135. *In*: Adey W. and Farrier D. (Eds). The Biology, Ecology, and Culture of *Mithrax spinosissimus* Utilizing Cultured Algal Turfs, Mericulture Institute, Washington, DC.

Walter Adey and Jeffrey Bannon (with Loveland K.) (1998). Dynamic Aquaria. Revised and expanded 2nd edition. p. 498. Academic Press.

Walter Adey and Jeffrey Bannon (with Loveland K.) (1991). Dynamic Aquaria: Building Living Ecosystems. Academic Press. p. 643.

Walter Adey and Jeffrey Bannon (with Finn M., Kangas P., Lange L., Luckett C. and Spoon D.) (1996). A Florida everglades mesocosm - model veracity after four years of self organization. *Ecological Engineering* **6(1-3)**: 171-224.

Walter Adey and Jeffrey Bannon (with Kangas P.) (2006). Ecosystems: Greenhouses. (*In*: Encyclopedia of Ecology, Elsevier (In Press).

Walter Adey and Jeffrey Bannon (with Kangas P. (*Eds*.)) (1996). Mesocosms and ecological engineering. *Special Issue of Ecological Engineering*, with editorial. **6(1-3)**: 1-6.

Walter Adey and Jeffrey Bannon (with Carpenter R. and Hackney J.) (1991). Measurements of primary productivity and nitrogenase activity of coral reef algae in a chamber incorporating oscillatory flow. *Limnol. Oceanog*. **36**: 40-49.

Walter Adey and Jeffrey Bannon (with Craggs R., Jessup B. and Oswald W.) (1996). A controlled stream mesocosm for tertiary and higher treatment of sewage. *Ecological Engineering*. 6(1-3): 149-170.

Walter Adey and Jeffrey Bannon (With Craggs R., Jensen K., St. John M., Green F.B. and Oswald W.) (1995). Phosphorus removal from wastewater using an algal turf scrubber. Proc. Int. Assoc. Water Quality. p. 11.

Walter Adey and Jeffrey Bannon (with Steneck R.) (1985). Highly productive eastern Caribbean reefs: synergistic effects of biological, chemical, physical and geological factors. *NOAA Symp. Ser. for Undersea Res*. **3(1)**: 163-187.

Walter Adey and Jeffrey Bannon (with Williams S.) (1983). *Thalassia testudinum* Banks ex Konig. Seedling success in a coral reef microcosm. *Aquatic Bot*. **16**: 181-188.

Walter Adey and Jeffrey Bannon (with Goertemiller T.) (1987). Coral reef algal turfs master producers in nutrient poor seas. *Phycologia*. **26**: 374-386.

Walter Adey and Jeffrey Bannon (with McConnaughey T. and Small A.) (2000). Optimizing calcification on the coral reef. *Limnology and Oceanography* **45**: 1667-1671.

Water purification system and apparatus – microcosm system integration with ATS. (1992). U.S. Patent no. 5, 097,795.

Water purification system and apparatus - Phosphorous precipitation in algal cell walls (1998). U.S. Patent no. 5,851,398.

Water Purification system and apparatus (1990). United States Patent 4,966,096:1-14, 3 drawings.

13

POTENTIAL IN COMPLIANCE AND VOLUNTARY CARBON MARKET FOR ALGAL BIODIESEL

Lokesh Chandra Dube[1*] and Ankit Kumar

Abstract

Algae are fast growing eukaryotes and contribute more than two third of the planet's photosynthesis. Concerns of sustainability and food security linked with the limited availability of land for cultivation of biodiesel crops and oil seed bearing trees have resulted in growing attention on algae based biodiesel production. Algae are the oldest and original oil producers on earth. Biodiesel produced by algae is considered as Green House Gas neutral. Oil producing macro and micro algae thus present promising Climate Change mitigation opportunities. The investment and other barriers could be overcome by carbon credit benefits coming from compliance and voluntary carbon market. Clean Development Mechanism (CDM) of the Kyoto Protocol is attractive option for implementation of algal biodiesel based Climate Change mitigation projects in developing countries. There are opportunities in voluntary carbon markets also driven by the corporate social responsibility to offset emission footprints of activities, products and services of individuals and companies. Climate Change mitigation opportunities in algal biodiesel sector include- Bio-sequestration in green algal biomass; Biodiesel production and consumption as a replacement of conventional fossil fuel in energy generation activities; and energy generation from de-oiled algal biomass. Eligibility of these options has been assessed in the paper under the CDM and VCM (Voluntary carbon market) from the methodological aspects. Based on the assumptions from past studies

1 Emergent Ventures India Pvt. Limited, Gurgaon, India

* *Corresponding author*: E-mail: lokesh@emergent-ventures.com

it is estimated that the algal biodiesel alone can generate 30 carbon credits per ha per year, offering the revenue of • 360 per hectare per year. This is an estimate and varies on actual performance of algae, biodiesel plant and power plant.

Key words: Carbon market, Algae Biofuel, Clean Development Mechanism (CDM), Kyoto Protocol, Green House Gases (GHG)

Algae for Biodiesel

Algae are the oldest and original oil producers on the Earth. Algae are diverse class of fast growing eukaryotes and contribute significantly to the planet's photosynthesis. Algae are very fast growing-faster than oil yielding plants that makes them competitive for biodiesel production. Microalgae commonly double their biomass within 24 h.

Since oil yielding algae are generally non edible, there are hardly any sustainability and food security issues associated. Biodiesel crops require considerable land area for their cultivation. There is limited availability of land for cultivation of biodiesel crops and oil seed bearing trees. Algae require less area to produce equal amount of biodiesel (Table 1).

Table 1. Oil yield from oil yielding crops vis a vis algae, on unit land area (Moholkar *et al.*, 2008)

Crop	***Oil in Liters per hectare***
Castor	1413
Sunflower	952
Palm	5950
Soybean	446
Jatropha and *Pongamia*	2000
Algae	100000

Biodiesel from sustainably cultivated Algal biomass sources are Green House Gas neutral. Oil producing macro and micro algae thus provide a good opportunity for climate change mitigation. Microalgae mass cultures could contribute to the mitigation of greenhouse gas (GHG) emissions by enabling the direct utilization of fossil CO_2 streams from concentrated sources and the production from the algal biomass of renewable bio-fuels. But, algae based biodiesel projects face typical implementation issues. These are, among others-

- Technical barrier- Know how of algal biodiesel production and impacts of utilization is still in infancy. Developing countries hardly have any technology of algal biodiesel production. It is not a common practice and is not under the spotlight. Some examples of such barriers are: Reliable source of radiant energy-If sunlight is the sole source of radiant energy, poor light penetration into the pond becomes a problem. Increased temperatures decreased the solubility of the gases in water, this may hamper algal growth. Algal harvesting techniques and optimizing transesterification conditions are also challenges for industrial production.
- Financial barrier – There are limited options of investment. Further the cost of biodiesel production from algae is higher than most of the vegetable oils and even petroleum diesel. Based on Hill's model Moholkar *et al.* (2008) compared (Table 2) the cost of production in USD/Liter. Authors have converted it in energy terms.

Table 2. Comparison of cost of production

Fuel	*Cost of production in USD/Liter*	*Cost of production in USD/MJ*
Algal Biodiesel	1.20	0.06
Crude palm oil	0.52	0.03
Petro-diesel	0.73	0.02

- Economics of producing microalgal biodiesel need to improve substantially to make it competitive with petrodiesel (Chisti, 2007), but the level of improvement necessary appears to be attainable with the help of CER revenues.
- Policy and Regulatory issues- The Aquatic species program, started by the US Department of energy in 1978 aimed to identify microalgae with potential for biodiesel production. Through screenings in many participating laboratories, 300 strains of potential biodiesel producers were identified. It was seen that lipid production increased under nutritional stress with a simultaneous decrease in the rate of cellular division. In order to genetically increase lipid content enzyme Acetyl CoA Carboxylase (ACCase) was isolated from diatom *Cyclotella cryptic* and then cloned and used to transform diatoms successfully, however, these experiments did not show a significant increase in lipid content. In 1995 the program was stopped due to lack of funding.

Carbon credit benefits from compliance and voluntary carbon market can help to overcome these barriers. USDOE (1998) also advocates that greenhouse gas mitigation credits would likely be of great considerations in any future externalities cost accounting.

Global Carbon Market

International carbon market can be categorized into two structures, first the Compliance market, also called regulatory market and second voluntary market. Compliance market is further categorized as Kyoto type that allows transactions under Kyoto Protocol to the United Nations Framework Convention on Climate Change (UNFCCC) in terms of well defined market based flexibility mechanisms like Clean Development Mechanism (CDM), Joint Implementation (JI) and International Emission Trade (IET); and Non Kyoto type that have been developed as many state legislations in the USA and as New South Wales GHG Abatement Scheme (NSW GHG AS) in Australia. Second structure of the carbon market is Voluntary market; segments of Voluntary Carbon Market (VCM) are legally binding ones like Chicago Climate Exchange (CCX) and the Over the Counter (OTC) market that includes the activities of companies and individuals who offset emission footprints of their activities, products or services. These companies or individuals often purchase Verified/Voluntary Emission Reductions (VERs) in small quantities. These VERs are usually not intended for compliance.

Voluntary Carbon Market has developed considerably in the past few years as an alternative to the Compliance Market. The Parties such as the USA, Canada and others have no legally binding targets as they have not ratified the Kyoto Protocol. Companies and individuals in these nations are realizing that they must take action to offset and reduce their emission footprints, as a sense of Corporate Social Responsibility or to gear up for possible future legislations. In comparison with compliance markets, voluntary markets are less mature and trade in lower volumes. There is no umbrella regulator in the voluntary carbon market.

Climate Change Mitigation Opportunities

Several types of renewable biofuels can be derived from algal biomass. These include; biodiesel derived from algal oil (Banerjee *et al.*, 2002; Gavrilescu and Chisti, 2005); methane produced by anaerobic digestion of the algal biomass (Spolaore *et al.*, 2006) and photobiologically produced biohydrogen (Ghirardi *et al.*, 2000; Kapdan and Kargi, 2006).

Bio-sequestration in green algal biomass

Algal biomass sequesters atmospheric carbon dioxide. But since this biomass is neither perennial nor woody, it stores carbon for very limited time and hence is hardly attractive for climate change mitigation. It has no place in carbon market. In controlled conditions where sustainable retention of biomass could be assured, this may be an option in future carbon markets.

Oceans are abode of algae, ocean carbon sequestration thus holds option for climate change mitigation, but this process is not a part of carbon market as of now. One possible way to increase the carbon sequestration potential of the ocean is to fertilize the surface waters with iron. The growth rate and amount of algae may depend upon iron. In general, the abundance and productivity of phytoplankton in the ocean depends a great deal on nutrients—especially nitrogen, phosphorous, and iron. Algae can be grown in manufactured ponds just about anywhere there is open land and plenty of sun light and it can be grown in fresh or salt water.

Biodiesel

Algae produce oil that can be transesterified to produce algal biodiesel. Algal biodiesel can be used as a replacement of conventional fossil fuel in energy generation activities for stationary and mobile combustion or power generation. Algal biodiesel holds good potential and is eligible under Clean Development Mechanism and Voluntary Carbon Market for generation of emission reductions. However no methodology for baseline and monitoring has been approved by CDM Executive board and any VCM registry/standard for algal biodiesel till date.

The Approved CDM Methodology III.T. Plant oil production and use for transport applications would also not be applicable for algal oils (non-transesterified), as it uses the term- "cultivation of oil seeds for production of plant oil". Similarly AM0047 is applicable for biodiesel from biogenic waste oil.

For development of new CDM methodology in this sector following points have to be taken care of (Dube, 2006):

1. Export of biodiesel produced to Annex I parties (developed country party to the Kyoto protocol) should be prevented either through applicability condition or by excluding the exported part from emission reduction calculations.

2. Blending proportion only over and above as mandated by host country specific policy/regulation would generate carbon credits.
3. Species selection, if included in methodology must clearly give the appropriate justifying reasons associated.
4. Combustion efficiency of engine with baseline fuel and project fuel (BDF) must be taken care of as well as the difference between heating values (NCVs) of the two fuels, this also affects the consumption of fuel per unit of distance travelled by vehicle (specific fuel consumption-SFC). Historic data is required for SFF calculation in the case of BLF (baseline fuel). SFC is more in the case of BDF.
5. The methodologies should cover explicit limitations and keep away from overlapping between procedures, steps, and processes to avoid possibilities of double counting.

Emission sources that may directly be attributed to the project activity and are measurable should be counted under the project activity emission. No business as usual or activity in common practice should be taken into account for such emission calculations Fig. 1.

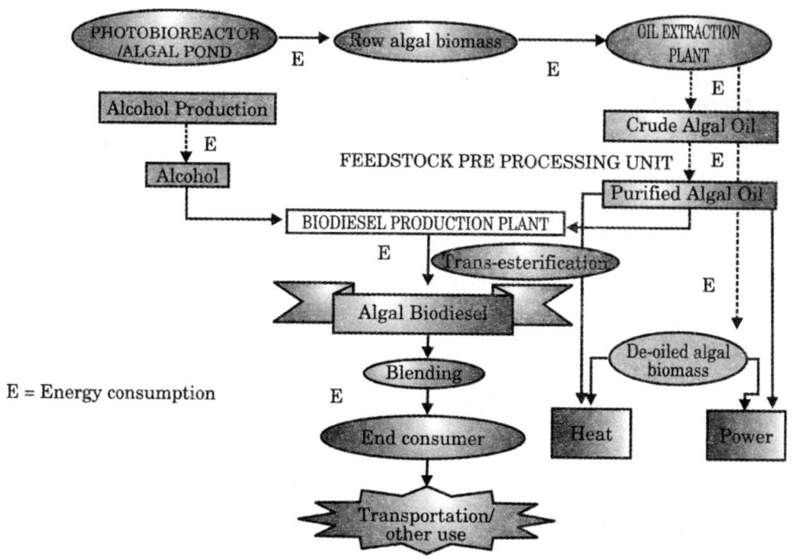

Fig. 1. Schematic representation of Biodiesel production by trans-esterification

Emissions from algal ponds and mechanical harvesting may include CO_2 emission from petroleum based fuel use in vehicles/farming appliances and electricity consumption. Fuel and electricity use in crushing plant contribute in emission from feedstock production stage, however the project proponent can also use the emissions from solvent use in the case of a chemical crushing plant. Transportation related emissions can be included in the project boundary arising from vehicles used inside the production plant (project boundary). However, emissions from transportation activities outside the project boundary, attributable to the project activity should be calculated as the leakage emissions.

Emissions from feedstock pre-processing include those arising from the use of electricity, hot water and steam in pre-treatment of feedstock mix and the purification of algal oil. Emissions from biodiesel production may include the energy consumption and production of steam in biodiesel producing plant.

Emission from biodiesel consumption is treated as carbon neutral, but in the process of trans-esterification some fossil carbon part from alcohol remains attached in the biodiesel, which is released upon combustion in engines, thus should be considered in calculations.

Wastewater treatment from the biodiesel plant can produce both, CO_2 from electricity use in aerobic processes and CH_4 from anaerobic treatment and should be calculated in Project activity emissions.

Algal Biomass

By-product oil cake can be used as manure, as fuel for boilers and also as a feed for bio-methanation process. Emission reduction can be claimed for fertilizer replacement also. If used as feed for digester and/or as fuel in boilers in biodiesel production plant, this could be adjusted in ER calculations.

Energy generation can also be done from de-oiled or dried algal biomass outside the biodiesel plant and is also an eligible activity for Carbon Credit generation. This can be done through direct burning, gasification or biogas production.

Approved methodologies for baseline and monitoring are available for this activity. These include:

ACM 0006	Consolidated methodology for grid-connected electricity generation from biomass residues
AMS-I.A.	Electricity generation by the user
AMS I.C.	Thermal energy for the user

Algae technology can extend the useful energy we get from coal combustion and reduce carbon emissions by recycling waste CO_2 from power plants into clean-burning biodiesel.

From 1978 to 1996, the U.S. Department of Energy's Office of Fuels Development funded a program to develop renewable transportation fuels from algae. The main focus of the program, know as the Aquatic Species Program (or ASP) was the production of biodiesel from high lipid-content algae grown in ponds, utilizing waste CO_2 from coal fired power plants. This may reduce the cost of CO_2 in photobioreactors.

Carbon dioxide from power plants can be captured and can be used to feed algal growth in oceans or in man made ponds (Fig. 2). The ocean carbon sequestration can be enhanced by Iron and other nutrient inputs. Studies suggest that microalgal cells develop high lipid contents in limited supply of Nitrogen (USDOE, 1998, 1999). Algal biomass and biodiesel can then be used for green house gas neutral combustion for stationary and mobile combustion in heat and/or power generation activities. The components of above process are shown in the Fig. 2.

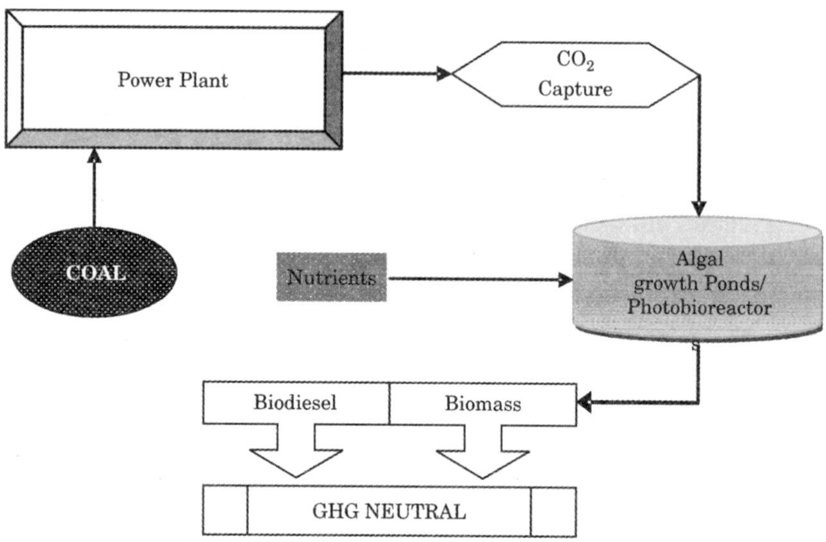

Fig. 2. Carbon dioxide capture from power plant and use in photo bioreactor/ algal pond

Residual biomass from biodiesel production processes can be used potentially as animal feed. Some of the residual biomass may be used to produce methane by anaerobic digestion, for generating the electrical power necessary for running the algal biomass production facility.

Mitigation Potential

Authors have estimated emission reduction potential in above identified algae based climate change mitigation options. Following assumptions were made based on literature reviews:

1. Algal biomass production rate: Biomass production rate differs from species to species and depends on rate of photosynthesis vis a vis rate of respiration (net primary productivity). We have assumed a production rate of 43.8 t/ha/year, which is a conservative estimate (Web 1).

   Oil content in algae: Oil content is also found varying in different species. Spoehr and Milner (1949); Metting (1996); Spolaore *et al.* (2006) reported oil contents as high as 80% of the dry weight.

   Following are given oil contents in percentage of dry weight in some major microalgae species (Rao, 2008).

Anabaena cylindrical	4–7%
Botryococcus braunii	25–75%
Chlamydomonas rheinhardii	21%
Chlorella sp.	28–32%
Neochloris oleoabundans	35–54%
Nitzschia sp.	45–47%
Schizochytrium sp.	50–77%

   Oil levels of 20–50% are quite common. We have assumed 50% oil content in algae for estimations.

2. Conversion to Biodiesel: Biodiesel conversion ratio refers to the ratio of total trans-esterified algal oil and oil consumed in conversion to biodiesel. We have assumed 95% conversion to biodiesel.

3. Heating Values and Emission Factors- IPCC Default values of Net Calorific Values and CO_2 Emission Factors have been used.

	NCV MJ/kg	***EF t C/TJ***	***Oxidation Factor***	***Emission Factort CO_2/t fuel***
Algal Biodiesels	27.0	19.3	1	1.911
Diesel Oil	43.0	20.2	1	3.185

4. Carbon credit value: Price of Carbon credits (CERs) is assumed to be 12 Euro/t CO_2.

Findings

- When used as diesel blend Biodiesel production and consumption stage would generate 30 carbon credits per ha per year, offering the revenue of •360 per hectare per year. This is an estimate and varies on actual performance of algae and biodiesel plant.
- Theoretical potential of power generation from de- oiled algal biomass is 21 carbon credits per ha per year, offering around • 250.
- Heat generation of de-oiled algal biomass would generate around 20 credits per ha per year, generating the revenue of •235.

References

Banerjee, A., Sharma, R., Chisti, Y. and Banerjee, U.C. (2002). *Botryococcus braunii*. A renewable source of hydrocarbons and other chemicals. *Crit Rev Biotechnol* **22**: 245-79.

Chisti, Yusuf (2007). Biodiesel from microalgae. *Biotechnology Advances* **25**: 294-306.

Dube, Lokesh Chandra (2006). Development of New Baseline and Monitoring Methodology and Project Design Document for a Biodiesel CDM project. M.Phil. Thesis, School of Energy and Environmental Studies, DAVV, Indore.

Gavrilescu, M. and Chisti, Y. (2005). Biotechnology—a sustainable alternative for chemical industry. *Biotechnol Adv.* **23**: 471-99.

Ghirardi, M.L., Zhang, J.P., Lee, J.W., Flynn, T., Seibert, M. and Greenbaum, E. (2000). Microalgae: a green source of renewable H_2. *Trends Biotechnol* **18**: 506-11.

Kapdan, I.K. and Kargi, F. (2006). Bio-hydrogen production from waste materials. *Enzyme Microb Technol.* **38**: 569-82.

Metting, F.B. (1996). Biodiversity and application of microalgae. *J Ind. Microbiol.* **17**: 477-89.

Moholkar, V.S., Purkait, M.K., Jana, S. and Amrita, A. (2008). Microalgal route to biodiesel: prospects, potential and process design. 5th International Biofuel Conference. Feb 7-8, 2008, New Delhi, India. p. 85.

Rao, B.V.S.K. (2008). Importance of Microalgal Sources in the Production of Biodiesel. 5th International Biofuel Conference. Feb 7-8, 2008, New Delhi, India. p. 187.

Spoehr, H.A. and Milner, H.W. (1949). "The chemical compostition of *Chlorella*; effect of environmental conditions." *Plant Physiol.* **24**: 120.

Spolaore, P., Joannis-Cassan, C., Duran, E. and Isambert, A. (2006). Commercial applications of microalgae. *J Biosci. Bioeng* **101**: 87-96.

USDOE (1998). John Sheehan, Terri Dunahay, John Benemann, Paul Roessler. A Look Back at the U.S. Department of Energy's Aquatic Species Program—Biodiesel from Algae. Prepared for: U.S. Department of Energy's Office of Fuels Development. Prepared by: the National Renewable Energy Laboratory. P. 256.

USDOE (1999). U.S. Department of Energy. Office of Science. Office of Fossil Energy. Carbon Sequestration: State of the Science. Retrieved from www.doe.gov.

Web 1: http://peswiki.com/index.php/Directory:Biodiesel_from_Algae_Oil, accessed on 30 July 2008.

14

CARBON NEUTRAL BIODIESEL PRODUCTION FROM MICROALGAE: PERSPECTIVES AND ADVANCES

Shakeel A. Khan[1*] and Rashmi[2]

Abstract

Continued use of petroleum-based fuels is now widely recognized as unsustainable because of depleting supplies / stocks and the contribution of these fuels to pollute the environment. There has been greater awareness on biodiesel in India in the recent times due to scarcity of petrodiesel and soaring prices. One possible alternative to fossil fuel is the use of oils of plant origin like vegetable oils, algal biomass and tree borne oil seeds. Usage of biodiesel will allow a balance to be sought between agriculture, economic development and the environment. Unfortunately, biodiesel from oil crops, waste cooking oil and animal fat cannot realistically satisfy even a small fraction of the existing demand for fuels. Recent research initiatives have proven that microalgae biomass appear to be the one of the promising source of renewable biodiesel which is capable of meeting the global demand for transport fuels. Microalgae are photosynthetic microorganisms which convert sunlight, water and carbon dioxide to sugars, from which biological macromolecules, such as lipids and triacylglycerol (TAG) can be obtained. The TAG content of some microalgae is so high and can be promising sustainable feedstock for biodiesel production. Using microalgae to produce biodiesel will not compromise production of food, fodder and other products derived from

[1] Division of Environmental Sciences, NRL Building, Pusa Campus, Indian Agricultural Research Institute, New Delhi - 110 012, India

[2] Department of Pharmaceutical Technology (Biotechnology), National Institute of Pharmaceutical and Education Research, S.A.S. Nagar, Mohali - 160 062, India

* *Corresponding author*: E-mail: shakeel_iari@yahoo.com, shakeel_env@iari.res.in

crops. Producing microalgal biomass is generally more expensive than growing crops but it can be reduced substantially by using a biorefinery based production strategy. Microalgal-based carbon sequestration technologies can, in principle, not only cover the cost of carbon capture and sequestration but also produce environment friendly biodiesel. So, Biodiesel from microalgae could be an environmentally benign solution for energy security & rural economy of India.

Key words: Biodiesel, Microalgae, Renewable Energy, Carbon Neutral, Triacylglycerol, Biorefinery

Introduction

The world has been confronted with an energy crisis due to depletion of finite resources of fossil fuel. Continued use of petroleum-based fuels is now widely recognized as unsustainable because of depleting supplies and contribution of these fuels to global warming and environmental pollution. Almost half a billion vehicles on the world's roads contribute to half of the global oil consumption and generate about 20 percent of the world's greenhouse gases, including carbon monoxide, nitrous oxides, and particulates. India meets nearly 75-85% of its total petroleum requirements through imports. With the increase in share of hydrocarbons in the energy supply/use, this share of imported energy is expected to in exceed 90% by 2030 (MoPNG, 2006). With 1.1 billion people, India is the world's second most populous country behind China and is expected to have the world's largest population by 2030. Further population increases and the country's continued economic growth are expected to increase India's energy demand from 537 million tons of oil equivalent (Mtoe) in 2005 to 770 Mtoe in 2015 and to 1,299 Mtoe by 2030 (WEO, 2007)[2]. According to annual report (2006-07) of Ministry of Petroleum and Natural Gas (MoPNG), India has imported about 99 MT of crude oil during the year 2005-06, causing a heavy burden of Rs. 171702 crore on foreign exchange (MoPNG, 2007). For the 2007-08, the crude oil import bill was 272699 Crores ($68 Billion), that was 28% of the total import of the country and we have more than 75% of oil import dependency (Fig. 1). The demand for crude oil increases dramatically year by year. Every one dollar rise in crude oil prices inflates the import bill by about 700 million dollars per year. The challenge, therefore, is to secure adequate energy supplies at the least possible cost. The increasing import bill has necessitated the search for liquid fuels as an alternative to diesel, which is being used in large quantities in transport of industrial and agricultural sector. India's transportation fuel requirements are unique in the world. According to an estimate, automobiles alone contribute to 70% of the total diesel consumptions.

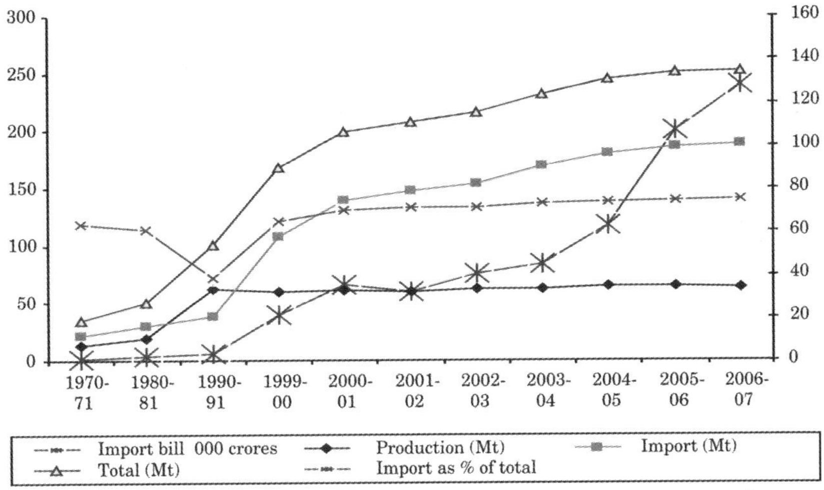

Fig. 1. Indian crude oil production and import

Source: Ministry of Petroleum and Natural Gas (MoPNG), Govt. of India. 2007

The country faces problems in regard to the fuel requirement for increased transportation demand. The demand for fuel has been growing continuously and it is expected that crude oil production will start declining from 2012. At the end of year 2007, the proved oil reserves of India was 0.7 thousand MT (5.5 thousand million barrel) and India share 0.4% of the world's total oil reserves with the reserve to production (R/P) ratio of 18.7 years (BP review, 2008). Proved reserves of oil, generally taken to be those quantities that geological and engineering information indicates with reasonable certainty can be recovered in the future from known reservoirs under existing economic and operating conditions. Reserves-to-production (R/P) ratio is the reserves remaining at the end of any year are divided by the production in that year; the result is the length of time that those remaining reserves would last if production were to continue at that rate. We consume almost six to seven times more diesel fuel than gasoline, whereas in the rest of the world, almost all the other countries use more gasoline than diesel fuel. Therefore, diesel is India's main liquid fuel; it burns roughly 64 million tonnes, or 450 million barrels, of the stuff a year, as opposed to about 84 million barrels of gasoline (MoPNG, 2007). Thus, our search for alternatives in India has to have a different emphasis. This is where biodiesel is comparatively much more important for us than for others. Many researchers have recommended the biodiesel as alternative fuels from vegetable oils. But unfortunately, biodiesel from oil crops, waste cooking oil and animal fat cannot realistically satisfy even a small fraction of the existing demand for fuels. In India, about 26 million hectares lands are used for the oilseed cultivation, mainly on marginal lands, dependent on monsoon rains (unirrigated) and with low levels of input usage. Yields are rather low at less than one ton per hectare (Mehta, 2005-06). With steady growth in population and personal income, Indian per capita consumption of edible oil has been growing steadily. Although, Indian vegetable oil economy is world's fourth largest after USA, China and Brazil. However, oilseeds output and in turn, vegetable oil production have been trailing consumption growth, necessitating imports to meet supply shortfall. So, for the country like us, which already depends on the imports of vegetable oil, have to see the other alternative for biodiesel production.

There is forceful need to explore the possibilities of producing biodiesel from microalgae, which would not compete with the land and cereal crops. The productivity of algal cultivation per area is 2 to 5 fold higher as compared with traditional agricultural crops and fast growing energy crops. While the mechanism of photosynthesis in microalgae is similar to that of higher plants, they are generally more efficient converters of solar energy because of their simple cellular structure. Recent research initiatives have proven that microalgae biomass appear to be the one of the promising source of renewable biodiesel which is capable of meeting the global demand for transport fuels. Being an agricultural country, requirement for diesel is more and we have to see the better alternative of diesel. Therefore, microalgal biodiesel is the only option to fulfill requirement in future.

Biodiesel Development in India

In the biodiesel sector, India has taken the initial steps toward commercial production. The focus of India's plans for biodiesel development and commercialization was the National Biodiesel Mission (NBM), formulated by the Planning Commission of the Government of India in April 2003. NBM had planned to be implemented in two phases. Phase I, has been carried out from 2003 to 2007 as demonstration project (PC report, 2003). The project involved the development of *Jatropha* oilseed nurseries, the cultivation of *Jatropha* on 400,000 hectares waste land, the setting up of seed collection and *Jatropha* oil expression centers, and the installation of 80,000 Mt/year transesterification to produce biodiesel from *Jatropha* oil. Phase II planned with a self sustaining expansion of the programme leading to the production of biodiesel to meet 20% of the country's diesel requirements by 2011-12. Based on extensive research carried out in agricultural research centers, it was decided to use *Jatropha curcas* oilseed as the major feedstock for India's biodiesel programme. The work accomplished so far includes developing high-yielding varieties of *Jatropha*, initiating *Jatropha* nurseries, setting up pilot-plants for biodiesel manufacture and testing biodiesel in public transport locomotives and buses. The lack of assured supplies of vegetable oil feedstock has foiled efforts by the private sector to set up biodiesel plants in India. So far only two firms, Naturol Bioenergy Limited (NBL) and Southern Online Biotechnologies, have embarked on biodiesel projects, both in the southern state of Andhra Pradesh. Naturol Bioenergy, a partnership between the Austrian biodiesel firm Energea Gmbh and the investment firm Fe Clean Energy, plans to install a 300 ton/day (90,000 t/year) biodiesel plant (Gonsalves, 2006). The State Government allocated 120,000 hectares of land for *Jatropha* cultivation to the firm but cultivation has not yet begun or in initial stage. The farmers were demanding that the market set the oilseed price, but NBL want the government to fix a price to reduce its risks in production. Southern Online Biotechnologies has a 30 ton/day (9000 t/year) project, which would require about 9500 t/year of oil. It was expected to get about 6000 t/year through cultivation of *Jatropha* and *Pongamia pinnata* oilseeds on wasteland, and plans to make up the balance through animal fats, but the results are not up to the level. So, there are many constraints for the biodiesel production in India and phase I of NBM has not given the expected results. Recent research and development on microalgae has exposed a new road ahead that can be promising sustainable feedstock for biodiesel production.

Microalgae as Emerging Feedstock for Biodiesel Production

Photosynthetic organisms that grew in aquatic environments include macroalgae, microalgae and emergents (Shechan *et al.*, 1998). Microalgae are, as the name suggests, microscopic photosynthetic organisms. These are primitive organism with a simple cellular structure and a large surface to volume body ratio, which gives them the possibility of large uptakes of nutrients. While the mechanism of photosynthesis in microalgae is similar to that of higher plants, they are generally more efficient converters of solar energy because of their simple cellular structure (Ghirardi *et al.*, 2000; Banerjee *et al.*, 2002; Melis, 2002; Metzger and Largeau, 2005; Singh *et al.*, 2005; Spolaore *et al.*, 2006; Walter *et al.*, 2005). In addition, because the cells grow in aqueous suspension, they have more efficient access to water, CO_2, and other nutrients. For these reasons, microalgae are capable of producing 30 times the amount of oil per unit area of land, compared to terrestrial oilseed crops. Algae, especially microalgae, appear to be a true solution, because of several advantages. Hu *et al.* (2008) described the following potential advantages of microalgae as feedstocks for biodiesel and biomaterials:

1. Synthesize and accumulate large quantities of neutral lipids/oil (20–50% dry cell weight).
2. Grow at high rates (*e.g.* 1–3 doublings per day).
3. Thrive in saline/brackish water/coastal seawater for which there are few competing demands
4. Tolerate marginal lands (*e.g.* desert, arid- and semi-arid lands) that are not suitable for conventional agriculture.
5. Utilize growth nutrients such as nitrogen and phosphorus from a variety of wastewater sources (*e.g.* agricultural run-off, concentrated animal feed operations, and industrial and municipal wastewaters), providing the additional benefit of wastewater bioremediation.
6. Sequester carbon dioxide from flue gases emitted from fossil fuel-fired power plants and other sources, thereby reducing emissions of a major greenhouse gas.
7. Produce value-added co-products or by-products (*e.g.* biopolymers, proteins, polysaccharides, pigments, animal feed and fertilizer).
8. Grow in suitable culture vessels (photo-bioreactors) throughout the year with annual biomass productivity on an area basis, exceeding that of terrestrial plants by approximately tenfold.

(a) Microalgae classification

Biologists have categorized microalgae in a variety of classes, mainly distinguished by their pigmentation, life cycle and basic cellular structure. The four most important (at least in terms of abundance) are:

1. *Diatoms (Bacillariophyceae)*: These algae dominate the phytoplankton of the oceans, but are also found in fresh and brackish water. Approximately 100,000 species are known to exist. All cells store carbon in a variety of forms. Diatoms store carbon in the form of natural oils or as a polymer of carbohydrates known as chyrsolaminarin.

2. *Green algae (Chlorophyceae)*: These are also quite abundant, especially in freshwater. Green algae are the evolutionary progenitors of modern plants. The main storage compound for green algae is starch, though oils can be produced under certain conditions.
3. *Blue-green algae (Cyanophyceae)*: Much closer to bacteria in structure and organization, these algae play an important role in fixing nitrogen from the atmosphere. There are approximately 2000 known species found in a variety of habitats.
4. *Golden algae (Chrysophyceae)*: This group of algae is similar to the diatoms. They have more complex pigment systems, and can appear yellow, brown or orange in color. Approximately 1000 species are known to exist, primarily in freshwater systems. They are similar to diatoms in pigmentation and biochemical composition. The golden algae produce natural oils and carbohydrates as storage compounds.

(b) Potential of microalgal biodiesel

Recent research initiatives have proven that microalgae biomass appear to be the one of the promising source of renewable biodiesel which is capable of meeting the global demand for transport fuels. Using microalgae to produce biodiesel will not compromise production of food, fodder and other products derived from crops because of its huge potential. A Comparison of some sources of crop oil is given in Table 1.

In view of Table 1, microalgae appear to be the only source of biodiesel that has the potential to completely displace fossil diesel. Unlike other oil crops, microalgae grow extremely rapidly and many are exceedingly rich in oil. Oil content in microalgae can exceed 80% by weight of dry biomass[18, 19]. Oil levels of 20-50% are quite common (Table 2) (Chisti, 2007). Oil productivity, that is the mass of oil produced per unit volume of the microalgal broth per day, depends on the algal growth rate and the oil content of the biomass. Oil content in microalgae can exceed 80% by weight of dry biomass. In addition, because the cells grow in aqueous suspension, they have more efficient access to water, CO_2, and other nutrients. For these reasons, microalgae are capable of producing 30 times the amount oil per unit area of land, compared to terrestrial oilseed crops.

(c) Composition of microalgal oil and transesterification

Several researchers have reported that algae produced more oil in stressed or unfavourable condition in comparison to optimal growth

Table 1. Comparison of some sources of crop oil for biofuel

Gallons of oil per Acre per Year	
Algae	***500-20000***
Oil Palm	635
Coconut	287
Jatropha	207
Rapeseed/Canola	127
Peanut	113
Sunflower	102
Safflower	83
Soybeans	48
Hemp	39
Corn	18

Table 2. Oil content of some microalgae

Microalgae	***Oil content (% dry wt)***
Botryococcus braunii	25–75
Chlorella sp.	28–32
Crypthecodinium cohnii	20
Cylindrotheca sp.	16–37
Dunaliella primolecta	23
Isochrysis sp.	25–33
Monallanthus salina	>20
Nannochloris sp.	20–35
Nannochloropsis sp.	31–68
Neochloris oleoabundans	35–54
Nitzschia sp.	45–47
Phaeodactylum tricornutum	20–30
Schizochytrium sp.	50–77
Tetraselmis sueica	15–23
Botryococcus braunii	25–75

condition. Under optimal conditions of growth, algae synthesize fatty acids principally for esterification into glycerol-based membrane lipids, which constitute about 5–20% of their dry cell weight (DCW). Fatty acids include medium-chain (C10– C14), long-chain (C16–18) and very-long-chain (≥C20) species and fatty acid derivatives. But under unfavourable environmental or stress conditions, many algae alter their lipid biosynthetic pathways towards the formation and accumulation of neutral lipids (20–50% DCW), mainly in the form of triacylglycerol (TAG). Unlike the glycerolipids found in membranes, TAGs do not perform a structural role but instead serve primarily as a storage form of carbon and energy. After being synthesized, TAGs are deposited in densely packed lipid bodies located in the cytoplasm of the algal cell, although formation and accumulation of lipid bodies also occur in the inter-thylakoid space of the chloroplast in certain green algae (Hu *et al.*, 2008). Oil production from microalgae can be performed by both photosynthesis and heterotrophy. The fatty acid composition of typical oil from microalgae has given in Table 3 (Meng *et al.*, 2009). It is mainly composed by a mixture of unsaturated fatty acids, as palmitoleic (16:1) oleic (18:1), linoleic (18:2) and linolenic acid (18:3). Saturated fatty acids, palmitic (16:0) and stearic (18:0) are also present to a small extent.

Table 3. Fatty acid composition of microalgal oil

Fatty acid	*Chain length: No. of double bonds*	*Oil composition (w/total lipid)*
Palmitic acid	16:0	12-21
Palmitoleic acid	16:1	55-57
Stearic acid	18:0	1-2
Oleic acid	18:1	58-60
Linoleic acid	18:2	4-20
Linolenic acid	18:3	14-30

TAGs consist of three long chains of fatty acids attached to a glycerol backbone. In 1980s, a chemical modification of TAG was introduced that helped to bring the viscosity of the oils within the range of current petroleum diesel. By reacting these TAGs with simple alcohols (a chemical reaction known as "transesterification"), we can create a chemical compound known as an alkyl ester, which is known more generically as biodiesel (Fig. 2). Its properties are very close to those of petroleum diesel. The reaction occurs stepwise: triglycerides are first converted to diglycerides, then to monoglycerides and finally to glycerol.

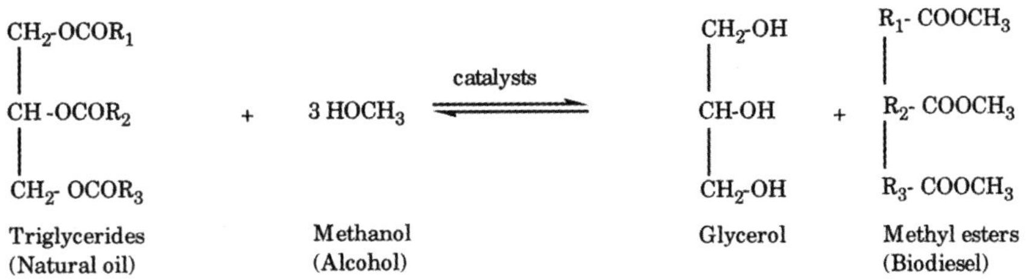

Fig. 2. Transesterification of triacylglycerols (TAGs)

Where R1, R2, and R3 are long chains of carbons and hydrogen atoms, sometimes called fatty acid chains. There are five types of chains that are common in soybean oil and animal fats (others are present in small amounts):

Palmitic	16:0	$R = -(CH_2)_{14} - CH_3$
Stearic	18:0	$R = -(CH_2)_{16} - CH_3$
Oleic	18:1	$R = -(CH_2)_7\ CH{=}CH\ (CH_2)_7CH_3$
Linoleic	18:2	$R = -(CH_2)_7CH{=}CH\text{-}CH_2\text{-}CH{=}CH\ (CH_2)_4CH_3$
Linolenic	18:3	$R = -(CH_2)_7\ CH{=}CH\text{-}CH_2\text{-}CH{=}CH\text{-}CH_2\text{-}CH{=}CH\text{-}CH_2\text{-}CH_3$

Transesterification requires 3 mol of alcohol for each mole of triglyceride to produce 1 mol of glycerol and 3 mol of methyl esters (Fig. 2). Industrial processes use 6 mol of methanol for each mole of triglyceride. This large excess of methanol ensures that the reaction is driven in the direction of methyl esters, *i.e.* towards biodiesel. Yield of methyl esters exceeds 98% on a weight basis (Fukuda *et al.*, 2001). Transesterification is catalyzed by acids; alkalis and lipase enzymes. Alkali-catalyzed transesterification is about 4000 times faster than the acid catalyzed reaction (Fukuda *et al.*, 2001). Consequently, alkalis such as sodium and potassium hydroxide are commonly used as commercial catalysts at a concentration of about 1% by weight of oil. Alkoxides such as sodium methoxide are even better catalysts than sodium hydroxide and are being increasingly used. Use of lipases offers important advantages, but is not currently feasible because of the relatively high cost of the catalyst. Alkali catalyzed transesterification is carried out at approximately 60°C under atmospheric pressure, as methanol boils at 65°C at atmospheric pressure. Under these conditions, reaction takes about 90 min to complete. A higher temperature can be used in combination with higher pressure, but this is expensive. Methanol and oil do not mix; hence the reaction mixture contains two liquid phases. Other alcohols can be used, but methanol is the least expensive. To prevent yield loss due to saponification reactions (*i.e.* soap formation), the oil and alcohol must be dry and the oil should have a minimum of free fatty acids. Biodiesel is recovered by repeated washing with water to remove glycerol and methanol.

(d) Microalgal diesel verses crop-derived biodiesel

It's becoming increasingly obvious that there isn't enough suitable land space to grow crops for food and feed as well as for biofuel, and to retain the forests and other land uses that sequester carbon in huge quantities. To replace worldwide petroleum use with biofuel we would have to consume 10.8 million square miles of farmland with the highest yielding biofuel crops, and there are only 5.8 million square miles of farmland on earth. Traditional raw materials for world-wide biodiesel production are mainly coming from four oil crops: rapeseed oil, sunflower oil, soybean oil and palm oil. In addition, some other oleaginous species such as *Jatropha curcas*, a plant that grows in harsh soils, and low value triacylglycerol feedstocks (restaurant grease, animal fat) are being used as an alternative for biodiesel production. Today biodiesel production is limited and not fully sustainable due to limited availability of raw materials that compete with food and other uses. Unfortunately, biodiesel from oil crops, waste cooking oil and animal fat cannot realistically satisfy even a small fraction of the existing demand of fuels.

Constant increases in the price of fossil fuels over the last few years have brought microalgae as alternative biofuel source back into the Research & Development (R&D) limelight. Since the demand for edible vegetable oil exceeds supply, the Government decided to use non-edible oil from *Jatropha curcas* oilseeds as biodiesel feedstock. But the main problems in getting the biodiesel programme rolling is the difficulty linked to initiating large-scale cultivation of Jatropha. Farmers do not yet consider *Jatropha* cultivation remunerative enough. For instance, sugarcane plantations yield 70 t/ha and fetch the farmer Rs. 70,000/ha at a sugarcane price of Rs. 1,000/t. In comparison, if the Jatropha farmer gets Rs. 5,000 per ton of oilseeds and if the yield is 3.75 t/ha, his income is only Rs. 18,750 per hectare (Gonsalves, 2006). The other main issue is the lack of seed collection and oil extraction infrastructure. In the absence of this infrastructure and available oilseeds, it will be difficult to persuade entrepreneurs to install transesterification plants. Across the world a debate started whether diversion of farm land for biofuel led to the rally and shortage in food grains. This year World Bank has reported that 33 countries may face unrest because of surging food costs and deepening poverty. UN's Food and Agriculture organization has also reported that Global food stocks are at their lowest since the 1980s. Many countries had raised the issue of how farm land was being increasingly used to produce biofuel in the developed world leading to a shortfall in grain production. Citing concerns about rising global grain prices, as well as potential land-grabs by large energy firms, a group of Indian ministers has quietly shelved the National Mission on Biodiesel. (The Economic Times, a leading business news paper of India has reported on 4th August 2008). So, again we are finding that microalgae are having promising result for the biodiesel industry of India.

Biologists have reported that the appropriate land for algae farms should be at +/- 37° of Equator and India is having the suitable climatic area for the enhanced production of algae biomass. Generally speaking low rainfall, high temperatures and sunshine hours are positive for algae growth. This means that geographically India have favourable conditions. In India, the total area of algae ponds required to produce biodiesel to replace all petroleum transportation fuels demand will be 13 million acres (5.4 million hectare). Equivalent acreage required if using *Jatropha* to produce same amount of biodiesel will be 952 million acres (384 million hectare), while India's total geographic area is 328 million hectare. Therefore, microalgae is having enormous potential while taking a very small piece of land (about 2%), although it is not easy to cultivate and harvest but further research efforts needed to develop innovative large scale culture system that will enable selected algal strains to achieve high and sustained growth rates and oil yields that is essential to developing an algal-based biofuel industry.

Biomass Production System

The vast bulk of microalgae cultivated today are grown in open ponds. Open ponds can be built and operated very economically and hence offer many advantages as long as the species for cultivation can be maintained [24]. Producing microalgal biomass is generally more costly than growing crops. To minimize expense, biodiesel production must rely on freely available sunlight, despite daily and seasonal variations in light levels which are easily available in India. In open pond method, fresh culture medium is fed at a constant rate and the same quantity of microalgal broth is withdrawn continuously. Feeding ceases during the night, but the mixing of broth continue to prevent settling of the biomass (Molina *et al.*, 1999). There are three potential and very common methods of large-scale production of microalgae:

1. Open ponds/Raceway ponds

Open ponds have a variety of shapes and sizes but the most commonly used design is the raceway pond. It is a closed loop of rectangular grid with recirculation channel. They usually operate at water depths of 15–20 cm, as at these depths biomass concentrations of 1 g dry weight L^{H1} and productivities of 60–100 mg L^{H1} day^{H1} (*i.e.* 10–25 gm^{H2} day^{H1}) can be obtained (Pulz, 2001). There is a paddle wheel, which mix and circulate the algal biomass (Fig. 3).

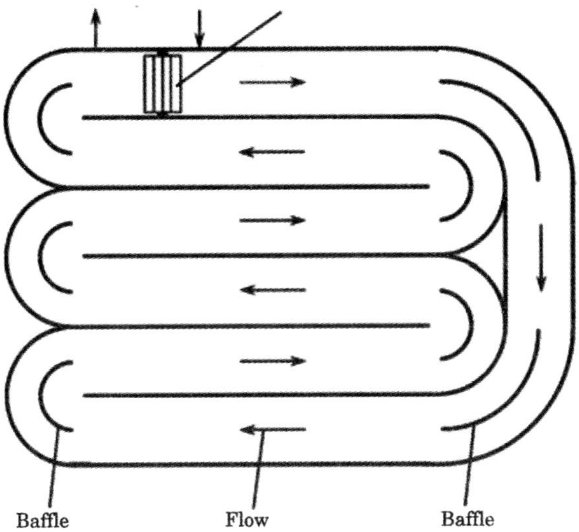

Fig. 3. A Raceway Pond

Flow is guided around bends by baffles positioned in the flow channel. Raceway channels are built in concrete or compacted earth, and generally lined with white plastic. During daylight, the culture is fed continuously in front of the paddlewheel where the flow begins. Broth is harvested behind the paddlewheel, on completion of the circulation loop. The paddlewheel operates all the time to prevent sedimentation. The main disadvantage of open systems is that by being open to the atmosphere, they loose water by evaporation at a rate similar to land crops and are also susceptible to contamination by unwanted species. From the aquatic species program collection of over 3000 photosynthetic organisms, none were found to be able to continually dominate an open pond and have desirable biofuel properties, *i.e.* high lipid contents (Shechan *et al.*, 1998). In practice open ponds are usually reported to be dominated by two to six species with a range of evolutionary advantages: rapid growth, resistance to predators, tolerance to high levels of dissolved oxygen, etc (Schenk *et al.*, 2008).

Raceways are perceived to be less expensive than photobioreactors, because they cost less to build and operate. Economically open pond system of biomass production is 10 time less costly in comparison to photobioreactors. Although raceways are low-cost, they have a low biomass productivity compared with photobioreactors. But the country like India, there is no need of photobioreactors because its climate is tropic *i.e.* ± 37° of Equator. There is need to approve the raceway pond system of biomass culture to achieve high and sustained growth rates and oil yields that is essential to developing an algal-based biofuel industry.

2. Closed bioreactors

Closed bioreactors permit essentially single-species culture of microalgae for prolonged durations. Most closed bioreactors are designed as tubular reactors, plate reactors, or bubble column reactors (Weissman and Goebel, 1988; Pulz, 2001). Other less common designs like semi-hollow-spheres have been reported to run successfully. The most common type of closed bioreactor is tubular photobioreactor. Tubular photobioreactor consists of an array of straight transparent tubes that are usually made of plastic or glass. The solar collector tubes are generally 0.1 m or less in diameter. Tube diameter is limited because light does not penetrate too deeply in the dense culture broth that is necessary for ensuring a high biomass productivity of the photobioreactor. Microalgal broth is circulated from a reservoir (*i.e.* the degassing column in Fig. 4.) to the solar collector and back to the reservoir (Chisti, 2007).

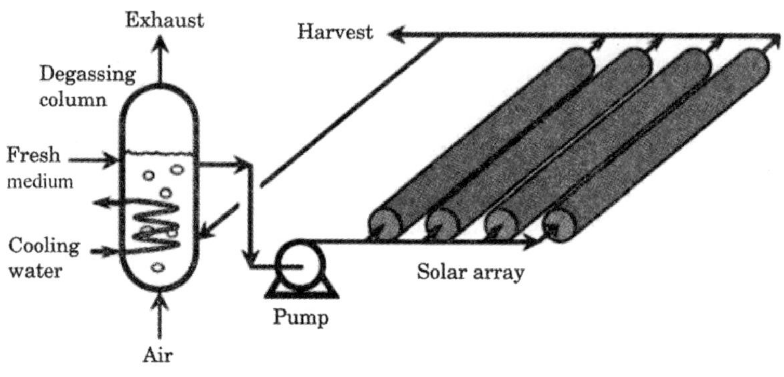

Fig. 4. Tubular photobioreactor with parallel run horizontal tubes

3. Hybrid systems

In this case both open ponds as well as closed bioreactor system used in combination to get the better results. Open ponds are a very proficient and lucrative method of cultivating algae, but they become contaminated with superfluous species very quickly. Photobioreactors are excellent for maintaining axenic cultures but setup costs are generally ten times higher than for open ponds. A combination of both systems is probably the most logical choice for cost-effective cultivation of high yielding strains for biofuels. Open ponds are inoculated with a desired strain that was invariably cultivated in a bioreactor, whether it be as simple as a plastic bag or a high tech fiber optic bioreactor. Importantly, the size of the inoculums needs to be large enough for the desired species to establish in the open system before an unwanted species. Therefore to minimize contamination issues, cleaning or flushing the ponds should be part of the aquaculture routine, and as such, open ponds can be considered batch cultures. This process has been demonstrated by Aquasearch (Hawaii, USA) cultivating *Haematococcus pluvialis* for the production of astaxanthin. Half of the Aquasearch facility was devoted to photobioreactors and half to open ponds. *H. pluvialis* is grown continuously in photobioreactors under nutrient sufficient conditions and then a portion is transferred to nutrient-limited open ponds to induce astaxanthin production (Schenk *et al.*, 2008).

Algal Biorefinery

Like a petroleum refinery, a biorefinery uses every component of the biomass raw material to produce useable products. Because all components of the biomass are used, the overall cost of producing desired product is lowered. Integrated biorefineries are already being operated in Canada, United States, and Germany for producing biofuels and other products from crops such as corn and soybean. This approach can be used to reduce the cost of making microalgal biodiesel. A conceptual

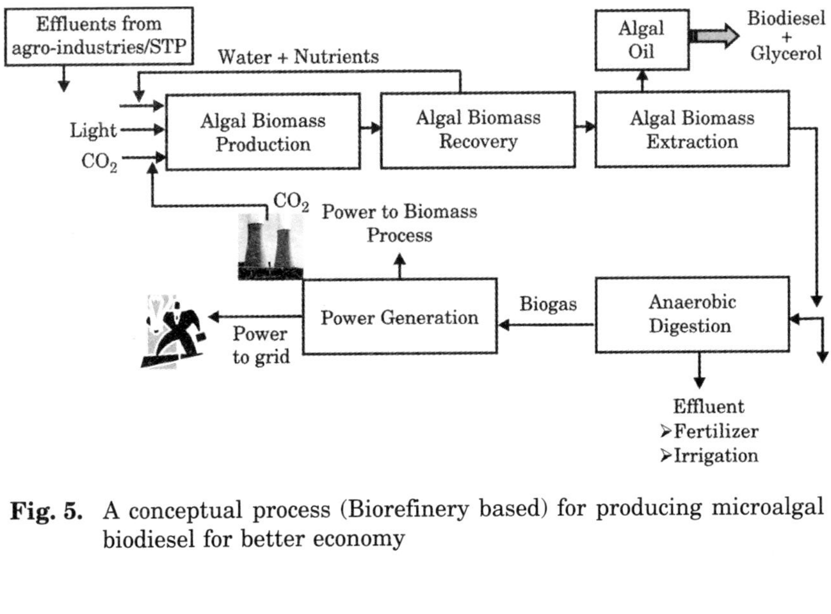

Fig. 5. A conceptual process (Biorefinery based) for producing microalgal biodiesel for better economy

process (biorefinery based) for producing microalgal biodiesel for better economy is given in Fig. 5.

In addition to oils, microalgal biomass contains significant quantities of proteins, carbohydrates and other nutrients (Sanchez Miron *et al.*, 2003). Therefore, the residual biomass from biodiesel production processes can be used potentially as animal and fisheries feed and after anaerobic digestion can be used as fertilizers (compost) and irrigation purposes (Fig. 5). Some of the residual biomass may be used to produce biogas by anaerobic digestion, for generating the electrical power necessary for running the microalgal biomass production facility. Excess power could be sold to defray the cost of producing biodiesel. The utilization of wastewater for microalgae cultivation will bring about remarkable advantages including the following:

1. Microalgae have been shown to be efficient in nitrogen and phosphorus removal, as well as in metal ion depletion, and combination of microalga cultivation with wastewater treatment will significantly enhance the environmental benefit of this strategy.
2. It will lead to savings in term of minimizing the use of chemicals such as sodium nitrate and potassium phosphorus as exogenous nutrients.
3. It will result in savings of the precious freshwater resources.

Algae as a Direct Green House Gas Mitigation Technology

CO_2 is recognized as the most important (at least in quantity) of the atmospheric pollutants that contribute to the "greenhouse effect," a term coined by the French mathematician Fourier in the mid-1800s to describe the trapping of heat in the Earth's atmosphere by gases capable of absorbing radiation. The burning of fossil fuels is the major source of the current build up of atmospheric CO_2. Global climate change requires immediate, substantial reductions in anthropogenic, Green House Gas (GHG) emissions, particularly fossil CO_2, and changes in the way we produce energy, in particular how we manage carbon. Carbon sequestration, capturing and storing carbon emitted from the global energy system, could be a major tool for reducing atmospheric CO_2 emissions from fossil fuel usage. Photosynthesis has long been recognized as a means, at least in theory, to sequester anthropogenic carbon dioxide. Biological CO_2 mitigation has attracted much attention as an alternative strategy because it leads to production of biomass energy in the process of CO_2 fixation through photosynthesis. Biological CO_2 mitigation can be carried out by plants and photosynthetic microorganisms. However, the potential for increased CO_2 capture in agriculture by plants has been estimated to contribute only 3–6% of fossil fuel emissions (Skjanes *et al.*, 2007), largely due to the slow growth rates of conventional terrestrial plants. On the other hand, microalgae, a group of fast-growing unicellular or simple multicellular microorganisms, have the ability to fix CO_2 while capturing solar energy with efficiency 10 to 50 times greater than that of terrestrial plants (Li *et al.*, 2008) and produced biomass for the subsequent production of biofuels. Microalgae have the advantages of efficient photosynthesis superior to C4 plants (those plants that form four carbon stable intermediates in the photosynthetic process; generally associated with agricultural and large terrestrial plants), fast proliferation rates, wide tolerance to extreme environments, and potential for intensive cultures. These advantages promise high performance in the reduction of carbon dioxide

On a worldwide basis, coal is, by far, the largest fossil energy resource available. To put this in perspective, consider the fact that, at current rates of consumption, coal reserves could last for over 200 years. Inevitably, the demand for electricity will have to be met by coal. Coal will remain the mainstay of world baseline electricity generation, accounting for half of electricity generation by the year 2015 (WEO, 2007). The long term demand for coal brings with it a demand for technologies that can mitigate the environmental problems associated with coal. While control technologies will be used to reduce air pollutants associated with acid rain, no technologies exist today, which address the problem of greenhouse gas emissions. Coal is the most carbon-intensive of the fossil fuels. In other words, for every unit of energy liberated by combustion, coal emits more CO_2 than either petroleum or natural gas. Flue gases from power plant are responsible for more than 7% of the total world CO_2 emissions (Sakai *et al.*, 1995). Carbon dioxide in flue gas is available at little or no cost. As estimated by the IPCC criteria, the CO_2 concentration of flue gas is up to 15% (Maeda *et al.*, 1995). Coal is the dominant fuel in India's energy mix, a condition that is expected to persist for at least the next 25 years. India has vast coal resources, but most are of low quality. Indigenous oil and gas reserves are in short supply while demand for oil almost quadrupled from 1980 to 2005. Oil imports are projected to increase even more going forward, leaving the country more vulnerable to international price spikes and potentially unreliable supplies. In 2005 India ranked fourth in energy consumption, after the United States, China, and Russia. By 2030, India is expected to surpass Russia and be the third largest energy consumer. India has an installed base of about

124,287 MW of electricity as of the year 2006, which includes about 66% thermal energy (85% of which is coal based) followed by hydro with 26%, nuclear with 3%, and renewable energy with 5% of the current total installed renewable energy base (WEO, 2007). According to World Energy outlook report of 2007, India's total CO_2 emission in 2007-08 was 1200 MT, of that 800 MT CO_2 (67%) came from coal combustion. We could solve the problem of this huge carbon emission from thermal power plant with the help of algal farm by recycling the carbon. The simple, direct method of GHG mitigation is the removal of CO_2 from stack gases, followed by long term sequestration of CO_2 by microalgae ponds. Thus, microalgae technology can extend the useful energy we get from coal combustion and reduce carbon emissions by recycling waste CO_2 from power plants into clean-burning biodiesel. Table 4 summarizes a few microalgal strains that have been studied for CO_2 biomitigation (Bei *et al.*, 2008). Some of these strains can tolerate high temperature and high CO_2 in the gas stream. CO_2 fixed through photosynthesis are converted to different

Table 4. Some microalgal strains studied for CO_2 biomitigation [adopted from Bei *et al.*, 2008]

Microalga	*CO_2 %*	*T °C*	*P g L^{-1} per day*	*P CO_2 g L^{-1} per day*	*References*
Chlorococcum littorale	40	30	N/A	1.0	[34, 35]
Chlorella kessleri	18	30	0.087	0.163[a]	[36]
Chlorella sp. UK001	15	35	N/A	>1	[35]
Chlorella vulgaris	15	N/A	0.624	[37]	
Chlorella vulgaris	Air	25	0.040	0.075 [a]	[38]
Chlorella vulgaris	Air	25	0.024	0.045 [a]	[38]
Chlorella sp.	40	42	N/A	1.0	[32]
Dunaliella	3	27	0.17	0.313 [a]	[39]
Haematococcus pluvialis	16–34	20	0.076	0.143	[40]
Scenedesmus obliquus	Air	–	0.009	0.016	[41]
Scenedesmus obliquus	Air	–	0.016	0.031	[41]
Botryococcus braunii	–	25–30	1.1	>1.0	[35]
Scenedesmus obliquus	18	30	0.14	0.26	[43]
[b]*Spirulina* sp.	12	30	0.22	0.413 [a]	[43]

[a] Calculated from the biomass productivity according to equation, CO_2 fixation rate (P_{CO2}) = 1.88 × biomass productivity (P), which is derived from the typical molecular formula of microalgal biomass, $CO_{0.48}H_{1.83}N_{0.11}P_{0.01}$ (Chisti, 2007).
[b]All species except *Spirulina* sp., which is a prokaryotic cyanobacteria (Cyanophyceae) species, are eukaryotic green algae (Chlorophyta) species (Bold and Wynne, 1985).

organic cell components including carbohydrates, lipids, proteins, and nucleic acids.

Microalgal-based carbon sequestration technologies can, in principle, not only cover the cost of carbon capture and sequestration but also produce environment friendly biodiesel. Carbon sequestration offers an opportunity for reducing green house gas emission that can complement the current strategies of improving the energy efficiency and increasing the user of non fossil energy resources. So, the microalgal biodiesel projects can qualify as CDM projects and thus bring in additional income through the sale of certified emission reductions.

Road Ahead for Algal Feedstock-Based Biodiesel

There is need for strenuous research on the biosynthesis of algal lipids, especially TAGs, if we are to better understand and manipulate algae for the production of biodiesel. While algae appear to provide the natural raw material in the form of a lipid-rich feedstock, our understanding of the details of lipid metabolism to enable manipulation of the process physiologically and genetically is lacking. Over 20 years ago, the Aquatic Species Programme (ASP) of the US Department of Energy (DOE) illustrated the potential of algae to provide liquid energy. To harvest the benefits of that potential will not only require critical engineering innovations and breakthroughs related to algal mass culture and downstream processing, but also focused research on essential biological questions related to regulation of lipid metabolism. Several biological and processing challenges and opportunities lie ahead.

Better understanding of the TAG and lipid synthesis pathways in microalgae

Available biochemical knowledge about fatty acid and TAG synthetic pathways in algae is still fragmentary. We need significant knowledge regarding both the regulatory and structural genes involved in these pathways and the potential interactions between pathways.

A critical evaluation of the relationship between the cell cycle and TAG production is needed

Understanding the control mechanisms underlying this phenomenon will enable genetic manipulation of selected algal strains that exhibit rapid growth and TAG accumulation simultaneously to ensure maximum sustainable oil production.

Enhancing algal biology by genetic engineering

Genetic and metabolic engineering are likely to have the greatest impact on improving the economics of production of microalgal diesel. So, there is need to do the metabolic engineering through genetic manipulation for enhancing the TAG production.

Developing large scale culturing condition and improved and innovative harvesting method for sustainable biomass production

The optimization of strain-specific cultivation conditions is of confronting complexity, with many interrelated factors that can each be limiting. These include temperature, mixing, fluid dynamics and hydrodynamic stress, gas bubble size and distribution, gas exchange, mass transfer, light cycle and intensity, water quality, pH, salinity, mineral and carbon regulation/bioavailability, cell fragility, cell density and growth inhibition. Algae typically have high water content and downstream harvesting and processing requires its removal. Harvest is considered to be an expensive and problematic part of industrial production of microalgal biomass due to low cell density. There is no single best method for harvesting microalgae and reducing their water content. In existing algal aquaculture the most common harvesting processes are flocculation, micro screening and centrifugation. Most importantly, cost-effective and energy-efficient harvesting methods are required to make the whole biofuels production process economical.

Coupling of a coal-fired power plant with an algae farm

The concept of coupling a coal-fired power plant with an algae farm provides an elegant approach to recycle of the CO_2 from coal combustion into a useable liquid fuel. Identification and cultivation of best microalga species which can grow in elevated CO_2 and high temperature for this process. Thus, microalgae technology can extend the useful energy we get from coal combustion and reduce carbon emissions by recycling waste CO_2 from power plants into clean-burning biodiesel.

Combination of wastewater treatment with microalgal CO_2 fixation

Combination of wastewater treatment and microalgal CO_2 fixation provides additional economic incentives due to the savings from chemicals (the nutrients) and the environment benefits. It provides a pathway for removing nitrogen, phosphorus, and metal from wastewater, and producing algal biomass, which can further be exploited for biofuel production, without using freshwater.

Conclusions

Biodiesel production from microalgae is a goal that still needs much research to be accomplished. Major R&D has to be done before the production of microalgae as a standalone biofuel project can play a competitive role in the fuel industry. Transition to Algal Biomass can deliver development & climate co-benefits. Use of the biorefinery concept and advances in raceway ponds engineering will further lower the cost of production. It is the only renewable bio-diesel that can potentially completely displace liquid fuels derived from petroleum. If the microalgal biorefinery concept can be adapted to a country like ours, it could become a highly distributed source of fuel oil and perhaps make us self- reliant and improve our economy many folds. In present scenario algal biomass is a key link between energy, local environment and climate change and further research are necessary to unlock full potential of algae.

References

Banerjee, A., Sharma, R., Chisti, Y. and Banerjee, U.C. (2002). *Botryococcus braunii*: a renewable source of hydrocarbons and other chemicals. *Crit Rev Biotechnol* **22**: 245-79.

Bei Wang, Yanqun Li, Nan Wu and Christopher Q. Lan (2008). CO_2 bio-mitigation using microalgae. *Appl. Microbial Biotechnology* **79**: 707-718.

Bold, H.C. and Wynne, M.J. (1985). Introduction to the Algae. 2nd Edn., Prentice-Hall, Inc., NJ, USA: Englewood Cliffs.

BP Statistical Review of the world energy, (2008).

Chisti, Y. (2007). Biodiesel from microalgae. *Biotechnol Adv* **25**: 294-306.

De Morais, M.G. and Costa, J.A.V. (2007). Biofixation of carbon dioxide by *Spirulina* sp. and *Scenedesmus obliquus* cultivated in a three stage serial tubular photobioreactor. *J Biotechnol* **129**: 439-445.

De Morais, M.G. and Costa, J.A.V. (2007). Isolation and selection of microalgae from coal fired thermoelectric power plant for biofixation of carbon dioxide. *Energy Convers Manag* **48**: 2169-2173.

Fukuda, H., Kondo, A. and Noda, H. (2001). Biodiesel fuel production by transesterification of oils. *J Biosci Bioeng* **92**: 405-16.

Ghirardi, M.L., Zhang, J.P., Lee, J.W., Flynn, T., Seibert, M. and Greenbaum, E. (2000). Microalgae: a green source of renewable H_2. *Trends Biotechnol* **18**: 506-11.

Gomez-Villa, H., Voltolina, D., Nieves, M. and Pina, P. (2005). Biomass production and nutrient budget in outdoor cultures of *Scenedesmus obliquus*

(Chlorophyceae) in artificial wastewater, under the winter and summer conditions of Mazatlan, Sinaloa, Mexico. *Vie et Milieu* **55**: 121-126.

Gonsalves, J.B. (2006). An assessment of the biofuels industry in India. United Nations conference on trade and development. UNCTAD/DITC/TED/2006/6.

Hu, Q., Sommerfeld, M., Jarvis, E., Ghirardi, M., Posewitz, M., Seibert, M. and Darzins, A. (2008). Microalgal triacylglycerols as feedstocks for biofuel production: perspectives and advances. *Plant J* **54**: 621-39.

Huntley, M.E. and Redalje, D.G. (2007). CO_2 mitigation and renewable oil from photosynthetic microbes: a new appraisal. *Mitig Adapt Strategies Glob Chang* **12**: 573-608.

Iwasaki, I., Hu, Q., Kurano, N. and Miyachi, S. (1998). Effect of extremely high-CO_2 stress on energy distribution between photosystem I and photosystem II in a 'high-CO_2' tolerant green alga, *Chlorococcum littorale* and the intolerant green alga *Stichococcus bacillaris*. *J Photochem Photobiol B* **44**: 184-190.

Kishimoto, M., Okakura, T., Nagashima, H., Minowa, T., Yokoyama, S.Y. and Yamaberi, K. (1994). CO_2 fixation and oil production using microalgae. *J Ferment Bioeng* **78**: 479-482.

Li, Y., Horsman, M., Wu, N., Lan, C.Q. and Dubois-Calero, N. (2008). *Biofuels from microalgae. Biotech Prog*. (in press) ASAP Article, DOI: 10.1021/bp0703 71kS8756-7938(07)00371-2.

Maeda, K., Owada, M., Kimura, N., Omata, K. and Karube, I. (1995). CO_2 fixation from the flue gas on coal-fired thermal power plant by microalgae. *Energy Convers Manag* **36**: 717-720.

Mehta, B.V. (2005-06). Overview of the Indian Oilseed Sector and Demand of edible Oils. Paper presented at Exchange Traded Commodities Outlook.

Melis, A. (2002). Green alga hydrogen production: progress, challenges and prospects. *Int J Hydrogen Energy* **27**: 1217-28.

Meng, X., Yang, J., Xu, X., Zhang, L., Nie, Q. and Xian, M. (2009). Biodiesel production from oleaginous microorganisms. *Renewable Energy* **34**: 1-5.

Metzger, P. and Largeau, C. (2005). *Botryococcus braunii*: a rich source for hydrocarbons and related ether lipids. *Appl Microbiol Biotechnol* **66**: 486-96.

Molina, Grima E., Acién, Fernández F.G., García, Camacho F. and Chisti, Y. (1999). Photobioreactors: light regime, mass transfer, and scaleup. *J Biotechnol* **70**: 231-47.

MoPNG, (2006). Report of the Working Group on Petroleum & Natural Gas Sector for the XI Plan (2007-2012).

MoPNG, Govt. of India (2007). Annual Report (2006-2007). Ministry of Petroleum and Natural Gas.

Murakami, M. and Ikenouchi, M. (1997). The biological CO_2 fixation and utilization project by RITE (2): screening and breeding of microalgae with high capability in fixing CO_2. *Energy Convers Manag* **38(Suppl 1)**: S493-S497.

Planning Commission, Government of India 2003. Report of the committee on Development of Bio-Fuel.

Pulz, O. (2001). Photobioreactors: production systems for phototrophic microorganisms. *Appl Microbiol Biotechnol* **57**: 287-293.

Sakai, N., Sakamoto, Y., Kishimoto, N., Chihara, M. and Karube, I. (1995). *Chlorella* strains from hot springs tolerant to high temperature and high CO_2. *Energy Convers Manag* **36**: 693-696.

Sánchez Mirón, A., Cerón García, M.-C., Contreras, Gómez A., García, Camacho F., Molina, Grima E. and Chisti, Y. (2003). Shear stress tolerance and biochemical characterization of *Phaeodactylum tricornutum* in quasi steady-state continuous culture in outdoor photobioreactors. *Biochem Eng J.* **16**: 287-97.

Schenk, P.M., Thomas-Hall, S.R., Stephens, E., Marx, U.C., Mussgnug, J.H. and Posten, C. (2008). Second generation biofuels: High-efficiency microalgae for biodiesel production. *Bioenerg Res* **1**: 20-43.

Scragg, A.H., Illman, A.M., Carden, A. and Shales, S.W. (2002). Growth of microalgae with increased calorific values in a tubular bioreactor. *Biomass Bioenergy* **23**: 67-73.

Sheehan, J., Dunahay, T., Benemann, J. and Roessler, P. (1998). A look back at the U.S. Department of Energy's Aquatic Species Program—biodiesel from algae. National Renewable Energy Laboratory, Golden, CO. Report NREL/TP-580-24190.

Singh, S., Kate, B.N. and Banerjee, U.C. (2005). Bioactive compounds from cyanobacteria and microalgae: an overview. *Crit Rev Biotechnol* **25**: 73-95.

Skjanes, K., Lindblad, P. and Muller, J. (2007). BioCO_2—a multidisciplinary, biological approach using solar energy to capture CO_2 while producing H_2 and high value products. *Biomol Eng* **24**: 405-413.

Spolaore, P., Joannis-Cassan, C., Duran, E. and Isambert, A. (2006). Commercial applications of microalgae. *J Biosci Bioeng* **101**: 87-96.

Van Gerpen, J. (2005). Biodiesel processing and production. *Fuel Process Technol.* **86**: 1097-107.

Walter, T.L., Purton, S., Becker, D.K. and Collet, C. (2005). Microalgae as bioreactor. *Plant Cell Rep* **24**: 629-41.

Weissman, J. and Goebel, R.P. (1988). Benemann, J.R. Photobioreactor design: mixing, carbon utilization, and oxygen accumulation. *Biotechnol Bioeng* **31**: 336-344.

World Energy outlook (2007). Published by The International Energy Agency (IEA).

Yun, Y.S., Lee, S.B., Park, J.M., Lee, C.I. and Yang, J.W. (1997). Carbon dioxide fixation by algal cultivation using wastewater nutrients. *J Chem Technol Biotechnol* **69**: 451-455.

15

OPPORTUNITIES AND ASSESSMENTS OF COMMODITY CHEMICAL FEEDSTOCKS FROM ALGAE

Steven J. Gluck[1] and Damien Perriman*

Abstract

Chemical feedstock from algal strains attracted the scientists world over. Non traditional agro-industries have emerged to fulfil the fuel need. Algae is a potential biological solution to the biofuel production because it can provide triglycerides or fatty acids, glycerol, sugars, alcohols and proteins. High-value nutraceuticals, pharmaceuticals, low-value commodities like bulk feed, oils and glycerol alongwith the energy feedstock can be obtained from algae.

Key words : Triglycerides, Algal feedstock, Biofuel, Nutraceuticals

The rising cost and unprecedented volatility of conventional energy has placed significant pressure on the chemicals industry to diversify to new sources for carbon feedstock (Lee *et al.*, 2003; Gluck and Wu, 2008; Pfeiffer *et al.*, 2006; Perriman, 2010). As a result, new opportunities are emerging for non-traditional agricultural industries to feed emerging biofuel and bioproduct pipelines with sources of renewable carbon. One opportunity is algae.

Algae have received a lot of attention as a potential biological option because of their simplicity to grow, independence from food production, and high productivity rates as well as viable other reasons. Algae can be grown virtually anywhere, providing the flexibility to meet the

[1] The Dow Chemical Company Environmental Technology Center, B 2009 Bldg, 2301 North Brazosport Boulevard, Freeport, TX 77541, USA

* *Corresponding author*: E-mail: dperriman@verdezyne.com

geographical needs of chemical supply chains that address market demand rather than petroleum supply.

Research is being invested to enhance plant biology with functionality tailored to the needs of the new value chain. New traits and technologies for developing optimized sources of biomass create the promise of a low cost raw material starting point. Algae hold the possibility to provide triglycerides or fatty acids, glycerol, sugars, replacement monomers, alcohols and proteins. Whether any of these materials will be realized in the near future depends on technical breakthroughs to achieve lower economics.

Algae are capable of producing high-value materials (nutraceuticals, pharmaceuticals), low-value commodities (*e.g.*, bulk feed, oils and glycerol) and energy feedstock (fuels). These outputs can be altered with breeding and genetic engineering tools, in conjunction with thermochemistry, to produce surfactants, lubricants, polymers, esters, olefins, acids and alcohols. This activity has created significant excitement within the investment community for advancing algae as a solution to high petroleum costs. Companies are advancing research with the scale and economics of algae for fuel as targets. Chemical products are commonly seen as a derivative of the fuels' value chain.

Unlike the historical petrochemical approach, algae operations have started with the low volume and high margin opportunities in nutraceuticals. A migration to the high volume and low margin fuel opportunity passes through the specialty and bulk chemicals opportunity. This intersecting platform provides an easier testing and development ground for algae. The economic and scale barriers for chemical feedstocks, although daunting, may be less of a challenge than those for providing a fuel.

As an example of the scale and economic challenge, consider that fuels are harvested for their ability to fit into an existing energy system and thus we see biodiesel, biogasoline and bioethanol opportunities. In 2005, only 10,000 mt of algae biomass was being produced globally. A single commodity chemical plant requires about 100,000 mt per year. One medium sized fuel refinery requires about 94 million mt per year, which is equivalent to providing feedstock for 94 commodity chemical plants or 9400 times the total 2005 global production of dried algal biomass.

In addition, similar to chemical feedstocks, bioproducts have highly controlled, and possibly consistent, compositions. Therefore, given the specificity of the product and the scale of the opportunity, it may make more sense to focus the algae industry effort towards chemical feedstocks before the fuels challenge is met. Indeed, there are significant barriers for algae as a newly introduced crop. Some of these barriers include:

- Growth of algae to commodity scale
- Land and water requirements
- Nutrient requirements
- Understanding the true CO_2 sequestration amounts
- Regulations
- Introduction of a new crop and a new way of growing the crop
- Energy return versus energy input
- Water and wastewater management
- Extraction, isolation and purification of materials
- Productivity and related growing area needs
- Process monitoring and control or automation
- Material supply chain
- Materials of construction
- Feedstock composition and downstream utility
- Management of growth rate limiting factors

Thus we have a long way to go, but there are good possibilities for new, large scale, algae opportunities. Scaling from low volume production through operations for chemical feedstocks to fuel may be the most achievable pathway.

References

Gluck, Steven J. and Wu, Qingzhong (2008). Strategic sustainable water solutions: what works and what doesn't proceedings of the water environment federation, WEFTEC session 11 through session **20**: 1279-1285(7) Publisher: Water Environment Federation.

Lee Max, M. , Tim, D. Logan, Kefu Sun, N. Spencer Hurley, Jr., Robert, A. Swatloski, and Steve, J. Gluck (2003). Direct injection GC method for measuring light hydrocarbon emissions from cooling-tower water. *Environmental Science and Technology* **37(24)**: 5721-5726.

Perriman Damien (2010). New biobased monomers. VII world congress on industrial biotechnology and bioprocessing. Washington.

Pfeiffer, Curtis D., Frawley, Nile N. , Peters, Thomas L., Savickas, Philip J., Albers, David R., Gluck, Steven J., Nicholson, Lawrence W., Esquivel, H. and Jose, B. (2006). Method for chemical analysis of biological material. IPC: G01n 30/88 (2006.01). G01 n 30/46 (2006.01) G01n 30/62 (2006.01), G01N 30/78 (2006.01)

16

FEASIBILITY OF ALGAL BIOMASS FOR BIODIESEL PRODUCTION

Sunil Pabbi[1*] and Dolly Wattal Dhar[1]

Abstract

Oil is an essential commodity and has brought numerous benefits to industrialized society, but the fossil fuel is non-renewable and it has given rise to environmental pollution, health hazards, global warming etc. The use of fossil fuels is widely accepted as unsustainable due to depleting resources and the accumulation of greenhouse gases in the environment which have already exceeded the dangerously high threshold levels. Thus, to eliminate the possibility of future energy shortage, efforts are directed for the development of alternative energy sources which are cheaper to produce, biodegradable, non-toxic and renewable. Currently, nearly all renewable energy sources (hydroelectric, solar, wind, tidal, geothermal) target the electricity market, while fuels make up a much larger share of global energy demand. Biofuels are therefore, rapidly being developed. Microalgae have the advantages that they can produce a wide range of feedstocks for the production of biodiesel, bioethanol, biomethane and biohydrogen. Biodiesel is currently produced from the oil synthesized by conventional fuel crops that harvest the sun's energy and store it as chemical energy. Producing biodiesel from algae is widely regarded as one of the most efficient ways of generating biofuels and appears to represent the only current renewable source of oil that could meet the global demand for transport fuels. Different microalgae having potential for the biofuel production are Scenedesmus, Pleurochrysis, Dunaliella, Spirulina, Chlorella etc. The main advantages of these are their a) Higher

[1] Centre for Conservation and Utilisation of Blue Green Algae, Indian Agricultural Research Institute, New Delhi - 110 012, India

* *Corresponding author*: E-mail: Sunil.pabbi@gmail.com

photosynthetic efficiency b) Faster growth and larger biomass c) Easy mass culture d) Easily controllable environmental conditions e) Round the year production d) Non polluting and environment friendly e) No net contribution to atmospheric CO_2 *levels f) Production reported to be 15000 gallons oil / acre / year g) Bio-oil from microalgae is rich in carbon and hydrogen as compared to wood but little less than fossil oil h) Algal oil contains No Sulphur and i) Microalgae also contain carbohydrates and protein. Microalgae contain lipids and fatty acids as membrane components, storage products, metabolites and sources of energy. Algal fatty acids and oils have a range of potential applications and algal oils possess characteristics similar to those of fish and vegetable oils, and can thus, are considered as potential substitutes for the products of fossil oil. After processing for biofuel, rest is good enough as animal feed or can be processed for ethanol production. Although micro algal biofuel production is relatively new and not very well explored, however, several aspects of this novel approach of using microalgae for biofuel is an interesting development. The unparalleled space efficiency of growing algae for biodiesel production gives it a major advantage over established biofuel production methods. With the ever increasing demand for arable land, not only because of biofuel demand, but also because of growing food consumption worldwide, it may become very hard to continue using cheap and well established oil crops such as corn and soy. In this respect, especially the potential of algae to be grown on non-arable land or on the sea is of great importance for energy production in the future. Taking all advantages together, including no fresh water requirements, carbon sequestration options, waste water treatment, algal biofuel may well be a prominent player of the future energy market. The new and broad field of microalgal biofuel production seems to be very promising approach for meeting ever increasing biofuel demands.*

Key words: Microalgae, Biodiesel, *Spirulina, Chlorella*, Bioethanol, Biomethane, Biohydrogen, *Scenedesmus, Pleurochrysis, Duneliella*

Introduction

Oil is an essential commodity and has brought numerous benefits to the industrialized society. But the fossil fuel is non-renewable and it has given rise to environmental pollution, health hazards and global warming etc. The use of fossil fuels is widely accepted as unsustainable due to depleting resources and the accumulation of green house gases in the environment which has already exceeded the dangerously high threshold levels. Thus, to eliminate the possibility of future energy shortage, efforts are directed for the development of alternative energy sources which are cheaper to produce, bio-degradable, non-toxic and renewable. The development of CO_2-neutral fuels is one of the most urgent challenges facing our society. This fact has been brought into sharp focus by the Stern Report 'The Economics of Climate Change' (Stern, 2006) and the Intergovernmental Panel on the Climate Change 'AR4 Synthesis report' (IPCC, 2007). These conclusions are leading governments to establish CO_2 emission reduction targets which lie in the range of 10-20% by 2020. Germany suggested a 30% cut by 2020 and these proposed emission reductions are not sufficient to stabilize CO_2 levels in an accepted 'safe zone' [*i.e.* below 450 ppm CO_2-e; (IPCC, 2007)]. Achieving a 60% CO_2 emission reduction by 2020 or even a 50-85% cut by 2050 is, however, an enormous global challenge. There is now a concerted effort to develop new biofuels, of which biodiesel and bioethanol are seen as being close to market options. The areas of biomethane, biomass-to-liquid (BTL)-diesel and biohydrogen are also developing rapidly (Schenk *et al.*, 2008).

Biofuels and Biodiesel in the Global Context

The most common concern related to the first generation biofuel systems is that as production capacities increase, so does their competition with agriculture for arable land used for food production. Current biodiesel supplies from oil-producing crops, supplemented with small amounts of animal fat and waste cooking oil account for an estimated 0.3% (approx. 12 million tons in 2007) of the current global oil consumption (BP statistics, 2007; Oilworld, 2007) and can not even come close to satisfying the existing and future demand for transport fuels. Currently, 8% of plant-based oil production is used as biodiesel (Oilworld, 2007) and this has already contributed to an increase of the price of oil crops over the last few years.

The biofuel producing countries have a small share of less than 6% amongst those in the European Union. The main producers are the USA and Brazil, whereas the share of Europe is rather small though it is the most important producer of biodiesel on the global market. From 1993 onwards, the European production level has increased by almost forty times, from 80.000 tons in 1993 to 780.000 tons in 2001, and finally to 3.184.000 tons in 2005. Most of the global biofuel production consists of ethanol and at present, all these countries are producing biodiesel from oil seed crops (Fig. 1).

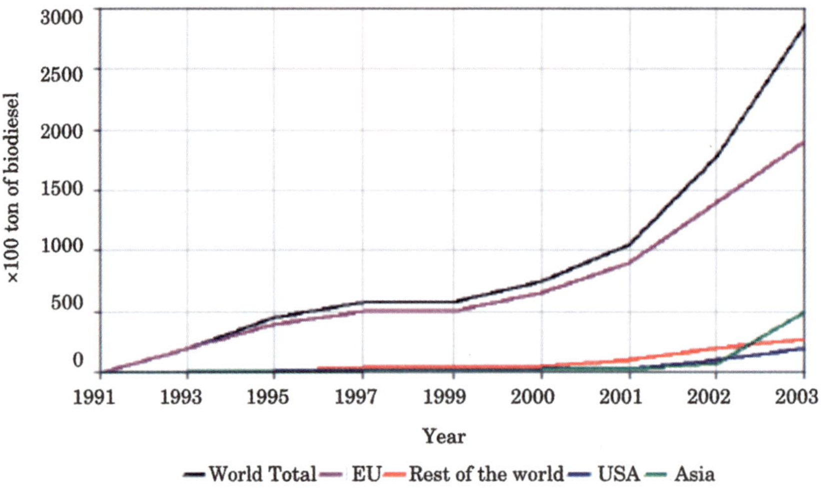

Fig. 1. World production of biodiesel (1991 till 2003)

Area Requirements for Biofuel Production

It is estimated that the surface of the Earth (510,072,000 km^2) on an average receives 170 W^{-2} of solar power (CIA World Fact book, 2007; World Energy Council, 2007). This is equivalent to 2.735 YJ of energy per year and corresponds to 5,600 times the global consumption of primary energy in 2005 (488 EJ; World Energy Council, 2007). Consequently, the solar energy required to produce biofuels is available in abundance. However, even if current oil-producing crops would be grown on all arable land (assuming 29.2% of the earth is land of which 13% are arable, energy conversion efficiencies of 1% from sunlight to biomass, and 20% yield as oil), these should be able to cover less than half of our energy demand today. The increased pressure on arable land has already resulted in considerable problems and unsustainable practices worldwide that have led to the coining of the phrase "peak soil". For example, rainforest regions in Brazil and South East Asia are currently being cleared at an unprecedented rate to make room for soybean and oil palm plantations for the production of biodiesel. Annual biodiesel production growth rates in Indonesia, Malaysia and Thailand are currently between 70% and 250% (USDA, 2007). The increased pressure on arable land currently used for food production could lead to severe food shortages, in particular for the developing world where already more than 800 million people suffer from hunger and malnutrition (FAO, 2007). In addition, the intensive use of land with high fertilizer and pesticide applications and water use can cause significant environmental problems.

Net Energy Balance

The NEB of first generation biofuels taking into consideration the energy required for farming, harvesting, processing, transport, etc. has recently been estimated to be 25% for corn ethanol and + 93% for soybean biodiesel (Hill *et al.*, 2006).

Carbon Balance

Biofuel production from crops is often estimated to be a near carbon-neutral process as nearly all carbon in conventional biofuel crops is derived from atmospheric CO_2, which is released on combustion. However, the overall CO_2 emissions balance for biofuel production, must also be evaluated on an individual production case basis and include energy-intensive fertilizer production, the use of machinery for cultivation and refinery, and transport which are processes that currently emit fossil fuel derived CO_2.

CO_2 Sequestration

One important development which is predicted not only to greatly improve the net CO_2 balance of biofuel processes, but actually contribute to atmospheric CO_2 reductions, is the coupling of biofuel production to CO_2 sequestration systems.

Second Generation Biofuel Systems

A range of second generation biofuel-production systems that will have much higher NEBs, are more water-efficient and require much less arable land are now under development. Of particular interest are lignocellulosic technologies and microalgae (Chisti, 2007; Hankamer, *et al.*, 2007; Schaub and Vetter, 2007). These have the potential to overcome many of the earlier limitations and target a newly emerging clean energy market which is predicted to expand rapidly to a value of US $500bn by 2050 (Stern, 2006), or more (current oil market value is US$2.65 Trillion). Microalgae are already reported to produce 15-300 times more oil for biodiesel production than traditional crops on an area basis These achieve a higher yield per hectare (potentially over 15-fold higher than oil palm and can be cultivated on non-arable land, thereby reducing the competition with food crops for land (Table 1, Chisti, 2007). Furthermore, compared with conventional crop plants which are usually harvested once or twice a year, microalgae have a very short harvesting cycle (1-10 days depending on the process), allowing multiple or continuous harvests with significantly increased yields. Greater light capture and conversion efficiencies ultimately lead to reduced fertilizer and nutrient inputs and so result in less waste and pollution. The use of water for algal cultivation is also a viable option. Furthermore, as the cultivation of microalgae for biofuel production can potentially be carried out on marginal or non-arable land, this could further reduce the competition for land and open up new economic opportunities for arid, drought or salinity-affected regions. In addition, while traditional biofuel crops require substantial amounts of fresh water, considerable savings in net water use can be achieved if closed bioreactor systems with minimal evaporation are

Table 1. Comparison of crop-dependent biodiesel production efficiencies from plant oils

Plant source	*Biodiesel* (L/ha/year)*	*Area to produce global oil demand (hectares x 10^6)*	*Area required as percent global land mass*	*Area as percent global arable land*
Cotton	325	15,002	100.7	756.9
Soybean	446	10,932	73.4	551.6
Mustard seed	572	8.524	57.2	430.1
Sunflower	952	5.121	34.4	258.4
Rapeseed/canola	1,190	4,097	27.5	206.7
Jatropha	1,892	2,577	17.3	130(0[a])
Oil palm	5,950	819	5.5	41.3
Algae (10 g m^{-2} day^{-1} at 30% TAG)	12,000	406	2.7	20.5 (0[a])
Algae (50 g m^{-2} day^{-1} at 50% TAG)	98,500	49	0.3	2.5(0[a])

* Presented yields are for peak performing crops, although Malaysia's average oil palm yield is actually about 4 tons/ha. Algae yield scenarios are based on existing production systems and their potential. Current algal production systems fall between these ranges (Seambiotic Israel (currently at 20 g m^{-2} day^{-1} at 8-40% Triacylglycerides (TAG). HR BioPetroleum Inc Hawaii (aims to achieve 50 g m^{-2} day^{-1} at 30% TAG).

[a] If algal ponds and bioreactors are situated on non-arable land; Jatropha is mainly grown on marginal land.

used for algal cultivation, especially using marine and halophilic strains of microalgae. The microalgal biomass generated in bioreactors can also be gasified or pyrolysed to produce a range of biofuels and Agri-char as part of CO_2 sequestration strategy. Other significant aspects of second generation microalgal systems are that they are amenable to highly innovative biotechnology approaches that have the potential for rapid improvements of algal strains. These aspects promise increases in yield by metabolic engineering. Recent advances have been made in many areas, and examples of increased photosynthetic efficiency and tolerance to grow in saline or waste water streams are described.

Potential of Microalgal Biodiesel

Replacing all the transport fuel consumed in the United States with biodiesel will require 0.53 billion m^3 of biodiesel annually at the current rate of consumption. Oil crops, waste cooking oil and animal fat can not realistically satisfy this demand. Microalgae appear to be only source of biodiesel that has the potential to completely displace fossil diesel. Unlike other oils crops, microalgae grow extremely rapidly and many are exceedingly rich in oil. These are sunlight-driven cell factories that convert carbon dioxide to potential biofuels, foods, feeds and high-value bioactives (Metting and Pyne, 1986; Kay, 1991; Shimizu, 2003; Singh *et al.*, 2005; Spolaore *et al.*, 2006). Microalgae can provide several different types of renewable biofuels. These include methane produced by anaerobic digestion of the algal biomass (Spolaore *et al.*, 2006); biodiesel derived from microalgal oil (Roessler *et al.*, 1994; Sawayama *et al.*, 1995; Dunahay *et al.*, 1996; Sheehan *et al.*, 1998; Banerjee *et al.*, 2002; Gavrilescu and Chisti, 2005); and photobiologically produced biodydrogen (Ghirardi *et al.*, 2000; Akkerman *et al.*, 2002; Melis, 2002; Fedorov *et al.*, 2005; Kapdan and Kargi, 2006). The idea of using microalgae as a source of fuel is not new (Chisti, 1980-81; Nagle and Lemke, 1990; Sawayama *et al.*, 1995), but it is now being taken seriously because of the escalating price of petroleum and, more significantly, the emerging concern about global warming that is associated with burning fossil fuels (Gavrilescu and Chisti, 2005). Microalgae commonly double their biomass within 24 h and biomass doubling time during exponential growth may be as short as 3.5 h. Oil content in microalgae can exceed 80% by weight of dry biomass (Metting, 1996; Spolaore *et al.*, 2006). Oil levels of 20-50% are quite common and production of methyl esters or biodiesel from microalgal oil has been demonstrated (Table 2, Belarbi *et al.*, 2000).

Table 2. Oil content (%dry wt) of some microalgae

Microalgae	***Oil content***
Botryococcus brauni	25–75
Chlorella sp.	28–32
Crypthecodinium cohni	20
Cylindrotheca sp.	16–37
Dunaliella primolecta	23
Isochrysis sp.	25–33
Monallanthus salina	>20
Nannochloris sp.	20–35
Nannochloropsis sp.	31–68
Neochloris oleoabundans	35–54
Nitzschia sp.	45–47
Phaeodactylum tricornutum	20–30
Schizochytrium sp.	50–70
Tetraselmis sueica	15–23

Mechanism of Biofuel Production in Microalgae

Microalgae exhibit oxygenic photosynthesis in which light is captured by specialized light harvesting complex proteins as LHCI and LHCII. These proteins bind the large proportions of photosynthetic pigments (chlorophyll and carotenoids) and play a key role in light capture and in the dissipation of excess energy. Excitation energy used to drive the photosynthetic reactions is funnelled to these photosynthetic reaction centres (PSI and PSII) via the highly coordinated network of pigments. Photosynthesis is the first step in the conversion of light energy to chemical energy which is ultimately responsible for driving the production of the feedstocks required for a wide range of fuel synthesis; protons and electrons (for biohydrogen), sugars and starch (for bioethanol), oils (for biodiesel) and biomass (for BTL, biomethane). Assimilatory powers (NADPH and ATP) generated during light reactions can be used to produce the sugars, starch, oils and other biomolecules that are required to produce bioethanol, biodiesel, biomethane and BTL-based biofuels. The Calvin cycle is an integral part of the photosynthetic process which uses ATP and NAD(P)H generated by the light reactions. Alternatively in some photosynthetic microorganisms like the green alga *Chlamydomonas reinhardtii* and also in blue green algae, the protons and electrons extracted from water (or starch) can be fed to the hydrogenase enzyme (*Hyd* A) via the electron transport chain to drive the direct photo-production of biohydrogen (Fig. 2).

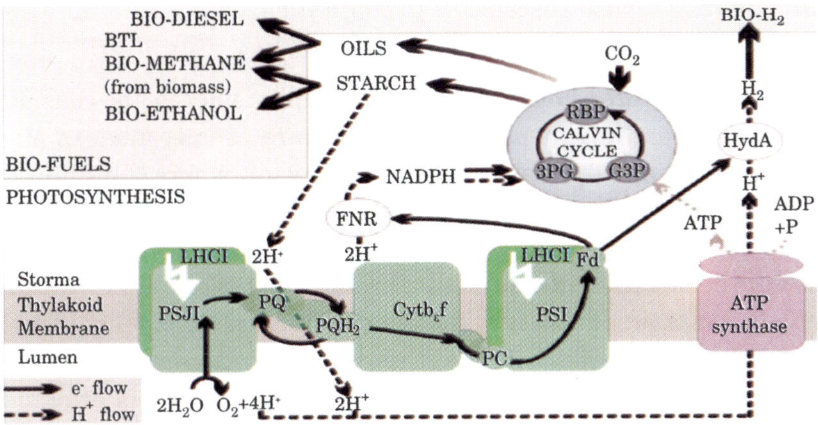

Fig. 2. The key processes to all biofuel production system in algae

Algal Biomass Production Systems

Efficient algal production systems should exhibit maximum growth and high cell density of organisms and therefore, a better understanding of the physical principles and bioreactor design is essential to achieve desired cell density. Optimal media formulation is also critical to ensure sufficient and stable supply of nutrients to attain maximal growth and cell density which would be ultimately used to produce biofuels at higher efficiencies. Algal production can be a multi-phased process with each step having independent optimal conditions for oil production and nitrogen and/or sulphur limitation or hydrogen evolution.

There are different types of cultivation systems which can be used for mass production of algae.

*1. **Autotrophic cultivation***: Cultivation systems for autotrophic growth provide mineral nutrients without organic carbon source in the presence of light for photosynthesis.

(a) *Open raceway pond system*: Open ponds have a variety of shapes and sizes but the most commonly used design is the raceway pond. Raceway pond area is divided into a rectangular grid, with each rectangle containing a channel in the oval shape and a paddle wheel which is used to drive water flow continuously around the circuit (Fig. 3). These systems usually operate at water depths of 15–20 cm, in which biomass concentrations of 1 g dry weight per litre and productivities of 60–100 mg per litre per day (10–25 g $m^{”2}$ $day^{”1}$) are possible. The main disadvantage of open systems is the loss of water by evaporation and such systems are also susceptible to contamination by other unwanted species.

(b) *Tubular photobioreactor*: A tubular photobioreactor consists of an array of straight transparent tubes that are usually made of plastic or glass. This tubular array is the solar collector, where the sunlight is captured (Fig. 3). The solar collector tubes are generally 0.1 m or less in diameter and the tube diameter is limited because light does not penetrate deeply in the dense culture broth that is necessary for ensuring a high biomass productivity of the photobioreactor. Micro algal broth is circulated from a reservoir (*i.e.* the degassing column) to the solar collector and back to the reservoir. The solar collector is oriented to maximize sunlight capture to facilitate high photosynthetic efficiency. In a typical arrangement, the solar tubes are placed oriented North-South, parallel to each other and flat above the ground and are sometime arranged like a fence to increase the number of tubes in a given area. The ground beneath the solar collector is often painted white, or covered with white sheets of plastic to increase reflectance which increases the total light received by the tube. Biomass sedimentation in tubes is prevented by maintaining highly turbulent flow that is produced using either a mechanical pump (Fig. 4), or an airlift pump. As the broth moves along a photobioreactor tube, pH increases because of consumption of carbon dioxide as it is fed in the degassing zone in response to a pH controller. Additional carbon dioxide may be injected at different at regular intervals along the tubes, to prevent carbon limitation and an excessive rise in pH. The main disadvantages of photobioreactor are photoinhibition due to photooxidative damage to algal cell.

Fig. 3. Typical commercial microalgae production facility (*Source*: Kona, Hawaii)

Fig. 4. Commercial photobioreactor for biodiesel production (*Source*: Dutch firm Bioking)

(c) *Comparison of photobioreactor and raceway pond*: The comparison of photobioreactor and raceway methods for an annual production level of 100 t of biomass shows that both production methods consume an identical amount of carbon dioxide if losses to atmosphere are not considered. Photobioreactors provide much greater oil yield per hectare compared with raceway ponds which is because the volumetric biomass productivity in a photobioreactor is more than 13-fold in comparison to that observed in raceway pond (Table 3).

Table 3. Comparison of Photobioreactor and Raceway methods for productivity

Variable	*Photobioreactor facility*	*Raceway pond*
Annual biomass production (kg)	100,000	1000,000
Volumetric productivity (kg/m^3 /day)	1.535	0.117
Areal productivity (kg/m^2 /day)	0.048[a] 0.072[c]	0.035[b]
Biomass concentration in broth (kg/m^3)	4.00	0.14
Dilution rate/day	0.384	0.250
Area needed (m^2)	5681	7828
Oil yield(m^3/ha)	139.9[d] 58.7[e]	99.4[d] 42.6[e]
Annual CO_2 consumption (kg)	183,333	183,333
System geometry	132 parallel tubes/ unit; 80 m long tubes; 0.06 m Tube diameter	978 m^2/pond; 12 m wide, 82 long, 0.30 m deep
Number of unit	6	8

a. Based on facility area.
b. Based on actual pond area
c. Projected area of photobioreactor tubes.
d. Based on 70% (by Wt.) oil in biomass.
e. 30% (by Wt.) oil in biomass

d) ***Hybrid system:*** Open ponds are very efficient and cost-effective method of cultivating algae, but these is prone to contamination with unwanted species. Photobioreactors are excellent for maintaining axenic cultures but the costs are generally ten times higher than that of open ponds. A combination of both the systems would probably be the most logical choice for cost-effective cultivation of high yielding strains of microalgae for biofuels. Open ponds are inoculated with a desired strain which can be cultivated in a bioreactor that can be as simple as a plastic bag or a technically refined optic bioreactor. Importantly, the size of the inoculum must be large enough for the desired species to establish in the open system before an unwanted species can contaminate. If the unwanted species contaminate and dominate in an open system, then the system is required to be cleaned and re-inoculated.

2. Heterotrophic cultivation: Cultivation systems for heterotrophic growth provide less quantity of mineral nutrients with organic source such as acetate in dark. In a study conducted in *Chlorella protothecoides,* the microalga in exponential period was inoculated (10%, v/v) in a liquid medium (Fig. 5). Heterotrophic cultivation of the microalgae can be initially carried out in a 500 mL Erlenmeyer flask containing 300 mL medium at 28 ± 1°C under continuous shaking (180 rpm) and air flowing in the dark (Miao and Wu, 2004). Further, heterotrophic cultivation can be performed in a 5L fermenter (Biostat Q, B.BRAUN, Germany) containing 3L medium, in which concentrated glucose solution can be

Fig. 5. A. Growth of of *C. protothecoides* under autotrophic (yellow) and heterotrophic (green yellow) culture condition. B and C are the cells of autotrophically and heterotrophically grown *C. protothecoides* under differential microscopy (Wu *et al.*, 1994)

batch-fed. Dissolved oxygen concentration is controlled by increasing agitation speed and airflow and aeration rate and the agitation speed can be variable and initially set at 0.5 vvm and 300 rpm respectively.

Harvesting

Algae have high water content, hence, downstream harvesting and processing requires the removal of water. The most common harvesting processes are flocculation, micro-screening and centrifugation. The harvesting method should be cost-effective and energy-efficient in order to make the biofuels production process economical. In this regard, strain selection is an important consideration since certain species are much easier to harvest than others. The long spiral shape of *Spirulina* (a blue green alga) can be harvested by microscreen harvesting method and large filamentous and colonial microalgae can be harvested by filtration. Flocculation, aggregation and sedimentation (or floatation) of algal biomass, is also a very common primary harvesting method used to concentrate algae. Inorganic chemicals such as alum, ferric chloride and lime are very effective flocculants but are considered to be too expensive for large scale operations. Flocculation increases particle size resulting in faster sedimentation. Nevertheless, addition of flocculants is currently not a method of choice for cheap and sustainable production. The algal chemical sludge can not be used for some downstream applications, such as animal feed supply or for anaerobic digestion. Organic cationic polyelectrolyte flocculants (cationic polymer Chitosan) are preferred options due to less amount required and the algae can be used in downstream process. Recent developments involve encouraging self-flocculation of the cells, which can occur during carbon limitation or pH shifts. Polyelectrolyte can also be used to assist spontaneous flocculation or bioflocculation which is likely to be the cheapest harvesting process. Certain species may flocculate naturally, while others flocculate in response to environmental stimuli, nitrogen stress, pH and level of dissolved oxygen. The problem, however, with spontaneous flocculation is the time taken which is not always reliable. An interesting variation of this can be the process of Co-flocculation presented by Ami Ben-Amotz. In this process, naturally flocculating alga *Skeletonema* was used to form flocs with high lipid producing strains of *Nannochloropsis.*

Oil Extraction Methods

There are a large number of methods used for the extraction of oil from microalgae (Fig. 6). These can be identified as under:

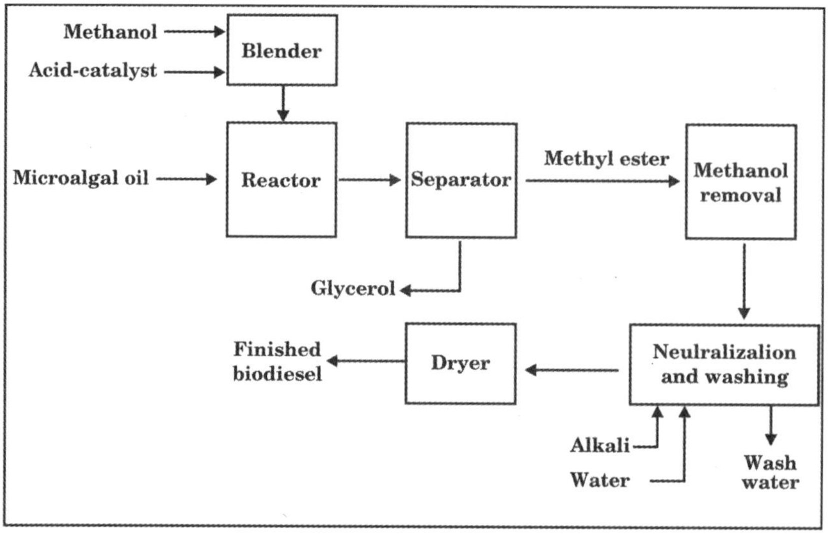

Fig. 6. Schematic representation of various steps involved in biodiesel production

I. *Expeller or press*: This involves the use of mechanical press for the algal oil extraction. Mostly a chemical solvent is applied in order to improve this process and efficiencies till 75% are feasible.

II. *Supercritical fluid extraction*: In this process, CO_2 is liquefied under pressure and heated in order to act as a solvent in extracting the oil. Efficiencies till almost 100% can be reached with the use of special machinery.

III. *Hexane solvent method*: Relatively cheap hexane can be used as a solvent in algal oil extraction. However, hexane is very inflammable; hence, its use can be very dangerous. Hexane can be used in combination with an expeller, to extract the last oil content from algae and an efficiency of 95% is feasible.

IV. *Osmotic shock*: Osmosis in a cell results in an internal pressure and when this is disturbed, the cells rupture and release oil content. The efficiency of this process is still under investigation.

V. *Enzymatic extraction*: Enzymes can be used to degrade the cell walls and the water can act as the solvent to stimulate fractionation. Efficiencies of this extraction method are not reported.

VI. *Ultrasonic-assisted extraction*: An ultrasonic reactor can produce ultrasonic waves which create bubbles in a solvent. The explosion of these bubbles near cell walls can result in cracks so that the oil can be extracted.

Important Fuel Properties of Biodiesel

Fuel properties of diesel and biodiesel from microalgae have been reported to be comparable (Table 4):

Table 4. Comparison of biodiesel from microalgae oil and diesel fuel

Properties	*Biodiesel from microalgae*	*Diesel*	*ASTM biodiesel standard*
Density (kg^{-1})	0.864	0.838	0.86–0.90
Viscosity (mm^{-2} s^{-1}, at 40°C)	5.2	1.9–4.1	3.5–5.0
Flash point (°C)	115	75	Min100
Solidifying point (°C)	–12	–50–10	–
Cold filter plugging point (°C)	–11	–3.0 (max –6.7)	Summer max 0 Winter max <–15
Acid value (mg KOH g^{-1})	0.374	Max 0.5	Max 0.5
Heating value (MJ kg^{-1})	41	40–45	Max 0.5
H/C ratio	1.81	1.81	–

1. *Cetane number*: As indicator of ignition quality, the CN is a prime indicator of fuel quality in the realm of diesel engines. It is conceptually similar to the octane number used for gasoline. Generally, a compound that has a high octane number tends to have a low CN and vice versa. The CN of a diesel fuel (DF) is related to the ignition delay (ID) time, *i.e.*, the time that passes between injection of the fuel into the cylinder and onset of ignition. The shorter the ID time, the higher the CN and vice versa.
2. *Oxidative stability*: Oxidative stability of biodiesel affects biodiesel primarily during extended storage. Generally factor such as presence of air elevated temperature or presence of metal facilitate oxidation. The reason for auto oxidation is the presence of double bond in the chain of many fatty compounds.
3. *Viscosity*: Viscosity affects the atomization of a fuel upon injection into the combustion chamber and thereby, ultimately the formation of engine deposits. The higher the viscosity, the greater the tendency of the fuel to cause such problems. The viscosity of transesterified oil, *i.e.*, biodiesel, is about an order of magnitude lower than that of the parent oil. High viscosity is the major fuel property which is why neat vegetable oils have been largely abandoned as alternative diesel fuel. Kinematic viscosity has been included in biodiesel standards (1.9–6.0 mm^2/s in ASTM D6751

and 3.5–5.0 mm^2/s in EN 14214). It can be determined by standards such as ASTM D445 or ISO 3104.

4. *Cold filter plugging point*: The temperature at which the fuel starts to form crystal /solidifies and blocks the filter of an engine.
5. *Heat of combustion*: - Gross heat of combustion (HG) is another fuel property indicating the suitability of fatty compounds as diesel fuel. The heats of combustion of fatty esters and triacylglycerols are in the range of approximately 1300 kg cal/mol to about 3500 kg cal/mol for C8–C22 fatty acids and esters. HG increases with chain length of fatty acids.
6. *Lubricity*: Desulfurization of conventional, petroleum derived diesel fuel reduces or eliminates the inherent lubricity of this fuel, which is essential for proper functioning of vital engine components such as fuel pumps and injectors. The lubricities of biodiesel or fatty compounds have shown the beneficial effect of these materials on the lubricity of conventional petroleum-derived diesel fuel, particularly low-sulfur petrodiesel fuel. Adding biodiesel at low levels (1–2%) restores the lubricity to low-sulfur petroleum-derived diesel fuel. An advantage of biodiesel compared to lubricity-enhancing additives is that biodiesel inherently possesses fuel properties competitive and compatible with conventional petroleum-based diesel fuel, which those additives usually do not possess.

Economic Feasibility of Microalgal Biodiesel

The potential of biofuel production is highlighted by the fact that even using current technology, Brazilian sugarcane-based and US corn-based bioethanol production are reported to be cost competitive at oil price of US $40 and US $60 per barrel, while the equivalent for biodiesel is US $80. Thus, since the oil price has recently reached US $100 per barrel, these fuels can be produced economically. Therefore, even modest increases in photosynthetic and process efficiency are expected to yield significant increases in economic competitiveness. Biodiesel is developing into one of the most important near-market biofuels as virtually all industrial vehicles used for farming, transport and trade are diesel-based. The increased demand for vegetable oils for the production of biodiesel has led to significant pressure on the vegetable oil market. Indeed the world's biodiesel industry is currently operating far below capacity due to a lack of feedstock. Second generation microalgal systems are increasingly predicted by international experts and policy makers to play a crucial role in a clean environmentally sustainable future.

A number of studies have attempted to calculate the cost of algal oil production from large scale farms. A comprehensive analysis (Benemann and Oswald, 1994) estimated the cost of algal oil to be in the range of US$39-$69/barrel (bbl), accounting for inflation to 2008 which would equate to US$52-$91/bbl. This estimate was based on 400 hectares of open ponds, using either pure CO_2 or flue gas from a coal-fired power station and productivity assumptions of 30-60 g m^{-2} day^{-1} with 50% algal lipid yield. Such high yields are theoretically possible but to date have not been demonstrated. A more recent analysis (Huntley and Redalje, 2006) estimated algae oil production costs to be US$84/bbl. This scenario was based on the infrastructure cost assumptions by Benemann and Oswald (1996) but utilized a hybrid system with an aerial productivity of 70.4 $gm^{-2}day^{-1}$ and 35% algal lipid yield. Companies commercially producing algae have not been as optimistic in the production costs. One such company that is close to achieving these values is Seambiotic Ltd. (Israel, Source of information). It estimates production costs of its dried algae to be at US$0.34/kg. The productivity was 20 $g.m^{-2}.day^{-1}$ and the total lipid content ranged from 8-40%. Assuming the average yield of 24% lipid, extracted at no further cost, then this equates to US$1.42/kg for lipids which equates to US$209/bbl. At the more optimistic 40% yield, lipids cost US$0.85/kg which equates to US$126/bbl. Calculations were based on the specific gravity of soybean oil [(0.925 kg/L; 1 kg = 1.08 L; 147.2 kg = 1 bbl; (Knothe *et al.*, 2005)]. Furthermore, industrial feasibility studies vary enormously on the precise parameters used. For example, incorporating improved efficiencies based on optimal bioreactor depth, or incorporating the economies of scale of bulk nutrient costs, have major effects on the calculated profitability of the plant. Simply doubling the operation of a plant from 5 to 10 years theoretically doubles the output at minimal additional capital cost. The conservative conclusion therefore, is that using existing low-tech approaches, microalgal biodiesel production system may already be economically viable.

Conclusions

Global atmospheric CO_2 increases and depletion of mineral oil reserves require the rapid development of carbon-neutral renewable alternatives. Biodiesel production from microalgae provides technical and economic feasibility that also has the potential for CO_2 sequestration and is therefore, likely to find wide acceptance. Algal biofuels appear to be the only current renewable source that could meet the global demand for transport fuels. Microalgal biofuels are also likely to have much lower impacts on the environment and the world's food supply than conventional biofuel-producing crops. The biggest challenge over the next few years in the biodiesel field will be to reduce costs for cultivation and to further improve the biology of oil production. New material and design for cultivation in closed bioreactors and the use of cutting-edge metabolic engineering and screening/selection techniques are thought to provide the biggest promises. In contrast to a pure genomics approach the integrated use of transcriptomics, proteomics and metabolomics/ metabonomics will allow insight into cellular processes. The identification of differentially expressed genes, proteins or metabolites given clues to rate-limiting processes in the cell, which can be backed up by the determination of metabolic flux. This systems biology approach will allow fine-tuning of algal properties by genetic or metabolic engineering.

References

Akkerman, I., Janssen, M., Rocha, J. and Wiffels, R.H. (2002). Photobiological hydrogen production: photochemical efficiency and bioreactor design. *Int J Hydrogen Energy*, **27**: 1195-1208.

Banerjee, A., Sharma, R., Chisti, Y. and Banerjee, U.C. (2002). *Botryococcus braunii*: a renewable source of hydrocarbons and other chemicals. *Crit Rev Biotechnol.* **22**: 245-79.

Belarbi, E.H., Molina, G.E. and Chisti, Y. (2000). A process for high yield and scaleable recovery of high purity eicosapentaenoic acid esters from microalgae and fish oil. *Enzyme Microb Technol.* **26**: 516-29.

Benemann, J. and Oswald, W. (1994). Systems and Economic Analysis of Microalgae Ponds for Conversion of CO_2 to Biomass, Quarterly Technical Project Report (Sept. 1993- Dec. 1993). US Department of Energy. Pittsburgh Energy Technology Center. p. 29.

Benemann, J. and Oswald, W. (1996). Systems and Economic Analysis of Microalgae Ponds for Conversion of CO_2 to Biomass. Final Report to the US Department of Energy, Pittsburgh Energy Technology Center.

BP statistics (2007). http://www.bp.com/statisticalreview. Cited 30 Nov 2007

Chisti, Y. (1980-81). An unusual hydrocarbon. *J Ramsay Soc.* **27-28**: 24-26.

Chisti, Y. (2007). Biodiesel from microalgae. *Biotech Advances*. **25**: 294-306.

CIA World Factbook. (2007). http://www.cia.gov/library/publications/the-world-factbook/index.html.Z Cited 6 Dec 2007.

Dunahay, T.G., Jarvis, E.E., Dais, S.S. and Roessler, P.G. (1996). Manipulation of microalgal lipid production using genetic engineering. *Appl Biochem Biotechnol.* **57-58**: 223-31.

Fedorov, A.S., Kosourov, S., Ghirardi, M.L. and Seibert, M. (2005). Continuous H_2 photoproduction by *Chlamydomonas reinhardtii* using a novel two-stage, sulfate-limited chemostat system. *Appl Biochem Biotechnol.* **121124**: 403-12.

Gavrilescu, M. and Chisti, Y. (2005). Biotechnology – a sustainable alternative for chemical industry. *Biotechnol Adv.* **23**: 471-99.

Ghirardi, M.L., Zhang, J.P., Lee, J.W., Flynn, T., Seibert, M. and Greenbaum, E. (2000). Microalgae; a green source of renewable H_2. *Trends Biotechnol.* **18**: 506-11.

Hankamer, B., Lehr, F., Rupprecht, J., Mussgnug, J.H., Posten, C. and Kruse, O. (2007). Photosynthetic biomass and H_2 production by green algae: from bioengineering to bioreactor scale-up. *Physiol. Plant.* **131**: 10-21.

Hill, J., Nelson, E., Tilman, D., Polasky, S. and Tiffany, D. (2006). Environmental, economic and energetic costs and benefits of biodiesel and ethanol biofuel. *Proc. Natl. Acad. Sci. USA.* **103**: 11206-11210.

Huntley, M.E. and Ridalje, D.J. (2006). CO_2 mitigation and renewable oil from photosynthetic microbes: A new appraisal. *In*: Report: Mitigation and Adoption strategies for global change. Univ. of Hawaii, Univ. of Mississippi.

IPCC. (2007). IPCC Fourth Assessment Report (Climate Change 2007).

Kapdan, I.K. and Kargi, F. (2006). Bio-hydrogen production from waste materials. *Enzyme Microb Technol.***38**:569-82.

Kay, R.A. (1991). Microalgae as food and supplement. *Crit Rev Food Sci Nutr.* **30**: 555-73.

Knothe, G., Krahl, J. and Van Gerpen, J. (*Eds.*) (2005). The Biodiesel Handbook. AOCS Press, Champaign, IL. USA.

Kruse, O. *et al.* (2005). Photosynthesis: A blueprint for solar energy capture and biohydrogen production technologies. *Photochem. Photobiol. Sci.* **12**: 957-970.

Melis, A. (2002). Green alga hydrogen production: progress, challenges and prospects. *Int J Hydrogen Energy* **27**: 1217-28.

Metting, B. and Pyne, J.W. (1986). Biologically active compounds from microalgae. *Enzyme Microb Technol.* **8**: 386-94.

Metting, F.B. (1996). Biodiversity and application of microalgae. *J Ind Microbiol.* **17**: 477-89.

Miao, X.L. and Wu, Q.Y. (2004). High yield bio-oil production from fast pyrolysis by metabolic controlling of *Chlorella protothecoides. J. Biotechnol.* **110**: 85-93.

Nagle, N. and Lemke, P. (1990). Production of methyl-ester fuel from microalgae. *Appl Biochem Biotechnol.* **24-5**: 355-61.

Oilworld. (2007). ISTA Mielke GmbH, Hamburg, Germany.

Roessler, P.G., Brown, L.M., Dunahay, T.G., Heacox, D.A., Jarvis, E.E. and Schneider, J.C. (1994). Genetic-engineering approaches for enhanced production of biodiesel fuel from microalgae. *ACS Symp Ser.* **566**: 255-70.

Sawayama, S., Inoue, S., Dote, Y. and Yokoyama, S.Y. (1995). CO_2 fixation and oil production through microalga. *Energy Convers Manag.* **36**: 729-31.

Schaub, G. and Vetter, A. (2007). Biofuels for Automobiles - An Overview. *Chemical Engineering & Technology* **31-5:** 721-729.

Schenk, P.M., Thoams-hall, S.R., Stephens, E., Marx, U.C., Mussgnug, J.H., Posten, C., Kruse, O., and Hankamer, B. (2008). Second generation biofuels: high-efficiency microalgae for biodiesel production, *Bioenergy Res.* **1**: 20-43.

Sheehan, J., Dunahay, T., Benemann, J. and Roessler, P. (1998). A look back at the U.S. Department of Energy's Aquatic Species Program – biodiesel from algae. National Renewable Energy Laboratory. Golden, CO; 1998. Report NREL/TP-580-24190.

Shimizu, Y. (2003).Microalgal metabolites. *Curr Opin Microbiol.* **6**: 236-43.

Singh, S., Kate, B.N. and Banerjee, U.C. (2005). Bioactive compounds from cyanobacteria and microalgae: an overview. *Crit Rev Biotechnol.* **25**: 73-95.

Spolaore, P., Joannis-Cassan, C., Duran, E. and Isambert, A. (2006). Commercial applications of microalgae. *J Biosci Bioeng.* **101**: 87-96.

Stern, N. (2006). *Stern review report on the economics of Climate Change.* HM treasury. http:/www.hmtreasury.gov.uk/independent_reviews/stern_review_economics_climate_change/stern_review_report.cfm

World Energy Council, (2007). Survey of Energy Resources. World Energy Council, London, UK. p. 600.

Wu, Q.Y., Yin, S., Sheng, G.Y. and Fu, J.M. (1994). New discoveries in study on hydrocarbons from thermal degradation of heterotrophically yellowing algae. *Sci. China (B)* **37**: 326-335.

17

CULTIVATION AND HARVESTING OF OIL PRODUCING ALGAE IN HYBRID POND SYSTEMS

S.K. Bhatnagar[1*] and Deepika Abrol[2]

Abstract

Algae is the potential feedstock for bio-fuel production as compared to other fodder crops. This group of microscopic and macroscopic autotroph has been given priority because of the eco-friendly nature, carbon sequestration tendency, production of high fatty acids content as raw material for biofuel industry, protein rich cells, farmer friendly cultivation and rapid multiplication. Developed countries are cultivating fatty acid rich algae in giant photo bioreactors which are quite expensive. The open pond cultivation of algae might lead to severe contamination and therefore the present authors has emphasized on the farmer friendly module of algae cultivation in a farmer dominated country like India by using hybrid pond system. Such ponds are easy to construct, operate and handle. This paper will facilitate the algal production by the smallest sectors of the society which shall be highly useful for mass production of bio-fuel by the industries.

Key words: Cultivation, Harvesting, Microalgae, Macroalgae, Open ponds, Hybrid ponds, Transesterification, Raceway ponds, Desert algae, Sewage algae.

Introduction

Indian subcontinent is profusely rich in microalgal and macroalgal flora because of high density of fresh water ponds, rivers, lakes and water

1 Department of Cell Biology, College of Biotechnology, Sardar Vallabhbhai Patel University of Agriculture & Technology, Meerut - 250 110, India

2 Department of Biotechnology, Maharishi Arvind Institute of Science and Technology, Jaipur, Rajasthan, India

* *Corresponding author*: E-mail : subodhbhatnagar18@gmail.com

reservoirs besides surrounding oceans. In India, the microalgal species had largely been exploited for biological nitrogen fixation and for their utilization as biofertilizer to improve soil fertility. Besides such microalgae, there are the forms which contains high amounts of the necessary fatty oils called triacylglycerols which can be used to make biofuels (Table 1). The necessity of green fuel has diverted the attention of the whole world towards the group of minutest microalgae due to the extreme potential of trapping solar energy and to utilize flue gases. Scientists at global level are confident that the day will come when the biodiesel from green algae will replace the fossil fuel completely (Table 2).

Table 1. Microalgae and their contents (after Becker, 1994)

S. No.	*Name of the algal strain*	*Protein*	*Carbo-hydrate*	*Lipids*	*Nucleic acids*
1	*Scenedesmus obliquus*	50–56	10–17	12–14	3–6
2	*Scenedesmus quadricauda*	47	–	1.9	–
3	*Scenedesmus dimorphus*	8–18	21–52	16–40	–
4	*Chlamydomonas rheinhardii*	48	17	21	–
5	*Chlorella vulgaris*	51–58	12–17	14–22	4–5
6	*Chlorella pyrenoidosa*	57	26	2	–
7	*Spirogyra sp.*	6–20	33–64	11–21	–
8	*Dunaliella bioculata*	49	4	8	–
9	*Dunaliella salina*	57	32	6	–
10	*Euglena gracilis*	39–61	14–18	14–20	–
11	*Prymnesium parvum*	28–45	25–33	22–38	1–2
12	*Tetraselmis maculata*	52	15	3	–
13	*Porphyridium cruentum*	28–39	40–57	9–14	–
14	*Spirulina platensis*	46–63	8–14	4–9	2–5
15	*Spirulina maxima*	60–71	13–16	6–7	3–4.5
16	*Synechoccus sp.*	63	15	11	5
17	*Anabaena cylindrica*	43–56	25–30	4–7	–

In Indian subcontinent, the population explosion has posed a threat to the biotechnologists and the person-land ratio has reduced. The use of fertile land for biofuel producing crops is just impossible unlike USA and other European countries where corn is being used for bioethanol production. Corn cultivation requires a lot of land, water and fertilizer which a country like India can not afford. On the contrary, algae requires relatively little space, uses clean and recyclable water and its use for fuel does not affect the cultivation of food crops.

Table 2. Algal species as oil producing crop

S. no.	*Name of algal species*	*Nature of algae*
1	*Neochloris oleobundans*	Microalga
2	*Scenedesmus dimorphus*	Unicellular
3	*Euglena gracilis*	Unicellular motile
4	*Phaeodactylum tricornutum*	Diatom
5	*Pleurochrysis carterae*	Unicellular coccolithophorid
6	*Prymnesium parvum*	Toxic algae
7	*Tetraselmis chui*	Marine unicellular
8	*Isochrysis galbana*	Microalgae
9	*Nannochloropsis salina*	Microalga
10	*Botryococcus braunii*	Siphonous alga
11	*Nannochloris*	Microalga

The most primitive forms of algae have a simple structure that also allows for easier conversion into biodiesel. Algae reproduce quickly and is nourished by carbon dioxide, the chief global warming gas and can rapidly recycle carbon dioxide emissions from the power plants and other industries. Fig. 1 indicates that algae has a higher energy yield than other biofuel sources.

Oil Source	Biomass (Mt/ha/yr)	Oil Content (% dry mass)	Biodiesel (Mt/ha/yr)	Energy Content (boe/1000ha/day)
Soya	1-2.5	20%	0.2-0.5	3-8
Rapeseed	3	40%	1-2	22
Palmoil	19	20%	1.2	63
Jtropha	7-5-10	30-50%	2.2-5.3	40-100
Microalgae	140-255	35-65%	50-100	1,150-2,000

mt = matric tons, ha = hectare, boe = barrel of oil equivalents

Fig. 1. Comparative chart showing that the algae has a higher energy yield than other biofuel sources. (*Courtesy* : HR BioPetroleum)

Companies like PetroSun, an Arizona-based diversified energy company has been researching the development of algae for jet fuels at an algae farm in Texas with plans to establish more algae farms and algae oil extraction plants in Louisiana, Arizona, Mexico and Central America. However, to date HR BioPetroleum has successfully moved algae production from the lab to the farm. Since 2004, HR BioPetroleum has operated a pilot facility in Kailua-Kona, Hawaii Island, to test its algae cultivation technology that uses marine algae to convert industrial carbon dioxide emissions into biomass feedstocks for use as biofuels.

Many algae farms use only one method for algae cultivation whereas HR BioPetroleum's technology uses two methods to cultivate algae for oil use. In the first stage, cultivation of the algae is done in a closed system which protects it from outside contaminants and also allows for maximum exposure to the sun. In the second phase, the algae are moved into open ponds to continue growth.

There are several reports on algae as oil producing feedstock, cultivation and harvesting of oil producing strains (Gilbert *et al.*, 1961; Bilanovie *et al.,* 1988; Guiry, 1991; John *et al.,* 1998; Bosma, 2003; Michael, 2004; Martin, 2005; Olivier, 2005; Gordana *et al.*, 2005; Afa, 2006; Mc Kenna, 2006; Chisti, 2007; Lee & Nam, 2009) from various countries but in India, the studies are at initial stage but due to very rich algal wealth, this country will provide a concrete solution to the problem of biofuel soon.

Methodology

After the algae are cultivated, it is then harvested and the algal oil is squeezed out, leaving a residue that can be used for animal feed and other high value products. To make biodiesel the algal oil is tread treated with sodium ethanolate and ethanol in a process called transesterification. By the end of the process, the algae's lipids are transformed into biodiesel (Chisti, 2007) and the non-fuel residual can be used for animal feed and other products. HR BioPetroleum's pilot facility inspired new methods for algae cultivation as a result, Royal Dutch Shell became partner with it in 2007 with the name Cellana. They are testing for large-scale commercial production of algae using open pond system.

The environmental impacts of algae farming will actually hold less contaminant than natural water and the microalgae used will be a common, non-genetically modified and naturally occurring species. As a promising source of sustainable and secure fuel, algae could play a vital role in providing clean and renewable energy for the entire world in future. Open pond cultivation and harvesting of algae can be studied under the following heads.

Cultivation

A. Cultivation of Algae in Open ponds

Cultivation of algae in open ponds has been extensively studied. Open ponds can be categorized into natural lakes, rivers, ponds and artificial ponds. The most commonly used systems include shallow water ponds, tanks and circular ponds. The open ponds are easier to construct and operate than the closed systems. However, major limitations in open ponds include poor light utilization by the cells, evaporative losses, diffusion of CO_2 to the atmosphere and requirement of large land area. Contamination by predators and other fast growing heterotrophs restrict the commercial production of algae in open culture systems. Due to inefficient stirring mechanisms in open cultivation system (Fig. 2), their mass transfer rate is very poor which leads to low biomass productivity.

Fig. 2. Cultivation of algae in an open pond

The ponds in which the algae are cultivated are usually called the raceway ponds (Figs. 3,4). In these ponds, the algae, water & nutrients circulate around a racetrack. Paddle wheels provide the flow and algae are kept suspended in the water, and are circulated back to the surface at regular intervals. The ponds are usually shallow because the algae need to be exposed to sunlight and the red and blue wavelengths of sunlight can only penetrate the pond water to a limited depth. The ponds are operated continuously with a regular supply of CO_2 and nutrients. The water containing algae is removed at the other end. The biggest

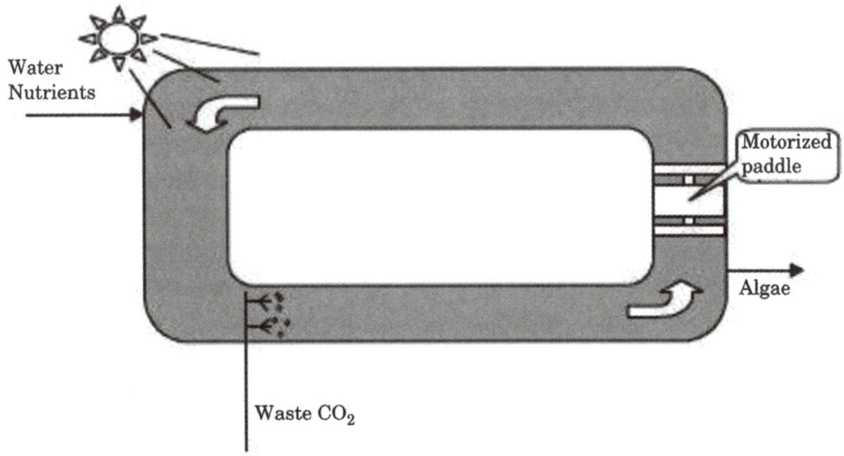

Fig. 3. Cultivation of algae in open pond system using raceway pond

advantage of these open ponds is its simplicity, low production cost and low operating cost.

Cultivating Algae for Liquid Fuel Production

With an increasing interest in biodiesel as an alternative to petrodiesel, many have looked at the possibilities of growing more oil seed crops as a solution to the problem of oil. There are two problems with this approach, first growing more oilseed crops would displace the food crops grown to feed mankind. Second, the traditional oilseed crops are not the most productive or efficient source of vegetable oil. Microalgae is the highest potential energy yield temperate vegetable oil crop. By using open, outdoor, racetrack ponds, only 15000 square miles could produce enough algae to meet the diesel requirement.

Fig. 4. A raceway pond showing rotators

Large Scale Algae Production

It is assumed that all gasoline powered vehicles can be replaced by biodiesel vehicles. The aquatic Species Program by National Renewable Energy Laboratory (NREL) indicated that one quad (7.5 billion gallons) of biodiesel could be produced on 200,000 ha (780 sq. miles) land thus for 19 quads (140.8 billion gallons) of biodiesel, nearly 15,000 sq. miles area will be needed.

Small Scale Algae Production

It will help the small farmers to produce their own fuels and will provide a farmer friendly module. It can be done in specially designed green houses in which two layers of green house plastic used. These two layers are separated by an air space which is inflated by a small fan to provide more rigidity to help the roof deflect wind and to shed rain and snow. Such green houses have two complete skins, one outside and one inside. During day time, this space may be filled with air but during night, the space is filled with soap bubbles.

Green house can be modified to produce algae throughout the year. Surface area limitation which applies to ponds can be overcome in a green house by adding a third layer of plastic inside the previously described two layers so that the water can flow in a thin film and receive enough solar radiation to grow algae. Thus the green house will allow better control of temperature and air to allow optimum growth of algae and will help in eliminating contamination with local algae. By small scale operations, local biodiesel processing plants can be constructed to convert raw oil into biofuel. It will not only increase the financial status of local farmers but will help in making them self sufficient for fuel requirements.

Drawbacks

While this is the simplest of all the techniques, it has some drawbacks. The environment in and around the pond is not under control. Bad weather can affect the algae growth adversely. Contamination from strains of bacteria or other outside organisms often results in the overflowing growth of undesirable species over the desired algae in the pond. The pond water has to be kept at a certain temperature for proper algal growth which may be difficult to maintain. Another drawback is the uneven light intensity and distribution within the pond. However, many companies today are trying out with the hybrid pond systems as compared to the expensive photobioreactors (Fig. 5).

Fig. 5. Hybrid enclosed system for cultivation of algae under semi controlled conditions

B. Cultivation of Algae in Deserts

Growing biofuel feedstock in the desert sounds good and the algae can be grown cheaply in saltwater ponds of the desert or even more efficiently in proprietary photobioreactors which will solve the problems encountered in open ponds. The photobioreactors could be placed in a desert environment but to keep water at a very consistent temperature of around 70 degrees Fahrenheit will be a challenge.

The primary inputs for growing algae are water, CO_2 and sunlight. This activity would be best accomplished in the deserts where seasonal sunlight levels and temperatures don't vary much. Another possible method to increase production would be to put the photobioreactors near a conventional coal-burning electric plant and harvest the significant amounts of CO_2 generated by the plant. But it is also true that the production of biodiesel shouldn't depend on the coal plant operating indefinitely.

Some of the algae strains suitable for the desert cultivation have been listed below.

- *Haematococcus pluvialis*
- *Microcoleus vaginatus*

- *Chlamydomonas perigranulata*
- *Synechocystis*

Valcent's HDVB Bioreactor requires very little water due to its closed circuit process, curtail labour cost and does not employ fossil fuel. Companies like Petrosun, Green Fuel Technologies and BioFields are launching large scale algae farms in desert regions.

C. Cultivation of Algae in Sewage Treatment Plant

The cultivation of Algae is useful primarily because it can extract nutrients from waste water and convert them to fats for the production of bio diesel and secondly, it can extract pollution from the atmosphere (Fig. 6).

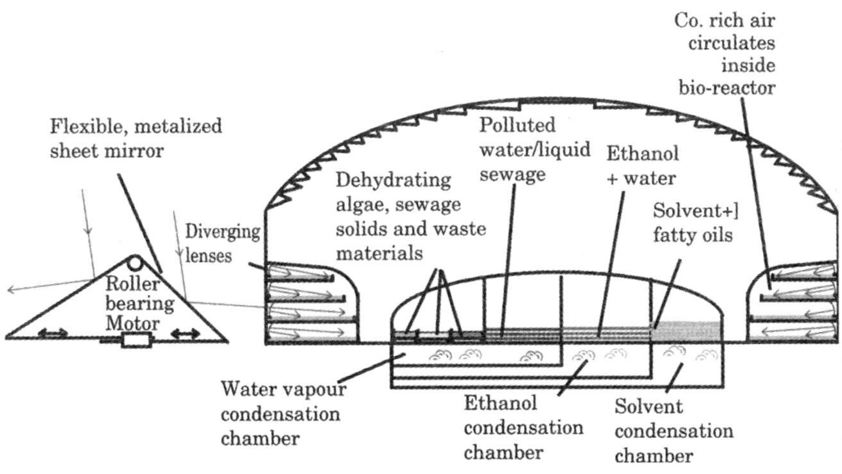

Fig. 6. Cultivation of algae in sewage water

Leaving aside, the expensive reactor systems, this approach relies on 'wild algae' *i.e.* the algae which already get colonized naturally in the sewage ponds. In New Zealand, Aqua Flow company has a small shed next to the sewage treatment plant which is producing just enough Biodiesel.

Harvesting of Algae

Algal strains such as *Botryococcus braunii* (Fig. 7) can produce long chain hydrocarbons representing almost 85% of its dry weight. Thousands of different algae can be grown naturally or under controlled conditions for harvesting and extraction of fuels, feedstocks, commercial and medical products. A comparative study shows the following projected potential gallons of oil per acre per year.

Corn	18
Soybeans	48
Safflower	83
Sunflower	102
Rapeseed	127
Oil Palm	635
Micro Algae	5000-15000

Separation of algae from the growing medium, drying, and delivering it either for end use or chemical processing are the initial steps of algae harvesting. For long distance delivery, the algae need to be packaged properly. Separating algae from its medium is also called harvesting algae. Algae can be harvested by using microscreens, centrifugation, flocculation and froth flotation.

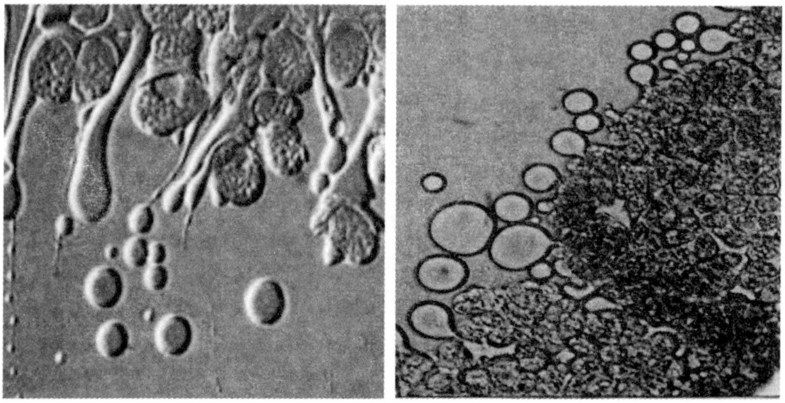

Fig. 7. *Botryococcus braunii*: A slow growing hydrocarbon producing microalgae with oil droplets

In froth flotation, the water and algae are aerated into a froth and the algae is then removed from water. Ultrasound based methods of algae harvesting are currently under development, and other additional methods are currently being developed.

A. Harvesting of Macro Algae

Macro-algae grow either on a solid substrate or free-floating in water. In the former case it is necessary to cut the algae into pieces. For free floating algae, harvesting can be done by simply rising a net installed in the pond with subsequent filtration for their separation (Fig. 8).

Fig. 8. Macroalgae

Earlier, the only way to collect the seaweeds was to pick them one by one from the nets which was a tedious and slow process. Today, petrol-driven rotary cutters are applied. One or two persons pull the net over the cutter while another manages the boat. The process can be adjusted so as to select only the larger fronds. Within a fortnight the remaining fronds and frond stumps grow sufficiently to allow a further crop cut. Usually the nets are harvested 3-4 times but the crop gets progressively smaller each time. At the end of the harvesting season, the nets are carefully cleaned with freshwater, dried and stored for the next season. Some harvesters use a machine with a vacuum pump to collect the cut plants. In harvesting the kelp are detached and collected in small boats and the plants are transported to land and dried under the sun.

B. Harvesting of Micro Algae

In most cases, it is unnecessary to separate micro-algae from the culture fluid. Excess and off-season production may, however, be concentrated and preserved. High-density algal cultures can be concentrated by either chemical flocculation or centrifugation. Products such as aluminum sulphate and ferric chloride cause cells to coagulate and precipitate to the bottom or float to the surface. Recovery of the Algal Biomass is done by siphoning off the supernatant.

Centrifugation of large volumes of algal culture is done by using a cream separator. Cells are deposited on the walls of the centrifuge head as a thick algal paste which is resuspended in a limited volume of water. The resulting slurry may be stored for 1-2 weeks in the refrigerator by using cryoprotective agents like glucose, dimethylsulfoxide etc.

C. Filtration

Filtration is carried out commonly on membranes of modified Cellulose (Fig. 9) with the help of a suction pump. This method is able to collect microalgae. However, filtration leads to the eventual blockage of the filter when vacuum is applied.

Fig. 9 . Manual filtration of algae

Various methods have been devised to avoid these problems which involve the use of a reverse-flow vacuum in which the pressure operates from above and avoid the packing of cells. Another process uses a direct vacuum but involves a stirring blade in the flask above the filter which prevents the settling of particles during the concentration process. They may be collected with the help of microstrainer but the original suspension may be light green.

D. Centrifugation

Centrifugation is a method of separating algae from the medium by using a centrifuge to cause the algae to settle to the bottom of a flask.

E. Centrifuge

A Centrifuge is a useful device for both biolipid extraction and chemical separation in biodiesel from algae. By using a homogenizer, biolipids and other useful materials can be separated from algae. Continuous-flow centrifugation with the classical rotor is efficient but the sensitive algal cells may be damaged by pelleting against the rotor wall and all the particles with higher sedimentation rate may be collected. Plankton including algae have been collected in sucrose gradients in the rotors (Fig. 10).

Fig. 10. Cell disruption and centrifugation

F. Flotation

Usually flotation is used in combination with flocculation for Algae Harvesting in waste water. It is a simple method by which algae can be made to float on the surface of the medium and removed as scum.

a. Dissolved Air Flotation

Dissolved Air Flotation (DAF) separates algae from its culture using features of both froth flotation and flocculation. It uses alum to flocculate an algae/air mixture with fine bubbles supplied by an air compressor. Alums are the trivalent sulfates of metal such as aluminum, chromium, or iron and a univalent metal such as potassium or sodium.

b. Froth Flotation

Froth Flotation is a method of separating algae from the medium by adjusting pH and passing air through a column to create a froth of algae that accumulates above the liquid and may be removed by suction. The pH required depends on algal species. Froth flotation and drying are comparatively expensive for commercial use although the cost may be reduced with further research.

G. Flocculation

Flocculation is a method of separating algae from the medium by using chemicals to form lumps of algae. The main drawback of this method is the removal of chemicals from the separated algae which makes it inadequate for commercial use. The technology for removal of these chemicals may be quite expensive for adapting commercially.

H. Flocculants

Flocculants, or flocculating agents are chemicals that promote flocculation by aggregating colloids and other suspended particles in liquids. Alum and ferric chloride are chemical flocculants which are commonly used to harvest algae. A commercial product called Chitosan (water purifier) can also be used as a flocculant but it is quite expensive. The shells of crustaceans are crushed into powder and processed to obtain chitin, a polysaccharide from which chitosan is derived by de-acetylation. Brackish and saline waters require additional chemical flocculant for causing flocculation.

Harvesting by chemical flocculation is too expensive and the interruption of carbon dioxide supply to an algal system may flocculate algae on its own (autoflocculation).

References

"Aphanizomenon Flos-Aquae Blue Green Algae". Energy for Life Wellness Center. http://www.sbgalgae.com/afa.htm. Retrieved 2006-08-29.

Bilanovic, D., Sukenik, A. and Shelef, G. (PDF) (1988). "Flocculation of microalgae with cationic polymers. Effects of medium salinity.". Elsevier Science Publishers Ltd, England. http://www.bemidjistate.edu/dbilanovic/Flocculation IonicStrength.pdf. Retrieved 2006-08-28.

Bosma, Rouke, Tramper, J. and Wijffels, R.H. (PDF) (2003). ULTRASOUND A new technique to harvest microalgae?. *Journal of Applied Phycology* **15(2–3)**: 143-153. Wageningen UR. http://www.springerlink.com/content/k03v3582334pt5j8/. Retrieved 2006-08-28.

Chisti, Y. (2007). Biodiesel from microalgae. *Biotechnology Advances* **25**: 294-306. doi:10.1016/j.biotechadv.2007.02.001.

Gilbert V. Levin, John R. Clendenning, Ahron Gibor and Frederick D. Bogar. (PDF) (1961). "Harvesting of Algae by Froth Flotation". Research Resources, Inc, Washington, D.C.http://www.pubmedcentral.gov/picrender.fcgi?artid =1057831&blobtype=pdf. Retrieved 2006-08-28.

Gordana Vunjak-Novakovic, Yoojeong Kim, Xiaoxi Wu, Isaac Berzin and Jose, C. Merchuk (2005). Air-Lift Bioreactors for Algal Growth on Flue Gas : Mathematical Modeling and Pilot-Plant Studies." *Ind. Eng. Chem. Res.*, **44(16)**, published on the web 6/28/2005. http://www.greenfuelonline.com/news/IECR.pdf

Guiry, M.D. and Blunden, G. (1991). Seaweed Resources in Europe: Uses and Potential. *John Wiley and Sons Ltd.*.

John Sheehan, Terri Dunahay, John Benemann and Paul Roessler (1998). A look back at the U.S. department of energy's aquatic species program: Bio-diesel from Algae, Closeout Report, July 1998, NREL/TP-580-24 190 http://www.nrel.gov/docs/legosti/fy98/24190 pdfJ

Lee, Keesoo and Nam, Paul K. (2009). Bioenergy from Algae: Cultivation and Harvest of locally-Isolated Microalgae Strains for Biomass Production. Missouri Energy Summit, University of Missouri, Columbia. Abst.

Martin LaMonica (2005). "Start-up drills for oil in algae," News.com, May 20, 2005 http://news.com.com/2102-1008_3-5714269.html?tag=st.util.print

McKenna, Phil (2006). "From smokestack to gas tank". *New Scientist* (Reed Business Information) **192(2572)**: 28-29. doi:10.1016/S0262-4079(06)60667-2. ISSN: 1032 1233.

Michael Briggs (2004). Wide scale Biodiesel production from algae, University of new Hampshire, Physics Department revised August 2004 http://www.unh.edu/p2/biodiesel/article_alge.html

Microalgae separator apparatus and method, United States Patent 6524486". United States Patent Department. http://www.freepatentsonline.com/6524486.html. Retrieved 2006-08-28.

Olivier, Danielo (2005). "An Algae Based Fuel," Biofutur, No. 255, May 2005 http://www.greenfuelonline.com/news/algaefuel.pdf

"Seaweeds and their Uses". *Methuen & Co. Ltd., London.*

18

HOW INDIA CAN ACHIEVE ENERGY INDEPENDENCE, COMBAT CLIMATE CHANGE & EMPOWER RURAL MASSES *VIA* BIOFUELS WITHIN JUST 20 YEARS?

Atul Saxena[1*]

Abstract

As we approach the end of first decade of new millennium, the world faces unprecedented environmental and social challenges. Devastating poverty, urban slums, unemployment, hopelessness, and terrorism proliferate as the human population swells beyond 6.5 billion and the gap between rich and poor continues to widen. At the same time, peak oil prices, choking mega-city pollution, loss of biodiversity, fresh water and food shortages, and the "inconvenient truth" of global climate change threaten ecosystem collapse as forests, soils, fisheries, and the atmosphere are pushed beyond their limits by over consumption and rapid industrialization. Political solutions to these social and environmental problems have not been forthcoming. The framework conditions needed for global governance have remained elusive; international aid and philanthropy have not been adequate to the challenge; and the use of force appears to create more problems than it solves. Economic globalization has shown promise, but thus far, it has not managed to reach the majority of humanity. Increasingly, people around the world are asking the question: Must capitalism's thirst for growth and profits serve only to exacerbate inequity and environmental deterioration? Clearly, the answer to this question must be an emphatic "no." The major challenge (and opportunity) of our

1 Founder & CEO, Growdiesel Mission
Mission Secretariat: Growdiesel Climate Care Council, B-196, Surajmal Vihar, Delhi -110 092, India

* *Corresponding author*: E-mail : info@growdieselmail.com

time is thus to create a form of commerce that uplifts the entire human community of 6.5 billion in a way that respects both natural systems and cultural diversity. *This is the only realistic and viable pathway to a sustainable world. And business can, and must, lead the way. Yet while these challenges call for bold innovation, most firms continue to focus on incremental strategies such as eco-efficiency, pollution prevention, product stewardship, and corporate social responsibility.*

Key words : Energy, climate change, biofuel, rural mass

The "next generation" of clean technologies hope to dramatically lower environmental impacts and promise to find innovative new ways to reach and include all of humanity.

In fact, there has been an explosion of sustainable innovation over the past few years: In 2006, for example, venture capitalists invested nearly $3 billion in "clean technology" deals. *Unfortunately, the vast majority of these investments are focused purely on sustainable technology development targeted at the "top of the pyramid."* Comparatively little attention is being paid to creative commercialization strategies or distribution models for clean technologies targeted at the "base of the pyramid,"

Mankind's current primary energy requirement is approximately 9.7 billion tons of crude oil equivalent units every year. Currently over 98% of liquid fuels used globally in transport sector come from fossil fuel that is also responsible for over 20% of all greenhouse gas emissions. Governments across the globe are currently focussed on three major tasks:

1. To lower the GHG emissions resulting from use of fossil fuel.
2. To secure energy independence
3. To develop sustainable renewable fuels with least harm to natural eco-system

The task at hand is to develop such biofuels that can help us to achieve all the aforementioned tasks. Today biofuels are produced almost entirely from food crops. The need of the hour is to focus on research and development leading to development of commercialization, sustainable production and creation of a value chain for algae, *Jatropha* and other waste biomass as a viable and cost effective feedstock for mass scale production of next generation biofuels that do not compete with food and also help in rapid carbon sequestration from atmosphere.

When it comes to biofuels we must do it right with sustainable, long-term solutions that can meet the triple bottom line criteria: *i.e.* while ensuring energy security, these biofuels must be working to improve the environment & empowerment of masses. Next Generation Biofuels like Growdiesel has the power to unleash a green economy that would provide a new engine of growth, putting the world on the road to prosperity again in a more sustainable way. The new, green economy would be based on developing the world's natural capital to provide lasting jobs and wealth that are needed, particularly for the poorest people on the planet. It makes more sense to invest in sustainable biofuels, which naturally absorbs carbon dioxide, than destroying the forests and then developing expensive technologies to solve the ensuing problems. The production of sustainable biofuels from Jatropha, algae and other wasted biomass can pave way for US, EU, India as well as the whole world to accrue great economic and environmental benefits.

Sources for Biofuels

Biofuels are limited in their ability to utilize natural resources for their production. It means, biofuels can only be produced from Biomass. Unlike other renewable energies, biofuels cannot be produced from solar, wind & hydro routes. There can be many approaches to production of biomass for biofuels. Replacing petrol and replacing diesel involve different technologies and markets. Since in India, the major fuel is Diesel, the focus here is principally on diesel replacement (Banerjee *et al.*, 2002; Carlsson *et al.*, 2007; Chinnasamy *et al.*, 2010; Chisti, 2008; Wikipedia, 2007).

We believe that a sustainable biofuel needs yield of at least 6-7 tons of biocrude equivalent per acre in the long run to meet the India's oil replacement needs. Chemical and water inputs and the effect on biodiversity should be minimal, if any. Cost should be below that of petroleum crude oil. Feedstock production should not materially increase the land under annual cultivation nor affect food security but should enhance energy security, reduce poverty and increase rural incomes. None of the "food/feed crop" based biofuels (corn or sugar based) or classic biodiesel sources (vegetable oils) come close to these targets. Is it really possible to achieve such a dream like situation? Fortunately the answer to this question is Yes!

Let us compare the competitiveness of biomass viz a viz fossil crude oil? A ton of biomass (on dry matter basis) contains about 2.5 times the energy content of a barrel of crude oil (14.5 million btu *vs.* 5.8 million btu). In other words $60/barrel oil could theoretically be competitive with $150/ton biomass. However, given the high cost and nascent nature of biomass industry today, we think a competitive feedstock cost delivered to the factory, has to be below $50/ton of dry biomass (plus or minus 25% depending upon feedstock type) to compete with $60/barrel oil. What are the price points needed for biomass to be profitable for farmers? It's worth noting that even at a $50 per ton price point, yields of as low as 7-8 tons/acre (which we are getting now) would allow farmers to be profitable.

As per the pricing estimates, we can focus on "multi-purpose energy crops" that can yield over 10 tons/acre in the mid-term effectively. A significant advantage of growing dedicated "multi-purpose energy crops" is that such crops shall be planned for FOOD & FUEL. This would create a win win situation wherein FUEL becomes a by-product of FOOD. Another area is to develop dedicated "multi-purpose energy crops" on waste/non-cultivable lands for example, non-food oil crops like Jatropha, algae etc.

The biomass obtained from such "multi-purpose energy crops" should be converted to "bio-oil/bio-crude" at small-scale localized BTL units. This "biocrude" can go into an existing refinery or used as feedstock in decentralized Bio refineries for production of transportation fuel like Growdiesel (Green Diesel), jet fuel and petrol (gasoline).

Waste Feedstocks

Promising waste feedstocks include municipal solid waste - the paper, wood, agro waste, plastics, rubber, hazardous waste etc. MSW which is now a big problem may soon become an opportunity! There is sufficient municipal waste to produce tens of billions of gallons of Biocrude. One of a fantastic possibility is to take all the waste carbon dioxide from flue gases of large utilities/industries and make Biofuel out of it by using algae as a converter.

The Way Ahead

The Indian government recently approved a National Biofuel Policy that predominantly aims at promoting biofuel sector in a big way. The policy aims at mainstreaming biofuels in the energy vision of India. It sets a target of achieving at least 20% blending of biofuels, both biodiesel and bioethanol, by 2017. The targeted 20% biofuel blending is expected to reduce imports by 10% by 2020. On environment front, petrol-driven vehicles are the major source of CO_2 emissions, contributing over 85% of total emissions, while diesel-driven vehicles are the major source of NOx, contributing over 90% of total emissions in India. The set target is expected to reduce the current vehicular emission by at least 20%. Hence the future of biofuels is secured & its advent poses great opportunities for farmers in India and across the globe. We firmly believe that with Advent of next generation biofuels, our transportation fuel needs can be met with fairly limited land usage.

For India, our calculations show that a laser sharp focussed vision, two decades of agricultural development, and process technology that is in pilots today put together on less than 2% of our total land (crop, timberlands & unused land) could more than supply our biocrude needs to replace all of our fossil crude oil imports by 2030. This can be the next revolution aptly called as Growdiesel Revolution.

References

Banerjee, A., Sharma, R., Chisti, Y. and Banerjee, U.C. (2002). *Botryococcus braunii*. A renewable source of hydrocarbons and other Chemicals. *Crit. Rev. Biotech* **22(3)**: 245-279.

Carlsson, A., Beilan van, J., Moller, R., Clayton, D. and Bowles, D.E. (2007). Micro and macroalgae-utility for industrial applications bioproducts. E.R.t.E.P.O.S.R. and Crops, f.N.f., CNAP, University of York. 86.

Chinnasamy, S., Bhatnagar, A., Claxton, Rand and Das, K.C. (2010). Biomass and bioenergy production potential of microalgae consortium in open and closed bioreactors using untreated carpet industry effluent as growth medium. *Biores. Technol.* **101**: 6751-6760.

Chisti , Y. (2008). Biodiesel from microalgae beats bioethanol. *Trends Biotechnolo.* **26(3)**: 126-131.

Wikipedia (2007). Algaculture. Retrieved 1 April 2007 from http://en.wikipedia.org/w/index.php?title=Algaculture & oldid=119402781.

19

BIO-ETHANOL FROM ALGAE FOR ENERGY SECURITY

R. Singh[1*] and S. Prasad

Abstract

Algae can play a significant role in energy security as it has the potential to produce bio-diesel and bio-ethanol. It can be grown in almost any enclosed space and it multiplies rapidly and requires very few inputs to flourish - mainly just sunlight, water and carbon dioxide and significant in the reduction of green house gases. Ethanol production from algae is better and efficient when compared with the generation from any other substrate because it is simple to ferment. Ethanol can also be produced by intracellular fermentation in algae. Algae can also be apart of integrated energy systems which can produce energy in different forms along with more useful by-products. Algae are very much sensitive towards the contamination that may hamper its growth. Genetic engineering can play an important role in developing the algae that can be used for the production of ethanol under the natural conditions.

Key words: Ethanol, fermentation, green house gases, integrated energy systems

Introduction

Biofuels are playing very important role in the development of sustainable supply of energy. The blending of ethanol with patrol and biodiesel with diesel up to some extend is loosening the strain imposed over the fossil fuels. In the future there is the endeavor to increase the

1 Division of Environmental Sciences, Indian Agricultural research Institute, New Delhi, India

* *Corresponding author*: E-mail : renu_icar@yahoo.com

sources and potential of biofuel. The sustainable production of biofuel requires high biodiversity and high carbon stock in the soil. Development of the new production strategies that have low energy input and more carbon binding potential should also be seek out for the biofuel production.

Normal alcohol one might encounter in daily life may consist of as little as 15% alcohol and 85% water. But the ethanol used as the fuel blend is 100% pure that is free from water. It is very clean source of energy that can replace gasoline. Ethanol can be obtained from any source of organic material that has carbohydrate. The fuels that mix bio-ethanol and gasoline are now being used around the world. The production of bio-ethanol is expected to increase to 58 billion liters by 2010 from 41 billion liters in 2006. The first generation bio-ethanol can be made from sugar and starch-based feedstocks such as sugarcane/ beet, corn, wheat, barley, rye, sorghum, and cassava. The second generation bio-ethanol is produced from the cellulosic biomass that is left after the crop harvesting or other sources i.e. stalks, leaves and husks of plants, wood chips, sawdust and algae, therefore need less input. The first and second generation reduces the carbon dioxide by between 20 and 50% and 90% respectively as compared to petroleum based fuels depending on the energy efficiency of the bio-ethanol (EU Commission, 2007). As, the first generation bio- ethanol depends upon food crops so it competes with the food security, leads to the higher food price, competition for the agricultural resources and effects the biodiversity. It may also accelerate deforestation because available agricultural land is overburdened already with the increasing food demand. In most of the cases, the crop developed for the biofuel production may not be grown with sustainability. Different type of food crops are used world wide as the biofuel substrate such as corn, wheat, barley (US and Canada); sugarcane and molasses (Central and South America); wheat corn barley, sugarbeet (European Countries); corn, wheat (China) and molasses (India). Corn price at Chicago trade has been increased from US$109 per ton (2005) to US$400 per ton (2008). According to World Bank bio-ethanol caused world food prices to increase by 75% and has helped worsen the conditions of 100 million people around the world living in poverty. The lignocellulosic based biofuel do not compete with the food crop but have some constraints in the mass production. The production costs are uncertain and vary with the different type composition of feedstock available. There is no clear candidate for "best technology pathway" between the competing biochemical and thermo-chemical routes.

As ethanol can be produced from organic matter having carbohydrates, algae could be a good source. Algae need primarily three components to grow: sunlight, carbon-dioxide and water. They use the sunlight to form all kind of substances like sugars, protein, fat, vitamins etc therefore available in most of the geographical areas. In principle, a biochemical process converts the energy of sunlight into a form of chemical energy. Algae needs low input and gives high yield up to 30-100 times more energy productivity and effectivity. It grows on the fresh or brackish water *i.e. Pleurochrysis carterae*; therefore do not compete with food crop. There are enough advancement in the open ponds and closed photo-bioreactors for the productivity and reliability of algae as the source of biomass for fuel. When physical conditions are stressful or certain nutrients are lacking, microalgae respond by producing more carbohydrates or lipids. Some algae species are known to produce high levels of carbohydrates (sugars and starch) instead of lipids (oils) as reserve polymers. Microalgae produce fermentable carbohydrates without containing lignin that make them an attractive feedstock for bio-ethanol production Algae produce a mixture of methanol, ethanol, propanol, and butanol. The idea of using algae for fuel production is over 60 years old and there has been extensive research on the algal biofuel regarding the algal strains, algae culture, harvesting and downstream process. Another advantage of algae cultivation for biofuel is its capacity to grow on low smoke stack gas. This technology allows low cost, higher industrial production of algae using flue gas from coal based thermal power plants having sulfur up to 60 ppm than the pure carbon dioxide.

Ethanol can be obtained from algae through two routes: saccharification and fermentation of algae and intracellular fermentation of algae. In the first route, the algae having good amount of starch (e.g. *Chlorella vulgaris,* 37%) is harvested and saccharified for the conversion of complex sugar into the fermentable sugars by using suitable enzymes. Then the simple sugars are subjected to the fermentation in the fermentor under the controlled conditions *i.e.* pH, temperature etc; 65% of the ethanol-conversion rate can be obtained as compared to the theoretical rate from starch. Therefore algal starch proved to be a good source of ethanol production using the conventional process. Center for Biorefining of the University of Minnesota has used algae and produced 5000 gallons of bio fuel per acre which is significantly higher that corn that yields 18 gallons, soybeans produce 48 gallons and palm trees yield 635 gallons per acre. The Korea Institute of Industrial Technology has succeeded in developing technology which can convert over 80% of cellulose from agar into sugar. Table 1. is showing the comparison of terrestrial and algal fermentable carbohydrate productivity. The percent fermentable carbohydrate is less in algae but due its high productivity, the fermentable carbohydrate productivity is far higher. The lack of lignin in algae is also making it suitable for the fuel production.

Table 1. Comparison of terrestrial and algal fermentable carbohydrate productivity

Feedstock	*Productivity (dry Mg/ ha/yr)*	*% fermen-table carbo-hydrate*	*% Lignin*	*fermen-table carbo-hydrate Producti-vity (dry Mg/ha/yr)*	*Lignin Producti-vity (dry Mg/ha/yr)*
Corn	7[a]	80[f]	15[f]	5.6	1.05
Switchrass	3.6–15[a]	76.4[f]	12[f]	2.8–11.5	0.4–1.8
Woody biomass	10–22[b]	70–85[f]	25–35[f]	7–18.7	4–7.7
Chlorella sp.	127.8–262.8[e]	33.4[c]	0[c]	42.7–87.8	0
Tetraselmis suecia	38e–139.4[d]	11–47[e]	0[e]	4.2–65.5	0
Anthrospira sp	27–70[e]	15–50[e]	0[e]	4.1–35	0

[a] Dismukes *et al.*, 2008; [b] Rgauskas *et al.*, 2006; [c] Kristensen, 1990; [d] Zittelli *et al.*, 2006; [e] Chisti, 2007; [f] Sachez, 2009

Another type of ethanol production process is the intracellular starch fermentation under dark and anaerobic conditions. Algal starch, which is photosynthetically accumulated in *Chlamydomonas* was observed to metabolize to low molecular weight compounds such as hydrogen, ethanol, glycerol and acetate under oxygen free, CO_2 atmosphere (Hirano *et al.*, 1997). During the logarithmic growth phase cells having starch get fermented and the degradation enhanced by 1.4 times due to the inhibition of hydrogenase enzyme that result to the higher production of ethanol and glycerol instead of hydrogen and acetate. It gives the ethanol yield up to 1(w/w)% that indicate that intracellular ethanol production is simpler and less energy intensive. Dark fermentation of green alga, *Chlorococcum littorale* having 27% cellular starch having maximum ethanol productivity of 450umol/g-dry weight at 30°C (Ueno *et al.*, 1998). Ethanol was formed from pyruvate by pyruvate decarboxylase and alcohol dehydrogenase. Algenol, an US based company, produce ethanol directly from algae which is removed without killing the algae. They claimed to generate 6000 gallons per acre per year of ethanol as compared with 400 gallons for corn.

There could be an integrated system where the algal biomass can be cultivated in the photo-bioreactor. In the primary reactor, hydrogen can be generated with the action of hydrogenase enzyme. At the end of the system, in the secondary reactor the biomass extracted may convert to ethanol by the action of *Saccharomyces*. Under another approach, the algal biomass can be first subjected to the extraction of oil content for the bio-diesel production then the leftover biomass having significant amount of starch can be fermented for ethanol fuel. In both of the pathways the waste generated can be utilized in biogas reactor, gasifiers for energy generation that may in turn give the power to the biofuel production system. More over waste which may have good content of the carbon source can be used as fodder.

Nowadays there is the development of photosynthetic microbial cathode half cell, employing micro algae species based such as *Chlorella vulgaris* that act as the direct electron acceptor. Then it is coupled with fermentative yeast anode, creating a complete microbial cell. This integrated system will generate power for use in the existing bio-ethanol plant. Moreover, the genetic manipulation techniques have been developed and applied for the increased production of biofuel in the algal systems. The progress in developing informed models of integrated algal metabolism and strides in algal transgenic have led to the better understanding of how fixed carbon is partitioned toward bio-energy related molecules (Powell and Hill, 2009).

There are some limitations with the handling of algae. The most common and important is the danger of contamination that enhances the competition with the target species (Zittelli *et al.*, 2006). It eventually reduces the productivity and the algae have to flourish under the extreme conditions. Some algal strains have the uniqueness regarding the conditions in which they can grow well in which most of the contaminants may not thrive well. Therefore these algae can be adopted for example *Arthrospira maxima* grow at very high pH (9.5-11), while *Tetraselmis* grow in extremely saline water (Dismukes *et al.*, 2008). Along with this, ethanol would need to be excreted into the growth medium at very high levels, to allow its recovery.

We have two major issues: environment and energy crisis and algae has the potential to reduce the use of fossil fuels, removal of CO_2 from the atmosphere and capturing and sequestration or utilize the CO_2 emitted by fossil fuel combustion before it enters the atmosphere.

As a country surrounded on two sides by water, India should consider ways to utilize marine microorganisms for bio-ethanol production suitable for the Indian climate. India should pay more attention to next generation bio-ethanol technologies which will develop rapidly. Bio-fuel is a comparatively juvenile market, leaving more opportunity for new entrants.

References

Chisti, Y. (2007). Biodiesel from microalgae. *Biotechnology Advances* **25**: 294-306.

Dismukes, G.C., Damian, C., Nicholas, B., Gennady, M.A., Matthew, C.P. (2008). Aquatic phototrophs: efficient alternatives to land-based crops for biofuels. *Current Opinion in Biotechnology* **19**: 235-240.

E., William, J.F. Jr., Jason, P.H., David, J.L., Charles, L.L., Jonathan, R.M., Richard, M., Richard, T. and Timothy, T. (2006). The path forward for biofuels and biomaterials. *Science* **311**: 484-489.

Hirano, A., Ryohei, U., Shin, H. and Yasuyuki, O. (1997). CO_2 fixation and ethanol production with microalgal photosynthesis and intracellular anaerobic fermentation. *Energy* **22(2/3)**: 137-142.

http://snrecmitigation.wordpress.com/2009/03/23/microalgae-from-biodiesel-to-bioethanol-and-beyond/

Kristensen, E. (1990). Characterization of biogenic organic matter by stepwise thermogravimetry(STG). *Biogeochemistry* **9(2)**: 135-159.

Maeda, I., Hikawa, H., Miyashiro, M., Yagi, K., Miura, Y., Miyasaka, H., Akano, T., Kiyohara, M., Matsumoto, H. and Ikuta, Y. (1994). Enhancement of starch degradation by CO_2 in a marine green alga, *Chlamydomonas* sp. MGA 161. *Journal of Fermentation Bioengineering* **78**: 383-385.

Powell, E.E. and Hill, G.A. (2009). Economic assessment of an integrated bioethanol–biodiesel–microbial fuel cell facility utilizing yeast and photosynthetic algae. *Chemical Engineering Research and Design* **87(9)**: 1340-1348.

Sánchez, C. (2009). Lignocellulosic residues: Biodegradation and bioconversion by fungi. *Biotechnology Advances* **27**: 185-194.

Ueno, Y., Kurano, N. and Miyachi, S. (1998). Ethanol production by dark fermentation in the marine green alga, *Chlorococcum littorale. Journal of Fermentation Bioengineering* **86**: 38-43.

Zittelli, G.C., Liliana, R., Natascia, B. and Mario, R.T. (2006). Productivity and photosynthetic efficiency of outdoor cultures of *Tetraselmis suecica* in annular columns. *Aquaculture* **261**: 932-943.

20

EVALUATION OF ALGAL CAKE AS DIETARY SUPPLEMENT

Raghvendar Singh[1*] and SK Bhatnagar[1]

Abstract

Algae are one of the potential sources of biofuels. Their residues supplement the body with protein, carbohydrates, carotenoids, amino acids, vitamins and trace minerals. It is especially advantageous for vegetarian animals and for the individuals suffering from malnutrition. Since algae stimulate the immune system, increase white blood cell count and promote the growth of healthy colonic flora, its supplements can be ideal for improving overall health. Algal supplements are also beneficial for treating anemia, infections, fatigue, obesity and toxicity. Present paper is an attempt to evaluate the effects of dietary supplementation with algal cake which contain high level of protein, carbohydrates, vitamins and trace minerals in order to improve overall quality and different productive traits like milk, meat, growth, subcutaneous fat quality and bio-energy of livestock. Bioactive factors of algal cake serve to modulate one or more processes such as immunological responses, circulation, cellular growth, cell repair and nutrients uptake. These bioactive factors are the topic of considerable researches, the world over and offer new opportunities for scientific investigation aimed at the characterization of their physiological effects and eventually for their commercial exploitation as high-value food or therapeutic agents for non-traditional markets. The integrated utility of algae need to be evaluated for making bio-fuel industry more economical and viable. Algal cake is extremely valuable, high protein by-product of our system that will be used for the production of nutritional food

1 College of Biotechnology, Sardar Vallabhbhai Patel University of Agriculture and Technology, Meerut - 250 110, India

* *Corresponding author*: E-mail: raghvendar@gmail.com

additives, vitamin supplements, animal/fish feed and other valuable consumer products.

Key words: Algal cake, Dietary supplement, Animal feed, Bioactive food

Introduction

Algae meal refers to the algal biomass with (deoiled cake) or without extracting the oil. When oil is removed from the algal biomass, the resulting cake does not have lipid content, and is primarily rich in protein and carbohydrates.

Algae cake is a source of nutrients for humans and animals, because the cake of many algal species has high protein content, sometimes as high as 50 to 60% of dry matter. Except for sulphur-containing amino acids (methionine and cystine), the essential amino acid content in many algal species is favourable for the nutrition of farm animals. Algae are also a rich source of carotene, vitamin C and K, and B-vitamins. Overall, the algal biomass comprises three main components – Carbohydrates, Proteins and Lipids. Once the lipids have been extracted the left-over cake is primarily composed of carbohydrates and proteins. Depending on the growth medium and the nutrients, algae meal could contain some substances such as lead, arsenic, mercury, and heavy metals. The exact composition of the algae meal depends on the algae species as well as the growth conditions. In addition, it also depends on the amount of oil that has been extracted.

Algal food production is one of the potential areas to supplement the food and simultaneously combating the challenges of fuel and environment in existing world scenario. Algal food products can create an abundance of food and energy thus reducing the demand for those food products which require extensive crop land, fresh water, fertilizers and fossil fuels. Food production that adds only oxygen to the atmosphere and does not pollute local ecosystems will provide a very positive net yield to the environment. Algal components are already integrated closely with our food, feed, cosmetics and medicines. A market basket test at Arizona State University concluded that nearly 70% of products purchased by the consumers from supermarkets contain algal components. Most people do not eat algae directly but enjoy the products made from algal components as algal flour in lieu of wheat, corn or soya flour. Low calorie, delicious algal chocolate enable consumers to eat their cakes without guilt rather than the high calories bakery products. Besides low fat and high nutrients in algal cakes as compared to the land based foods, researchers in Russia and Japan suggests that the algae may alter enzyme activity in the liver and thus control the metabolism of fatty acids, resulting in to reduced levels of fat, cholesterol and triglycerides in the blood. Algae may be humanity's best friend as it can provide sustainable and affordable food, fuel as well as effective solutions to the environmental problems (algaeindustrymagazine.com).

Animal Feed

Algae produce high-protein, low-cost, nutritious animal feed with numerous vitamins, minerals and nutrients. Local algal production in villages would feed millions of animals and save 20 million acres of forests and grasslands per annum from desertification due to animal forage (drmetrics@cox.net).

Blue green animal food, a blend of coarse grade alpha and Omega algae is potential diet for strengthening the immune system of animals *i.e.* dogs, cats, horses, or rabbits, birds, creatures at home or on the farm and the term "ultimate super-food" for animals was coined (http://www.bhealthe.com/ct2.htm). The effect of dietary supplementation with a marine algae product (MA) containing a high level of docosahexaenoic acid (DHA) on heavy pig production performances (including meat and subcutaneous fat quality) was evaluated and significant differences were observed with respect to pH value, meat colour, loin composition and the iodine number of subcutaneous fat. Pigs on MA diet showed a significant ($P < 0.01$) improvement in DHA levels (Fig. 1) both in loin and in subcutaneous fat (Sardi *et al.,* 2006).

Fig. 1. Dietary supplementation with a marine algae product (MA) containing a high level of docosahexaenoic acid (DHA) for heavy pig production

After the bulk algal biomass is processed for oil and astaxanthin, the rema.ining dry algal cake serves as a nutritious and easily digestible feed for livestock. When algal cake is mixed with trace amounts of astaxanthin, livestock show noticeable increase in fertility rate and other health benefits in poultry and cattle (Fig. 2) such as (i) increase in calcium content and strength of the shell, (ii) increase in the number of eggs per year, (iii) increase in egg laying period from 14 months to 24 months, (iv) decrease mortality rate from 10% to less than 2%, (v) prevent, reduce or eliminate osteoporosis and degenerative conditions, (vi) increase quantity of healthier and better tasting milk (vii) easy for cattle to digest, (viii) increase 'good' butter fat percentage, (ix) significantly enhances milking period (x) prevent, reduce or eliminate osteoporosis and degenerative conditions and (xi) prevent, reduce and speed up repair of physiological and physical mastitis (http://www.algae2omega.com/index).

Fig. 2. Use of algal cake for improved reproductive and productive traits of cattle

The effect of sun-dried *Spirulina platensis* in poultry diet was studied in a 12-week feeding trial by replacing either fishmeal (FM) or groundnut cake (GC) from a commercial diet with algae at isonitrogenous concentrations of 140 g/kg and 170 g/kg respectively. Additional vitamins/minerals were omitted from the algal diets because *Spirulina* is rich in vitamins and minerals (Fig. 3). The efficiency of food utilization, protein efficiency ratio and dressing percentage indicated that the

Fig. 3. Commercial product of *Spirulina* for different dietary applications

substitution of FM or GC by algae does not affect the performance of broilers. None of the diets affected the weights, compositions and histopathology of various organs of the chicks and the meat quality remained unchanged except for an intense colour in the case of birds fed on the alga containing diets (Venkataraman *et al.*, 1994).

The possibility of using unicellular algae (*Chlorella* and *Scenedesmus*) as feed for cattle has been studied. Inclusion of algal suspension did not improve total metabolizable energy (ME) or crude protein (N x 6.25) intake but an increase in daily gain and feed conversion efficiency was 6.2 and 7.4 g live weight gain per MJ ME intake for the oil cake and algae groups respectively. Crude fiber digestibility was significantly ($P < 0.01$) higher in case of algal diet (Chowdhury *et al.*, 1995).

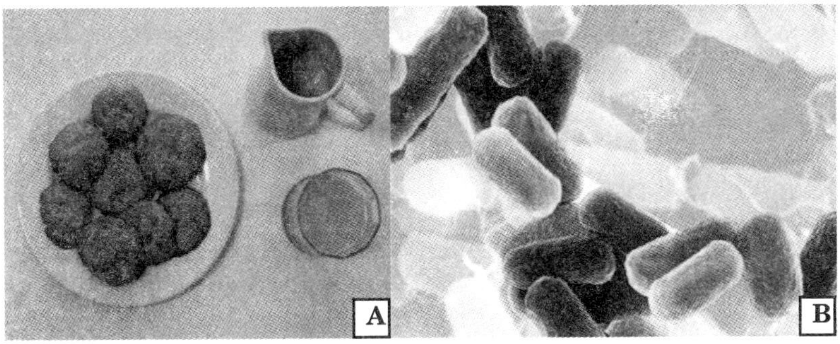

Fig. 4 A, B. Different algal cake health food products

One of the major constraints to livestock production in most of the tropical and developing countries is the shortage of protein (Preston and Leng 1987; Tareque and Saadullah 1988; Huque *et al.*, 1992). In the early fifties, Japanese and German workers explored the possibility of utilizing fast growing unicellular algae as a source of nutrients (Fig. 4a, b) for humans and animals (Halama, 1990). Of the different algal species, *Chlorella* and *Scenedesmus* are among the most widely studied genera. Their protein content ranges from 50 to 60% of dry matter. Except for sulphur-containing amino acids (methionine and cystine), the essential amino acid contents are favourable for the nutrition of farm animals (Halama, 1990). Algae are also a rich source of carotene, vitamin C and K, and B-vitamins (Halama, 1990). Because of the cellulosic cell membrane, *Chlorella* and *Scenedesmus* are not utilized more efficiently by monogastric animals. Pregastric fermentation by microbes enable the ruminants to utilize cellulose much efficiently.

Algal cake of many species is a source of nutrients for humans and animals because of high protein content. Besides this, algae are also a very rich source of carotene, vitamin C and K, and B-vitamins. The exact composition of the algal meal depends on the algal species as well as their growth conditions. In addition, it also depends on the amount of oil that has been extracted

Major advantage with the blue-green algae (cyanobacteria) is the abundance of minerals, which is important because of their synergistic relationship with vitamins. Vitamins and minerals promote the assimilation of one another. Minerals are also necessary to form body fluids, blood and bone besides maintaining a healthy nerve function. Blue-green algae contains almost all organic minerals in chelated form, meaning that each is attached to a protein molecule that transports it to the bloodstream thus enhancing absorption. Amino acids also form protein molecules called neuropeptides, which transmit information between the brain and nerve cells and throughout the body. Any weakness or deficiency in these 'electric signals' may promote disease. Amino acids help in the repair, rebuilding and strengthening of neuropeptides so that the brain's neurons can function optimally (http:/ /www.immunesupport.com).

Spirulina extract inhibits HIV replication in human T-cells, peripheral blood mononuclear cells (PBMC) and the Langerhans cells (Ayehunie *et al.,* 1998). According to Barmejo-Bescós *et al.* (2008), *Spirulina* might possess iron chelating properties. In an experiment, human neuroblastoma cells were treated with a toxic amount of iron and a subsequent treatment of *Spirulina* was given. It was observed that the iron-induced oxidative stress was reduced by *Spirulina.*

Similarly, the study with diabetic rats concluded that *Spirulina maxima* was effective in correcting the abnormal carbohydrate and lipid metabolisms. *Spirulin* helps in preventing heart damage caused by chemotherapy using Doxorubicin without interfering with its anti-tumor activity (Khan *et al.,* 2005). It also improves weight-gain.

It has been reported that *Spirulina* is beneficial against high fever and reduces the inflammation involved in arthritis in geriatric patients by stimulating the secretion of interleukin-2 which helps in regulating the inflammatory response (Mao *et al.,* 2005; Park *et al.,* 2008).

Torres *et al.* (2007) studied on 36 volunteers taking 4.5 grams of *Spirulina* per day, over a six week period, exhibited significant changes in cholesterol and blood pressure. They concluded that *Spirulina* improved the antioxidant potential of many geriatric patients who were administered *Spirulina* for 16 weeks. The plasma of these patients showed a measured increased level of total antioxidant status (Hsieh *et al.,* 2007; Kalafati *et al.,* 2010) concluding a positive effect although the mechanism was not well understood.

Bioactive Potential

Proteins from algae are well balanced in their amino acid composition and hence are quite important dietary sources. Seaweeds are a rich source of iodine, phlorotannins, glutathione, fucoxanthin and also carbohydrates such as alginates. Much interest has also been expressed in algae because of epidemiological evidence has linked habitual consumption of seaweed to reduced risk of a number of chronic diseases in Japanese and Chinese population. Fish skin may also be used for production of collagen as well as gelatin, the latter for special nutritional, religious and nutraceutical purposes. Meanwhile, unexploited aquatic resources from the oceans offer many opportunities for future developments (Shahidi, 2007).

Innovative global approaches for commercial applications:

- *Fabrics*: Algal carbohydrates are similar to wood and may be made into textiles, paper and building materials. Algal paper and building materials save forests. Fabrics save crop land for food crops and provide warmth. Algae may be made into biodegradable plastics, biofuels or other refined products.
- *Foreign aid*: American foreign aid provides subsidized U.S. food, undermines or destroys local food production because farmers cannot compete with U.S. subsidized food. Gifting food fails to address the root cause of hunger and poverty. Algaculture foreign aid transfer knowledge and some start-up materials to grow algal foods, fuels, fodder, fertilizer and medicines locally.
- *Famine and disaster relief*: Algae, with its rich set of vitamins and minerals, activates the immune system and wards off starvation while providing fuel, fodder, fabrics, fertilizers and fine medicines. Disaster relief with local algaculture production may prevent community starvation for millions. Local algal production solves the critical problem of food distribution.
- *Fine medicines*: High-quality, affordable medicines, vaccines and pharmaceuticals may be made from algal co-products or grown in algae bioengineered to produce advanced compounds such as antibiotics, vitamins, nutraceuticals and vaccines. These

compounds are grown today in land plants and animals so algae offer significantly faster and lower cost production.

- Designer algae grown locally in villages could save millions of lives by providing low cost vaccines or other medicines that need no packaging or distribution. Fine medicines, especially personalized drugs tailored to an individual, may offer more value than all other algal co-products combined.

References

Algaeindustrymagazine.com *(2010).*What's it all about Algae? Algae 101 (Part 20).

Ayehunie, S. *et al.* (1998). Inhibition of HIV-1 Replication by an aqueous extract of *Spirulina platensis* (*Arthrospira platensis*). *Journal of Acquired Immune Deficiency Syndromes & Human Retrovirology* **18(1)**: 7-12.

Barmejo-Bescós, P., Piñero-Estrada, E. and Villar del Fresno, A. (2008). Neuroprotection by *Spirulina platensis* protean extract and phycocyanin against iron-induced toxicity in SH-SY5Y neuroblastoma cells. *Toxicology in vitro* **22**: 1496-1502. doi:10.1016/j.tiv.2008.05.004

Chowdhury, S.A., Huque, K.S., Khatun, M. and Quamrun, Nahar (1995). Study on the use of algae as a substitute for oil cake for growing calves. *Livestock Research for Rural Development*, **6(3):** 22-26.

Edwards, Mark (2009). *Green Solar Gardens: Algae's Promise to end Hunger. http:/ /www.*drmetrics@cox.net

Halama K. (1990). Single Cell Protein. *In*: Non-conventional Feed Stuffs in the Nutrition of Farm Animals (*Editor*: Kolman Boda). Elsevier, pp. 34-49.

Huque K.S., Huque, Q.M.E. and Jalil, M.A. (1992). Energy: most critical nutrient for the ruminant in Bangladesh. *In*: The Proceedings of the Fourth National Conference of Bangladesh Animal Husbandry Association. pp. 141-148.

Lu, H.K., Hsieh, C.C., Hsu, J.J., Yang, Y.K. and Chou, H.N. (2007). Preventative effects of *Spirulina platensis* on skeletal muscle damage under exercise induced oxidative stress. *European Journal of Applied Physiology* **98**: 220-226.

http://www.algae2omega.com/index (2010). Algae to omega. Advanced green technologies, 4040 Galt Ocean Drive, Suite L10 Fort Lauderdale, FL 33308.

http://www.immunesupport.com/news/94wtr001.htm.Blue-Green Algae: Nature's Perfect Food.

Kevin Flatt (2008). Diet Rich in Algae Oil Can Lower Triglycerides. *In*: Marcia Wood, Agricultural Research Service Information Staff. USDA, ARS. (http:// www.naturalremediescures.com)

Kalafati, Maria, Jamurtas, A.Z., Nikolaidis, M.G., Paschalis, V., Theodorou, A.A., Sakellariou, G.K., Koutedakis, Y. and Kouretas, D. (2010). Ergogenic and antioxidant effects of *Spirulina. Supplementation in Humans, Medicine and Science in Sports and Exercise* **42(1)**: 142-151.

Khan, M. *et al.* (2005). "Protective effect of *Spirulina* against doxorubicin-induced cardiotoxicity. *Phytotherapy Research* **19(12)**: 1030-7.

Mao, T.K. *et al.* (2005). "Effects of a *Spirulina* based dietary supplement on cytokine production from allergic rhinitis patients. *Journal of Medicinal Food* **8(1)**: 27-30.

Park, H., Lee, Y., Ryu, H., Kim, M., Chung, H. and Kim, W. (2008). A randomized double-blind, placebo-controlled study to establish the effects of *Spirulina* in elderly Koreans. *Annals of Nutrition & Metabolism* **52(4)**: 322-328. doi:10.1159/000151486. PMID 18714150.

References and further reading may be available for this article. To view references and further reading you must purchase this article.

Preston, T.R. and Leng, R.A. (1987). Matching Ruminant Production Systems with Available Resources in the Tropics and Sub-Tropics. Australia, Penambul Books, Armidale.

Sardi, L., Martelli, G., Lambertini, L., Parisini, P. and Mordenti, A. (2006). Effects of a dietary supplement of DHA-rich marine algae on Italian heavy pig production parameters. *Livestock Science*, **103(1-2)**: 95-103.

Shahidi, Fereidoon, (2007). Nutraceuticals and healthful products from aquatic resources. *Feeding and Healing Humans* **2(2)**: 36-48

Tareque, A.M.M. and Saadullah, M. (1988). Feed availability, requirements for animals and current pattern of feed utilization in Bangladesh. *In*: Non-conventional Feed Resources and Fibrous agricultural Residues. Strategies for Expanded Utilization. (*Editor*: Devendra, C.). International Development Research Center and Indian Council of Agricultural Research. pp. 116-130.

Torres-Dran, Ferreira-Hermosillo and Juarez-Oropeza (2007). Antihyperlipemic and antihypertensive effects of *Spirulina maxima* in an open sample of mexican population: A preliminary report. *Lipids in Health and Disease*. **6**: 33

Venkataraman, L.V., Somasekaran, T. and Becker, E.W. (1994). Replacement value of blue-green alga (Spirulina platensis) for fishmeal and a vitamin-mineral premix for broiler chicks. *Br Poult Sci.*, **35(3)**: 373-381.

Further Readings

AOAC (1984). Official Methods of Analysis of the Association of Official Analytical Chemist (Centennial Edition). Arlington, Virginia 22209 USA

ARC (1980). Nutrient Requirements of the Ruminants Livestock. Commonwealth Agricultural Bureaux, Slough, England.

ARC (1984). Nutrient Requirements of the Ruminants Livestock. Supplement No. 1. Commonwealth Agricultural Bureaux, Slough, England

Boda, K. (1990). Basic information on unconventional feed stuffs. *In*: Non-conventional Feed Stuffs in the Nutrition of Farm Animals (Kolman Boda, Editor). Elseivier pp. 11-29

http://www.bhealthe.com/ct2.htm (2005). A superior blue green algae ´s most one of nature nutritious foods.

http://www.algaecake.com (2008). Algae cake Technologies Corporation.

Leng, R.A. (1990). Factors effecting the utilization of poor quality forages by ruminants particularly under tropical conditions. *Nutrition Research Reviews* **3**: 277-303

Rahman, M.M., Islam, M.R., Rahman, M.M., Akhtaruzzaman, M. and Malek, M.A. (1990). Study on livestock feeds and fodder and feeding practices in Bangladesh and their nutritive evaluation. Report presented to the Workshop on Contract Research Programme held on October, 9, 1990. Bangladesh Agricultural Research Council, Dhaka, pp. 22.

Saadullah, M. (1984). Supplementing Ammoniated Rice Straw for Native Cattle in Bangladesh. PhD Thesis. The Royal Veterinary and Agricultural University, Institute of Animal Science, Denmark

Silva, A.T., Greenhalgh, J.F.D. and Orskov, E.R. (1989). Influence of ammonia treatment and supplementation on the intake, digestibility and weight gain of sheep and cattle on barley straw diets. *Animal Production* **48**: 99-108

Webster, A.J.F. (1989). Bioenergetics, bioengineering and growth. *Animal Production*, **48**: 81-91.

21

CURBING EUTROPHICATION BY GROWING MIXOTROPHIC ALGAE FOR BIOFUEL FEEDSTOCKS

Monica Bhatnagar, Ashish Bhatnagar[1], Senthil Chinnasamy[2] and K.C. Das[3]

Abstract

In view of the dwindling water resources curbing eutrophication needs immediate attention from the governments at all levels on one hand while growing fuel needs, global warming and impending petroleum depletion are forcing search for safe and renewable energy sources. Cultivation of algae is one such way that would generate clean water, algal biomass for utilization and new opportunities for enterprise and employment. Mixotrophic algae have immense potential to suit to the requirements of organically polluted waters. However, they need to be prospected locally to meet the demands for local environments and people.

Key words: Eutrophication, Mixotrophic algae, Biofuel, Algae

Introduction

Fresh water is just three per cent of earth's water. Surface water and ground water total up to 30.4 per cent of this pool. Inability to contain the precipitated water, pollution of surface water and withdrawal of groundwater at a rate higher than its recharge are pushing the countries into severe water stress situation. In 2004, a study of 61 major lakes,

1 Algae Biofuels & Biomolecules Initiative, Department of Microbiology, Maharshi Dayanand Saraswati University, Ajmer - 305 009 Rajasthan, India

2 Naruveli Ventures, No.1 Karunanidhi Second Street, Kottur, Chennai - 600 085, India

3 Biorefining and Carbon Cycling Program, Faculty of Biological and Agricultural Engineering, The University of Georgia, Athens, GA 30605, USA

pond and tanks of India showed ranges of DO, BOD and fecal coliforms as 2.4 to 10.4 mg L^{-1} (mean 6.5); 0.3 to 25 mg L^{-1} (mean 7.3) and 2 to 390833 MPN 100 m L^{-1} (mean 8936), nitrites 0.004 to 2.9 mg L^{-1} (mean 0.298), nitrates 0.03 to 21.98 mg L^{-1} (mean 2.299) and ammoniacal nitrogen 0.01 to 5 mg L^{-1} (mean 0.8305), respectively (www.cpcb.nic.in/oldwebsite/Water/Lakes-Ponds) indicating that most of them were polluted. Malin Falkenmark and associates at the Stockholm International Water Institute estimated that on average, each person on the earth needs a minimum of 1,000 m^3 of water per year for drinking, hygiene and growing food for sustenance a warning to the acute water crisis due to unrestricted increase in the human population. Globally we generate >3 billion ton of domestic wastewater every year (Fahm, 1980), which is the major source of eutrophication of water bodies. Thus used water can be a big source to fulfill human needs other than those of drinking and growing food. Growing and harvesting algal biomass as biofuel feedstock is one such way.

Eutrophication

High amounts of nutrients may arise naturally in water bodies from rainfall, sediment, soil in contact and through erosion in some lakes. Human interference additionally loads nutrients by way of agriculture, wastewater and an enhanced rate of soil erosion. The domestic sewage and agricultural runoff carry heavy loads of nutrients and cause eutrophication when added to the water bodies. High nutrient levels in natural waters generate extremely high levels of algae and water plants that cover the surface, block light and diffusion of gases and in turn asphyxiate the life below. The oxygen requiring organisms die and further add nutrients to the lake and promote more growth of phytoplankton. Anaerobic conditions lead to acidity and incomplete digestion and hence generate bad odor and toxic molecules. Such an environment also allows survival to pathogens. Nutrient excess even increases infestation by parasites and pathogens that affect human beings, fishes and corals (McKenzie and Townsend, 2007).

Eutrophication may severely alter the recreational value of water bodies due to odor (*e.g.* geosmine), reduced transparency, increased incidence of stinging insects and swimmer's itch and impair related activities such as swimming and fishing. It affects potability of water due to odor, persistent and unpleasant taste, enhanced availability of manganese, iron and ammonia giving rise to color or other disturbances; risk of increased bacterial growth due to the fouling of the distribution networks and the nutrient content. In water supply lines filters get rapidly clogged by algae and flocculation treatment by organic substances gets disturbed. Furthermore, these waters are often extensively chlorinated during treatment. High levels of both chlorine and organic substances lead to formation of organochlorines that adversely affect health and are considered to be carcinogenic. Thus eutrophication needs to be prevented or cured. Other than the domestic wastewater, runoff and percolation of water from agricultural fields is another reason for eutrophication.

Agriculture

Organized animal husbandry in developed countries lead to point pollution conditions that can be managed better than that of the developing world, however the scale of waste generation may require huge infrastructure. Agriculture in the former situation asks for heavy input of resources causing higher levels of pollution. Agricultural scenario in the developing world is typified by average application of fertilizers, large cattle load, large scale flood plain farming and farming in catchment area and dried areas within ponds/lakes (Agrawal, 1999). It is known that depending on soil, crop, cropping practices and amount and duration of rainfall, broadcast application of granular fertilizers leads to upto 60 per cent losses of the fertilizer nitrogen including volatilization. Less than 40 per cent N recovery from fertilizers in rice crop is a cause of concern for Asian farmers (Freney *et al.,* at http://www.agnet.org/library/eb/414/), while sugarcane with broadcasted urea is reported to cause 45-59 per cent N-losses (Prasertsak *et al.*, 2002). Baker and Timmons (1994) reported leaching losses of nitrogen from corn crops to be in the range of 4.8 to 16.1 per cent. Leached nitrogen reaches water bodies on the surface or below the ground. In the present context nitrogen deposition with critical loads for eutrophication has increased by 7-18 per cent in the global natural ecosystems (Bouwman *et al.*, 2002).

Mineral P fertilizer and/or animal manures may be a significant nonpoint source of pollution as a result of P leaching, especially in intensive dairy pastoral systems (Toor *et al.*, 2004b). Number of reports (Hooda *et al.*, 2000; Toor *et al.*, 2004a) show that P loss by surface pathways in grassland systems seldom exceeds 2 kg ha^{-1} yr^{-1} yet this amount may play significant impact on water quality and use (Toor *et al.*, 2003; Condron, 2004). P availability in water is also affected by the redox potential (Eh). Below the compensation point (the depth at which oxygen produced (photosynthesis) is equal to oxygen consumed (respiration)), Eh decreases with depth in water column and sediment. When it reaches ~120mv, both soluble ferrous iron and soluble phosphate accumulate. At <-75mV Eh, sulfate is reduced to sulfide that removes iron and produces ferrous sulfide (Kortmann and Rich, 1994). When FeS is converted to pyrite (FeS_2) it is no longer susceptible to reduction to ferric iron even on the return of the aerobic conditions. The merit of algicide (copper sulfate) and alum (aluminium sulfate) treatment thus needs to be assessed in this perspective on a long term basis.

Phosphorus may enter a water body through the inflows, precipitation, dry fallout and sediments, while it is removed by sedimentation and/or through the outflow. For several lakes of United States, Holdren and Armstrong (1980) reported sediment phosphorus release rates as 0-13 mg P m^{-2} d^{-1} under aerobic conditions and 0-50 [upto 150] mg P m^{-2} d^{-1} under anaerobic conditions. In addition to the inputs and outputs described for phosphorus, nitrogen can enter and leave a water body in the form of free nitrogen gas through atmospheric exchange. Carbon dioxide diffuses into water at rates sufficient to meet the needs of photosynthesis. Therefore conventionally phosphorus is the obvious choice for eutrophication control since it cycles between living and nonliving particulate forms and the dissolved form. Further, the reduction of nitrogen input creating low N: P ratio, favors blooms of nitrogen fixing nuisance algae with insufficient reduction in the productivity.

Controlling Eutrophication

The main strategy to control eutrophication is reduction of the external load. Control of point sources of pollution from municipalities and industries is usually given priority as it is generally the most cost-effective measure. Diffuse source control is more difficult to achieve. Yet, in many cases, effective prevention of eutrophication or restoration of eutrophied waters cannot be achieved without such control. A possible strategy is to modify agricultural practices with reference to the control of waste from intensive animal husbandry, the dose, period and methods of fertilizer application in order to achieve minimal loss and optimum up-take by crops, erosion and run-off from tillage land and from forestry operations and over-irrigation (fertilizer leaching to ground waters) (Daniel *et al.*, 1994).

Herbicides and algicides cannot really control eutrophication as they present toxic hazard. The sulfated pesticides used to kill algae might ultimately release more of P by generating low redox potential conditions.

A number of strategies have been used for restoration of eutrophied water bodies with varying rates of success. They include, sediment removal (Peterson, 1981), circular canalization (OECD, 1982), siphoning of hypolimnetic water (OECD, 1982), hypolimnetic aeration (Fast, 1977; Olem and Flock, 1990), dilution/flushing technique (Welch, 1981), artificial circulation (Olem and Flock, 1990), chemical precipitation in the lake (OECD, 1982; Olem and Flock, 1990), lime treatment (Murphy and Prepas, 1990; Murphy *et al.*, 1991) and shading and sediment covers (Olem and Flock, 1990). But none can succeed for long unless preventive steps are taken up. Phosphorus control in wastewater treatment (OECD, 1982; Olem and Flock, 1990) by Activated Sludge Process (ASP) and controlled eutrophication process (CEP) (Benemann *et al.*, 2002) using partitioned aquaculture system (PAS) (Brune *et al.*, 2001) to manage point and non point pollution are the common preventive modes that have been tried.

Controlled eutrophication process involves utilization of raceway type High Rate Algal Ponds (HRAPs) that differs from Activated Sludge Process (ASP) in its dual utilization of the activity of both bacteria and algae (including cyanobacteria wherever used in this article). Algae generate 20 kg O_2 kwh^{-1} as compared to only 3.5 kg O_2 kwh^{-1} in the aerated systems, since in HRAP the entire energy bill is footed through the process of photosynthesis (Fallowfield and Garrett, 1985). They also help in raising the pH (to 8-9.5) by utilizing dissolved carbon dioxide.

The only drawback of HRAP in comparison to ASP is that it requires larger land area, a costly commodity today. However, the advantages with HRAP are many being relatively hassle free with low inputs for management. It adds ten times more oxygen than any of the conventional mechanical means. To treat 1000 m^3d^{-1}, HRAP requires just 60 kwh of energy to run as against 724 kwh for an ASP (Fallowfield and Garrett, 1985). Except sludge, the ASP and Trickling Filter systems do not provide any co-product worth utilization and even disposal of sludge is a problem. In case of HRAP however sludge formation is nil or negligible and the process can be standardized to produce a number of co-products so as to obtain financial gains along with clean water (Fallowfield and Garrett, 1985). Based on local environmental conditions and the suitable strains for the region, the process may yield high value oils, biodiesel, high protein animal/fish feed, methane and compost while earning green house gas credits being a carbon neutral process. The system does not need arable land.

Microalgae contain high amounts of N and P (10 and 1 per cent w/w respectively). Algae also show luxurious utilization of phosphorus that is, after meeting cellular demand the uptake continues to store large amounts of extra phosphorus. A phenomenon discovered way back in 1953 by Mackereth in *Asterionella formosa* that stored 25 times the minimum cell content of phosphorus before the spring outburst of growth especially under low nutrient conditions. They tend to allow significant variability in N: P ratios (3 to 30N for each P) (Benemann *et al.*, 2002). Algae grown in typical high rate ponds (paddle wheel mixed raceway) after passing the domestic sewage through a conventional anaerobic pond. The solid sediments of the anaerobic pond may be recycled on land as organic fertilizer after anaerobic digestion to produce methane. A settling pond after HRAPs allows harvesting algae for either oil or starch as the feedstock material for biofuels. Alternatively it may be sent to a polishing pond to culture fishes turning algal biomass into settleable form that may be turned into biogas (Benemann *et al.*, 2002).

We compared different cultivation systems to grow algae in carpet industry waste waters (Chinnasamy *et al.*, 2010) using raceway ponds, vertical tank reactors and polybags. The raceway ponds (1.52 m wide x 2.44 m long x 0.6 m deep) were made of opaque plastic. Acrylic bodied vertical tube reactors had 0.45 m diameter and 100 L working volume while low density polyethylene was used to fabricate 95 cm deep polybag reactors having 15 cm diameter and 20L working volume. 5-6% carbon dioxide-air mixture was bubbled into all reactors. Areal biomass productivity was the best in Polybags (21.1 g m^{-2} d^{-1}) followed by vertical tube reactors (8.1 g m^{-2} d^{-1}) and raceways (5.9 g m^{-2} d^{-1}). Biomass productivity was estimated to be 51 tons ha^{-1} $year^{-1}$ using 20 L capacity Polybags arranged in a cassette of triple rows of such bags. However for an economically feasible production, the costs needed to be reduced to \$10 m^{-2}. Since the algal biomass was grown in the wastewater from the carpet industry, economic feasibility must be considered in view of its remediation potential as well, thus it would be appropriate to mention that a productivity of 51 tons ha^{-1} $year^{-1}$ would remove ~5 tons of nitrogen and ~0.5 tons of phosphorus from the wastewater.

A novel reactor design called Algal Turf Scrubber (ATS) was developed by Walter H. Adey (1980). It utilizes multi-species assemblages dominated by mostly attached (benthic) algae. Studies on attached algae cultivation system by Johnson and Wen (2009) showed higher biomass productivity and effective waste water treatment with improved reactor designs. They used animal manure as growth medium in their studies. Nevertheless, the effect of high strength industrial wastewaters rich in nitrogen (N) and phosphorus (P) in algal turf scrubber system has not been vividly studied and there is little information available on appropriate reactor system for attached- algae cultivation and their cost economics in with a perspective of biofuel production. Our continuing studies with algae in a modified version of algal turf scrubber allowed a mixed culture productivity of 23 g m^{-2} d^{-1} during summer months in carpet industry wastewater at the University of Georgia, Athens.

The Ideal Alga

An ideal alga for cultivation in wastewater for biofuel feedstock shall be a halophile if saline water body eutrophication is to be tackled and community pollution tolerant if freshwater body is concerned. It must be capable of dominating the environment, fast growing, copiotrophic, capable to survive or live in anaerobic conditions, acidity tolerant, osmotolerant, r-selected (tolerant to wide range of environmental fluctuations) and easy to harvest (Azov *et al.*, 1980; Bhatnagar and Bhatnagar, 1998, 2001). Additionally for community wastewater additions it may require attributes of facultative heterotrophy or mixotrophy (Bhatnagar *et al.*, 2010), antibiotic activity, surfactant producing or emulsifier producing capabilities.

Studies on *Chlorella minutissima* (Bhatnagar *et al.*, 2010) showed the possibility to find in nature such an organism that carries most of these attributes listed above. It was obtained from dark colored, turbid sewage (BOD_5 720.15 ppm and D.O. nil) of Wazirabad Oxidation Pond System, Delhi. It also dominated the flora through all stages of treatment with an average cell count of 7 x 10^3 ml^{-1} with 50-85 per cent cells being bottom dwellers. A preliminary screening was performed using anaerobic BG 11 (0.5 per cent sodium sulfite and 1 per cent glucose) in sealed completely filled tubes to create anaerobic conditions. The alga could grow both in absence and presence of light. Presence of light however did not create completely anaerobic environment as photosynthesis might have generated oxygen to create safe heavens.

Growth of *C. minutissima* was better in sterilized 75 per cent filtered sewage (diluted with tap water) containing 95.9 ppm N and 7 ppm P than in BG 11 medium. However, the optimum growth occurred in 25 per cent sewage. It could grow in presence of as low as 150 lux illuminance. The alga was capable to grow in dark when the medium was amended with 1 per cent glucose. A minimum of 11h 44min generation time could be obtained at 2200 lux in presence of glucose with a minimum of 19h 12min at the same illuminance in absence of glucose. The growth in terms of chlorophyll a in presence of both light and glucose was ~1.5 times more than the sum of the growth in absence of glucose in light and in presence of glucose in dark thus indicating that the alga was not a facultative heterotroph but a mixotroph.

C. minutissima had a wide substrate profile for N and C sources. It preferred ammonium-N over nitrate-N showing 2.2 times better growth. Over urea however growth was inhibited. In terms of carbon sources, other than carbon dioxide and glucose, it could also utilize citrate, acetate and sucrose and in presence of light the growth increased albeit less than BG 11 on acetate and slightly higher than BG 11 over citrate and sucrose.

Its versatility was also proven in terms of tolerance to pH from 4 to 8 though the optimum was near 8. In BG 11,80 per cent cells settled down on 7 day incubation but when amended with glucose only 53 per cent cells settled. Thus in wastewater when organic carbon is present a large number of cells remain floating and are able to grow mixotrophically, while in clear water not having organic carbon, the number of suspended cells gets reduced.

Mixotrophic growth allowed upto 15 per cent of lipid accumulation in the cells of *C. minutissima.* It was estimated that approximately 10.3 ton biomass and 1.2 ton of lipid per hectare per year could be obtained through it.

Our study on carpet industry was done using a consortium of two to three mixotrophic algae namely *Chlorella minutissima, Chlamydomonas globosa* and *Scenedesmus bijuga* (Chinnasamy *et al.*, 2010). This consortium was rich in protein (~53.8 per cent) and low in carbohydrates and lipids thus could be considered to produce methane instead of alcohol or biodiesel. Our estimates showed a theoretical productivity of methane to the tune of 12.13 m^3 per hectare per year.

Tackling Eutrophication and Growing Algae *Awareness*

Tackling eutrophication has to be a multipronged approach. There is need to create awareness amongst people towards their diminishing water bodies. Clubs at the level of schools and colleges along with NGOs particularly those involving women and senior citizens may prove to be vital in stressing the need of preventing pollution of the water bodies and preserving their quality by way of conducting activity camps. People need to understand that the lakes and ponds must be open systems having both entry as well as exit of sufficient amount of water annually. One way to achieve this is to harvest roof top water of the city and use it to recharge the pond and at the exit of the pond the sewage may be diluted with pond water and treated using HRAP or turf scrubbers to grow algae while removing nutrients that lead to eutrophication.

Agriculture

Based on the soil analysis and crops being grown, the farmers need to be informed about the exact need of fertilizers and irrigation in their fields. Besides there is need to orient agricultural research to achieve minimal loss and optimum utilization of nutrients especially phosphorus. Though in developing countries like India animal husbandry is not as intensively managed as in developed world, yet the wastes need to be managed wisely. They may be a great source of nutrients for the cultivation of algae reducing competition with crops for fertilizers. The roof top water may be harvested and high rate algal ponds may be erected at the points of entry of water in the water body thus treating the agricultural runoff and recharging the water body.

Forestry

Plantation and wind breaks play major role in checking soil erosion that too contributes to the enrichment of nutrients in the water bodies. Thus plantation to check erosion is a must, but the forester must acquaint himself with the local flora that may be utilized for the purpose.

Algology/Microbiology

Research needs to be focused on making wastewater remediation from an economically feasible alternative to a profit making industry. In each climatic region, one needs to make pilot scale studies to compare algae cultivation with ASP systems. Even HRAPs can be used to polish the ASP treated wastewater that may still have amounts of phosphorus that may cause environmental concern. Similarly for these areas, some local strains need to be searched, isolated and prospected for their utility for removal of both nutrients as well as generating some co-products. Since not all algae will be easy to mass cultivate, this area also needs to be targeted while dealing with the selection of algal strains. Once such systems get *in vogue* a number of other unforeseen problems will emerge and therefore would require an ongoing process of research and innovation.

Role of Local Bodies

Local governments need to be involved in erecting algae based treatment systems at the points through which water enters the water bodies. For maintenance and running the system, a group of trained people shall be available. If due to co-products the system becomes profitable then it may be leased to these persons, thus making an earning for them besides providing profits to the local governing body.

References

Agrawal, G.D. (1999). Diffuse agricultural water pollution in India. *Wat. Sci. Tech.* **39(3)**: 33-47.

Azov, Y., Shelef, G., Moraine, R. and Levy, A. (1980). Controlling algal genera in high rate oxidation ponds. *In*: Algae biomass production and use. Shelef, G. and Soeder, C.J. (*Eds.*). pp. 245-254. Elsevier/North Holland Biomedical Press, Amsterdam.

Baker, J.L. and Timmons, D.R. (1994). Effect of fertilizer management on leaching of labeled nitrogen for no-till corn in field lysimeters. *J. Environ. Qual.* **23**: 305-310.

Benemann, J.R., van Olst, J.C., Massingill, M.J., Weissman, J.C. and Brune, D.E. (2002). The controlled eutrophication process: Using microalgae for CO_2 utilization and agricultural fertilizer recycling. Working paper. I5-2. "http:// www.unh.edu/p2/ biodiesel/pdf/algae_salton_sea. pdf".

Bhatnagar, A. and Bhatnagar, M. (1998). Algae based wastewater remediation systems. *In*: Advances in Phycology. Verma, B.N., Kargupta, A.N. and Goyal, S.K. (*Eds.*). pp. 375-392. APC Publ. Pvt. Ltd., New Delhi.

Bhatnagar A and Bhatnagar M. (2001). Strategies to employ algae and cyanobacteria for wastewater remediation. *In*: Innovative Approaches in Microbiology. Maheshwari, D.K. and Dubey, R.C. (*Eds.*). pp. 379-403.

Bhatnagar, A., Bhatnagar, M., Chinnasamy, C. and Das, K.C. (2010). *Chlorella minutissima*—A promising fuel alga for cultivation in municipal wastewaters. *Appl. Biochem. Biotechnol.* **161**: 523-536.

Bouwman, A.F., Van Vuuren, D.P., Derwent, R.G. and Posch, M. (2002). A global analysis of acidification and eutrophication of terrestrial ecosystems. *Wat. Air Soil Pollution.* **141**: 349-382.

Brune, D.E., Collier, J.A. and Schwedler, T.E. (2001). Partitioned Aquaculture System, US Patent No. 6, 192, 833.

Chinnasamy, S., Bhatnagar, A., Claxton, R. and Das, K.C. (2010). Biomass and bioenergy production potential of microalgae consortium in open and closed bioreactors using untreated carpet industry effluent as growth medium. *Biores. Technol.* **101**: 6751-6760.

Condron, L.M. (2004). Phosphorus-Surplus and deficiency. *In*: Managing soil quality-Challenges in modern agriculture. Schjonning *et al.* (*Ed.*) pp. 69-84. CAB International, Wallingford, UK.

Daniel, T.C., Edwards, D.R., Lemunyon, J.L., Sharpley, A.N. and Wedepohl, R. (1994). Minimizing surface water eutrophication from agriculture by phosphorus management. *J. Soil Wat. Conservation Suppl.* **49**: 30-38.

Fahm, L. (1980). The waste of nations: The economic utilization of human waste in agriculture. Montclair, New Jersey, USA: Allenhand, Osmun & Co. Pub, Inc.

Fallowfield, H.J. and Garrett, M.K. (1985). The treatment of wastes by algal culture. *J. Appl. Bacteriol. Symp. Suppl.* **59**: 187S-205S.

Fast, A.W. (1977). Artificial aeration as a lake restoration technique. *In*: Recovery of Damaged Ecosystems. Cairns, J. Jr., Dickson, K.L. and Herricks, E.E. (*Eds*.). pp. 135-165. University Press of Virginia, Charlottesville, Virginia.

Freney, J.R., Peoples, M.B. and Mosier, A.R. Efficient use of fertilizer nitrogen by crops. http://www.agnet.org/library/eb/414/.

Holdren, G.C. and Armstrong, D.E. (1980). Factors affecting phosphorus release from intact sediment cores. *Environ. Sci. Technol.* **14(1)**: 79-87.

Hooda, P.S., Edwards, A.C., Anderson, H.A. and Miller, A. (2000). A review of water quality concerns in livestock farming areas. *Sci. Total Environ.* **250**: 143-167.

Kortmann, R.W. and Rich, P.H. (1994). Lake ecosystem energetic: The missing management link. Lake and Reserv. *Manage* **8(2)**: 77-97.

Mackereth, F.Y.H. (1953). Phosphorus utilization by *Asterionella formosa* Haas. *J. Exp. Bot.* **4**: 296-313.

McKenzie, V.J. and Townsend, A.R. (2007). Parasitic and infectious disease responses to changing global nutrient cycles. *Eco Health* **4**: 384-396.

Murphy, T.P. and Prepas, E.E. (1990). Lime treatment of hardwater lakes to reduce eutrophication. *Verh. Internat. Verein. Limnol.* **24**: 327-334.

Murphy, T.P., Prepas, E.E. and Babin, J. (1991). Limnology of Figure Eight Lake in 1988. Effects of 1986 and 1987 lime treatments on water quality. National Water Research Institute Report No. 91-13. p. 52.

OECD. (1982). Eutrophication of waters, monitoring, assessment and control. Organization for Economic Cooperation and Development, Paris.

Olem, H. and Flock, G. (*Eds*.) (1990). Lake and reservoir restoration guidance manual. 2/e. USEPA.

Peterson, S.A. (1981). Lake restoration by sediment removal. *J. Amer. Wat. Resour. Asso.* **18(3)**: 423-436.

Prasertsak, P., Freney, J.R., Denmead, O.T., Saffigna, P.G., Prove, B.G. and Reghenzani, J.R. (2002). Effect of fertilizer placement on nitrogen loss from sugarcane in tropical Queensland. *Nutr. Cycl. Agroecosys.* **62(3)**: 229-239.

Toor, G.S., Condron, L.M., Di, H.J., Cameron, K.C. and Cade-Menun, B.J. (2003). Characterization of organic phosphorus in leachate from a grassland soil. *Soil Biol. Biochem.* **35**: 1317-1323.

Toor, G.S., Condron, L.M., Di, H.J. and Cameron, K.C. (2004a). Seasonal fluctuations in phosphorus loss by leaching from a grassland soil. *Soil Sci. Soc. Am. J.* **68**: 1429-1436.

Toor, G.S., Condron, L.M., Di, H.J., Cameron, K.C. and Sims, J.T. (2004b). Assessment of phosphorus leaching losses from a free draining grassland soil. *Nut. Cycling Agroecosys.* **69**: 169-184.

Welch, E.B. (1981). The dilution/flushing technique in lake restoration. *Wat. Resour. Bull.* **17**: 558-564.

www.cpcb.nic.in/oldwebsite/Water/Lakes-Ponds (2004). Water quality of lakes-ponds and tanks (Statewise) in India

SUBJECT INDEX

E

F

G

H

I

J

K

L

M

N

O

P

U

V

W

X

Z